IET CONTROL, ROBOTICS AND SENSORS SERIES 123

Data-Driven Modeling, Filtering and Control

Other volumes in this series:

Data-Driven Modeling, Filtering and Control

Methods and applications

Edited by
Carlo Novara and Simone Formentin

The Institution of Engineering and Technology

Published by The Institution of Engineering and Technology, London, United Kingdom

The Institution of Engineering and Technology is registered as a Charity in England & Wales (no. 211014) and Scotland (no. SC038698).

First published 2019

The Institution of Engineering and Technology
Michael Faraday House
Six Hills Way, Stevenage
Herts, SG1 2AY, United Kingdom

www.theiet.org

British Library Cataloguing in Publication Data
A catalogue record for this product is available from the British Library

ISBN 978-1-78561-712-6 (hardback)
ISBN 978-1-78561-713-3 (PDF)

Typeset in India by MPS Limited
Printed in the UK by CPI Group (UK) Ltd, Croydon

Contents

Chapter 1
Introduction

Simone Formentin[1] and Carlo Novara[2]

"In theory, there is no difference between theory and practice. In practice, there is."
Yogi Berra

1.1 Introduction

One of the main challenges of modern engineering applications is how to effectively handle the increasingly large amount of data and information available in today's complex dynamical systems [1]. Data can not only be used for mathematical modeling but also for monitoring, filtering and control.

For instance, it is often estimated that more than 75% of the costs associated to an advanced control project goes into accurate mathematical modeling [2]. For some real plants, some work might be dedicated to devise a model of the physics underlying the process, but nowadays most of the efforts are usually devoted to data-driven modeling, that is, experiment design and implementation, model learning from data and data-driven validation.

Within this framework, *system identification*, namely, the science of deriving dynamical models from input/output measurements has played a key role throughout the last 50 years. The traditional objective of system identification has been that of finding the model which best fits the data, so that model-based filters and controllers could be designed standing on the assumption that the identified mathematical description coincides with the real system (the so-called certainty equivalence principle).

Application-oriented system identification. Since—quoting G. Box—"all models are wrong, some are useful" [3], the above approach may lead to detrimental effects when applied to real world setups. In fact, since modeling usually represents only the intermediate phase of a two-step design procedure, modeling errors may seriously affect the performance achieved with the system of interest, e.g., a model-based controller or a virtual sensor. For this reason, robust approaches have been developed,

[1]Dipartimento di Elettronica, Informazione e Bioingegneria, Politecnico di Milano, Milan, Italy
[2]Dipartimento di Elettronica e Telecomunicazioni, Politecnico di Torino, Torino, Italy

which take into consideration not only a nominal model but also a description of the uncertainty region [4,5].

In any case, the model (and the related uncertainty region) which best fits the data might be a bad choice for model-based design, in which the loss function accounted for during the identification phase (typically, a norm of the data-fitting error) does not usually coincide with the final application-oriented cost to minimize (e.g., in control it is a measure of the mismatch between the desired closed-loop behavior and the achieved one). Nonetheless, robust design is safe but may lead in some cases to very conservative performance, depending on the size of the uncertainty set. Also, deriving the required uncertainty region model may be a difficult task, especially in the case of nonlinear system. Some additional knowledge about the specific domain or the model application is then *necessary* to deal with real-world systems in an effective way.

Identification for control (I4C). Based on the above observations, in the 1990s, a few leading experts in system identification started to analyze the interplay between identification and control, see, e.g., the plenary [6] at the 1991 IFAC Symposium on System Identification. The new field originated from such studies, called *I4C*, proposes a change of perspective, in which "modeling" is seen as *a design problem*. In other words, one could think of designing the identification process in order to find a model (and an uncertainty set) which does not necessarily fit the data at best but which obtains satisfactory closed-loop performance when the (robust) model-based controller is applied to the real system, by—at the same time—minimizing the level of conservativeness.

An alternative yet appealing approach is to map the data directly onto the controller parameters. From now on, this approach will be referred to as *direct data-driven control* (D^3C) [7]. D^3C can be seen as a subfield of I4C, in which the identification process is designed so that the system to identify from data is directly the controller. An interesting feature of this approach is that the control design stage is automatically incorporated into the modeling part, thus simplifying the procedure.

The overall I4C rationale is depicted in Figure 1.1 (solid lines). Unlike standard identification and control, the modeling step is "control-oriented," in which data are

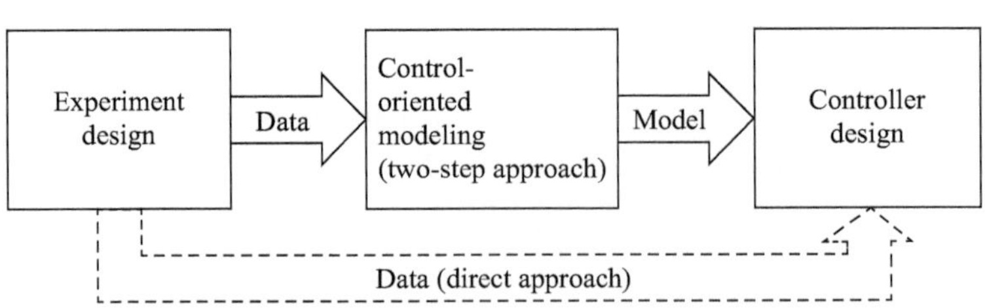

Figure 1.1 The I4C design procedure based on experimental data (solid lines). First, data are mapped onto a model of the system, by taking the control objective into account. Then, the controller is designed based on the identified model and, possibly, an estimate of the uncertainty set. If a D^3C approach is adapted, the data are directly mapped onto the control parameters (dashed lines)

mapped onto the model which is best suited for the control design. In the same figure, the D^3C alternative is illustrated using dashed lines.

Identification for filtering (I4F). Similar to what happens in I4C, also in the context of *I4F*, two main approaches can be distinguished: the two-step approach, where first a model of the system of interest is identified from data and then a filter/observer is designed from the identified model; the direct approach, where the filter/observer is directly designed from the available data. As discussed in [8], the two-step approach is in general not optimal, due to the following reasons: (1) only approximated models can be identified from measured data, and a filter which is optimal for some identified model may display a large estimation error when applied to the real system; (2) in the case of nonlinear systems, designing a computationally tractable optimal filter is in general difficult, and often only approximate filters can be derived, whose accuracy and/or stability are not guaranteed. Evaluating how these two sources of approximation affect the filter estimation accuracy is a largely open problem. Note that robust filtering does not provide at present an efficient solution to the filter design problem. Indeed, the design of a robust filter is based on the knowledge of an uncertainty model (e.g., a nominal model plus a description of the parametric uncertainty). However, identifying reliable uncertainty models from experimental data is an open problem, especially when nonlinear systems are involved. Moreover, in the case of nonlinear systems, designing a computationally tractable robust filter is in general hard (see [9,10]) and approximate filters are used [11,12].

The direct approach can overcome these issues, allowing the design of optimal filters for linear and nonlinear systems [8,13]. Another advantage of the direct approach is that it is in general simpler.

In summary, in I4F, we can say that the direct approach may be more convenient than the two-step approach in many situations. A similar statement does not hold in I4C, where the direct approach may present some advantages with respect to the two-step approach, but, in some situations, the latter may be more convenient (see the discussion in Section 1.2).

The overall I4F rationale is depicted in Figure 1.2.

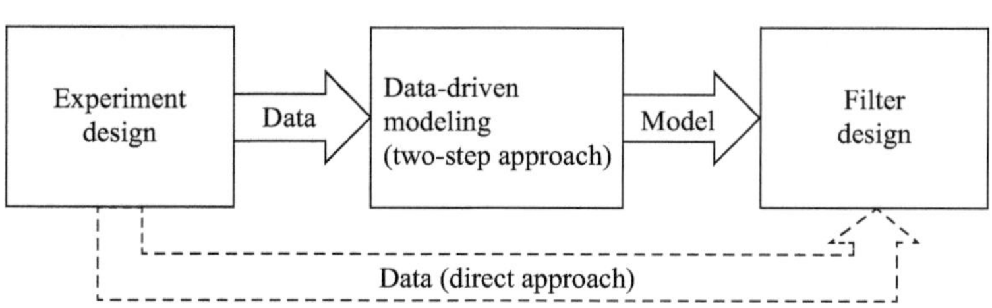

Figure 1.2 *The I4F design procedure based on experimental data (solid lines). First, data are mapped onto a model of the system, by taking the control objective into account. Then, the filter/observer is designed based on the identified model and, possibly, an estimate of the uncertainty set. If a direct approach is adapted, the data are directly mapped onto the filter/observer (dashed lines)*

1.2 State-of-the-art

System identification is the science of learning dynamical models from experimental data. A thorough review of the state-of-the-art for this discipline would take too much space and would anyway sound sketchy and incomplete. Therefore, after a brief introduction on the topic, we will hereafter focus on the interplay between system identification and model-based design applications, to introduce the key technical difficulties which still limit the performance of many real-world filters and controllers.

The first steps. To start with, consider that system identification goes as far back as the work of Gauss and Legendre in the late eighteenth and early nineteenth century. Until the late 1950s, the field had been mainly developed by mathematicians, statisticians, time-series analysts and econometricians. The main reason why control engineers had not been attracted by the discipline was that control design was mostly based on Bode and Nyquist plots, Ziegler–Nichols charts and other graphical design methods.

In the 1960s, thanks to the work by Kalman [14,15], graphical design techniques were rapidly replaced by more effective model-based (certainty equivalence) control design. Parametric models became the central focus of attention, and the development of parametric identification techniques naturally followed that of model based control design.

The founding year of modern system identification is definitely the year 1965, with two different milestones. The Ho–Kalman paper [16] provided the first solution to the determination of a minimal state-space representation from impulse response data and paved the way for the development of subspace identification [17]. The Åström–Bohlin paper [18] introduced the maximum likelihood framework developed by time series analysts into the control community and gave rise to the successful prediction error minimization (PEM) framework [19].

For more than 20 years, system identification has been at the service of control engineers but developed as a science strictly distinct from control theory. The aim of identification was only that of finding a good model of the plant: the better the model (i.e., the closer to the "real" system description), the better the final control performance (see again Figure 1.1). All problems related to inaccurate modeling had to be handled in the control design phase, e.g., by employing robust techniques.

Identification as a design process. It was in the early 1990s that, within the identification community, the faith that the true system could be modeled almost perfectly started to waver. In those years, providing a description of the model uncertainty became at least as important as computing the "nominal" (i.e., most likely) model: people started to see identification as something to *design* with a specific model application in mind [6]. The main rationale behind this change of perspective was that, if model uncertainty is unavoidable, it is smart to *shape* the error in such a way that its effect is practically negligible for the closed-loop system. This objective can be achieved by playing, e.g., with the experimental conditions, the selection of the identification cost and/or the model complexity [2]. All the approaches with these features were put inside the set of I4C methods.

It must be said that the first analyses of the interplay between identification and control were already provided by Fel'dbaum in the 1960s. Using the concept of "dual" control, Fel'dbaum showed that, when controlling a system whose parameters are unknown, the control effort must pursue the dual goal of "investigating," i.e., obtaining information about the system, and "directing," i.e., piloting the system toward the best achievable state [20]. Unfortunately, the solution of the optimization problem trading-off investigation and directioning proved to be computationally intractable, even in the simplest cases. Moreover, the whole dual control theory was essentially developed for the unrealistic case where the structure of the true system is known, and the system is in the model set. The above problems made dual control very hard to use in real applications. The same drawbacks were shared by the early attempts in adaptive control [21], in which the model parameters are adjusted at every time instant and the analysis of closed-loop properties is even more complex.

Conversely, the studies carried out in the 1990s led to the development of simpler design schemes, in which the model is found by minimizing some control-oriented cost and the controller is derived accordingly. In such studies, it was also observed that, since any "control-oriented" criterion implicitly requires the knowledge of the controller that will be implemented on the real system, any design scheme of this sort can only be iterative: namely, at each iteration a model is identified from the data obtained from the real plant on which the most recent controller is acting; the controller parameters are in turn computed from the most recent model. As a result, the succession of nominal models has a bias error distribution that is tuned for control design. This means that the (typically low order) nominal models have a model error that is small in the frequency areas where it needs to be small for the design of a better controller, typically around the present cross-over frequency [2].

Within the I4C field, many observations were made which turn out to be very useful for everyday engineering, e.g.: a closed-loop identification experiment may be more effective than an open loop one [22], identification methods delivering accurate error bounds may provide models not well suited for robust control design and many others.

Despite the success of I4C in a number of real-world process control applications, it was shown in [23] that, even with the simplest control performance criterion, such iterative schemes are not guaranteed to converge to a minimum of the control performance criterion over all models in the chosen set. This may even cause the iterative parameter adjustment scheme to drift to a controller making the closed loop system unstable. Therefore, model validation for control and controller validation criteria were introduced to guarantee the stability of the closed-loop [24]. After that, most of the efforts have been devoted to model-error modeling, see, e.g. [25,26], and, above all, to an (renewed) interest in optimal experiment design, aimed to take the most from the available data from a control-oriented perspective, see, e.g. [27,28].

Notwithstanding all the above valuable efforts, after more than 20 years from the foundation of I4C (say, the plenary [6] at the 1991 IFAC Symposium on System Identification), identification and control are far from being processes synergically designed. On the one hand, robust control is still mainly based on hard bounds on model uncertainty [5], which are different from the soft ones provided by the most widely used (i.e., stochastic) identification methods. On the other hand, identification people have not yet provided an automatic modeling procedure able to give guarantees for the real-world closed-loop systems. In fact, some interesting validation procedures have been proposed for model-based controllers, but no algorithm is currently able to embed the problem specifications inside the design procedure.

Moreover, as already mentioned in Section 1.1, limitations are given in terms of admissible systems (linear time-invariant (LTI)), identification approaches (PEM), validity of the results (infinite dataset), convergence of the iterative algorithms (not guaranteed) and systematic use of the prior knowledge. For all the above reasons, I4C cannot be considered yet a solid methodology for automatic data-driven controller design.

Controller identification. As an alternative to control-oriented model identification, within the past 20 years, I4C people have also produced a number of methods which, instead of considering the control objective inside the identification phase, directly map the data onto the controller parameters (D^3C [7]). In this way, modeling does not need to be designed to optimize the control performance, because it is actually fully skipped.

Some of these methods already existed before the 1990s and were adaptive design methods; in other words, in these techniques, the controller parameters are adjusted at each time instant, depending on the new measurements coming from the closed-loop system. The major difference between these algorithms and the more classical model-based adaptive control is that the controller parameters update rule does not require a parameterization of the plant [29]. However, being adaptive schemes, the analysis problems of classical adaptive control were shared also by these approaches. The same holds for more recent D^3C methods like the unfalsified control approach [30], which anyway can be considered as a big leap towards modern direct design of feedback controllers from data.

Instead, the methods proposed as alternatives to iterative identification and control after the foundation of I4C are *off-line methods*: the controller is designed based on the information contained in one or more bunches of experimental data, but the controller is fixed while the closed-loop system is operated. This feature allows a simpler analysis of the closed-loop and avoids the typical problems of adaptive control. Among these methods, iterative feedback tuning (IFT, [31]) involves iterative optimization of the parameters of a fixed-order controller according to a data-driven estimate of the gradient of the control performance criterion. In iterative correlation-based tuning [32], the controller parameters are instead tuned iteratively either to decorrelate the closed-loop output error between the desired and the achieved closed-loop systems with the external reference signal (decorrelation procedure) or to reduce this correlation (correlation reduction).

Both the above methods require several experiments (especially in the multivariable case for IFT [33]) to update the controller parameters. Instead, in virtual reference feedback tuning (VRFT, [34]) and noniterative correlation-based tuning [35], the controller is designed in "one-shot," by employing a single set of data. Since a single set of data is also the starting point of many model-based control design methods, this feature allowed I4C researchers to carry out a fair comparison between D^3C and certainty equivalence design (namely, classical system identification plus model-based control).

Such a comparative analysis showed that "going direct" certainly gives some advantages in fixed-order controller tuning, like avoiding model parameterization and model error modeling. At the same time, some drawbacks can be listed, among which the major ones are the following: (1) only sufficient (and overly conservative) conditions are currently available to guarantee closed-loop stability [36], (2) direct design is statistically less efficient than model-based design with correct model parameterization [37,38], (3) if the controller can be freely chosen, the problem of model parameterization is not skipped, but simply transformed into that of controller selection (the aim, in this case, is to select a controller class including the controller achieving the desired closed-loop properties).

It should be stressed that, unlike more classical I4C, in D^3C, the fact that the achieved closed-loop is uncertain due to the randomness of the dataset has never been analyzed in detail (uncertainty assessment in D^3C is not needed for control design). Nevertheless, as an advantage over more classical I4C, control of nonlinear plants has been more deeply investigated, see, e.g., use of IFT with nonlinear systems [39], nonlinear VRFT [40], iVRFT [41], direct control of linear parameter-varying (LPV) systems [42] and Direct FeedbacK (DFK) control [43]. Also in the nonlinear setting, the methods lack some important features. For instance, stability guarantees are given only in DFK, which however has been developed within a set-membership framework and therefore relies on bounded noise hypothesis without exploiting any prior knowledge on the noise distribution. Moreover, the DFK method can be used only in the restrictive case where the full state is measurable.

Therefore, despite several successful practical applications (see, e.g., [44–49]), also in D^3C a lot of research work still needs to be done to provide a general data-driven methodology with reliable performance guarantees.

Filter identification. As discussed in Section 1.1, and analogously to what happens in I4C, two main approaches to I4F can be found in the literature: the two-step approach, based on model identification from data and subsequent filter design, and the direct approach, where the filter is directly identified from data, avoiding the model identification step.

The two-step approach is in general not optimal, especially in the case of nonlinear system, where designing an effective filter is in general a hard task [8]. Conversely, the direct approach may allow the design of filters with suitable optimality properties, even for complex nonlinear systems. The direct approach is based on the simple idea that a filter/observer is a dynamic system whose inputs are the input and output of the system of interest and the output is an estimate of the system state variables (or a function of them). Hence, it is more efficient to identify this filter/observer system directly from data using a suitable identification method (linear or nonlinear), rather than designing it on the basis of a model identified from the same data, whose uncertainties and/or nonlinearities may strongly affect the filter estimation accuracy.

Filters/observers are often called "virtual sensors" or "soft sensors," since they can be used in place of (or in addition to) the more traditional physical sensors, yielding relevant advantages such as cost reduction, redundancy, increased robustness and safety, and space saving.

In the literature, several works on direct filter identification for LTI systems [50–54] and LPV systems [55,56] can be found. However, the majority of works regard nonlinear systems. Indeed, the direct filter-identification approach has proven to be a key enabler to solve filter design problems for nonlinear systems that cannot be solved using standard techniques. In the nonlinear context, several state-of-the-art approaches can be found: neural network approaches [57–59], set membership approaches [13,60], piecewise affine approaches [61], approaches based on radial basis functions [62] and gray-box approaches [63]. A limited number of these works provide theoretical/methodological analyses/results [8,13,50,52–54,60,61,64–67]. Most of works are concerned with numerical analyses and/or applications of practical interest in different fields of science and technology, see, among many others [57–59,62,63,68–79].

1.3 Goals and structure of the book

Although very appealing, the above data-driven approaches suffer from some important limitations, which make them not yet competitive with state-of-the-art model-based design. At the same time, the cutting-edge identification technologies, like nonparametric identification, are not yet ready to be applied for modeling of real-world systems. More specifically, the major limitations of current state-of-the-art, also following the discussion of the previous section, can be listed as follows:

- Recent nonparametric-identification approaches have been mainly developed for LTI plants, while most real-world systems exhibit an LPV behavior or specific nonlinearities (e.g., Wiener systems).
- When dealing with real-world applications, the science of learning and identification usually mixes with the art of domain-expert people. This should reflect into practical hints that are often missing in the scientific papers.
- D^3C design has been mainly developed in the linear SISO case, while modern systems are usually characterized by coupled nonlinear dynamics.
- Many specific design problems like that of dynamic sensor and fault-detection usually rely on a (uncertain) model of the plant.
- Data-driven design problems may rely on the knowledge of some system parameters, which is milder than the knowledge of the full system dynamics. In those cases, it would be suitable to directly identify those parameters from data. This is the case, e.g., of the H_∞ norm usually employed to characterize a bound on the modeling error.
- Although many data-driven techniques have been proposed, theoretical and experimental analyses are missing in the literature with the aim to compare the eventual performance of different methods on challenging engineering applications.

Facing the above issues, this book proposes a number of contributions that try to overcome the current limitations of the state-of-the-art to make the direct data-driven design technology competitive with model-based alternatives. Moreover, challenging modeling problems in modern engineering systems are addressed in a novel way to show both the methods and the implementation tricks that potentially make an application successful.

In particular, the book is structure in two parts. The first part (Chapters 2–5) is dedicated to *application-oriented system identification* and proposes four contributions in which existing techniques are complemented with additional tools to deal with challenging real-world problems. The second part (Chapters 6–13) is devoted to *data-driven design* and deals with the estimation of application-oriented parameters as well as direct identification of controllers and filters within complex scenarios.

Part I is entitled **data-driven modeling** and is composed as follows:

- **Chapter 2** extends nonparametric identification to Wiener systems. A tractable convex relaxation of the original problem is derived to handle cases involving systems with high-dimensional outputs. Such a theoretical extension is motivated by (and validated on) a classification problem using video data.
- **Chapter 3** deals with the wing-flutter-identification problem in a situation of poor information. In particular, an LPV approach is proposed and theoretically analyzed.
- **Chapter 4** proposes LPV identification of a web-winding machine using multi-local models. Experimental issues when working with real-world data are discussed in detail.
- In **Chapter 5**, an electrochemical impedance spectroscopy for Li-ion batteries, based on system identification techniques, is discussed together with the main implementation issues.

The title of Part II is **Data-driven filtering and control**. The following chapters are included:

- **Chapter 6** considers the problem of dynamic measurements in metrology, that is, the fact that sensors can be characterized as dynamical systems and, as such, affect the quality of the measurement. In this chapter, the measurement is then dynamically compensated from data without relying on the full model of the sensor.
- **Chapter 7** reviews the latest developments of iterative learning control (ILC) by outlining a range of design approaches for multivariable ILC with a focus on cases with limited model knowledge.
- **Chapter 8** deals with the data-driven estimation of the H_∞ norm of an LTI system. This is an important issue in real-world problems in which a bound on the modeling errors should be provided together with the nominal model (e.g., for robust control purposes).
- **Chapter 9** compares two D^3C design approaches, developed within a statistical and a deterministic setting, respectively: VRFT and set-membership controller tuning.
- **Chapter 10** addresses the difficult problem of direct Fisher matrix approximation from data. Such a matrix is very useful in system identification for benchmarking

the achievable performance, as it provides a lower bound for the variance of maximum likelihood estimates.

- In **Chapter 11**, a method to design feedback controllers from data for linear constrained systems is proposed. The method is based on a novel application of receding horizon policies within a model-free scenario.
- **Chapter 12** presents an approach to fault detection for nonlinear dynamic systems, based on the recently introduced quasi-local set membership-identification method.
- In **Chapter 13**, a data-driven controller design methodology is developed, that guarantees $\mathcal{H}_\infty$ performance and closed-loop stability for linear systems that are subject to nonlinear distortions.

References

[1] R.M. Murray, K.J. Åström, S.P. Boyd, R.W. Brockett, and G. Stein. Future directions in control in an information-rich world. *IEEE Control Systems*, 23(2):20–33, 2003.

[2] M. Gevers. Identification for control: from the early achievements to the revival of experiment design. *European Journal of Control*, 11(4):335–352, 2005.

[3] G.E.P. Box. Robustness in the strategy of scientific model building. *Robustness in Statistics*, 1:201–236, 1979.

[4] M. Milanese and A. Vicino. Optimal algorithms estimation theory for dynamic systems with set membership uncertainty: an overview. *Automatica*, 27: 997–1009, 1991.

[5] I.R. Petersen and R. Tempo. Robust control of uncertain systems: classical results and recent developments. *Automatica*, 50(5):1315–1335, 2014.

[6] M. Gevers. Connecting identification and robust control. In *The Modeling of Uncertainty in Control Systems*, pages 35–37. Berlin Heidelberg: Springer, 1994.

[7] A.S. Bazanella, L. Campestrini, and D. Eckhard. *Data-Driven Controller Design: The H_2 Approach*. Netherlands: Springer, 2011.

[8] C. Novara, M. Milanese, E. Bitar, and K. Poolla. The filter design from data (FD2) problem: parametric-statistical approach. *International Journal of Robust and Nonlinear Control*, 22:1853–1872, 2012.

[9] S.K. Nguang and M. Fu. Robust nonlinear H_∞ filtering. *Automatica*, 32(8):1195–1199, 1996.

[10] W.M. McEneaney. Robust/H_∞ filtering for nonlinear systems. *Systems and Control Letters*, 33(5):315–325, 1998.

[11] J. Seo, M.-J. Yu, C. G. Park, and J. G. Lee. An extended robust h_∞ filter for nonlinear constrained uncertain systems. *IEEE Transactions on Signal Processing*, 54(11):4471–4475, 2006.

[12] B.-S. Chen, W.-H. Chen, and H.-L. Wu. Robust h_2/h_∞ global linearization filter design for nonlinear stochastic systems. *IEEE Transactions on Circuits and Systems I: Regular Papers*, 56(7):1441–1454, 2009.

[13] C. Novara, F. Ruiz, and M. Milanese. Direct filtering: a new approach to optimal filter design for nonlinear systems. *IEEE Transactions on Automatic Control*, 58(1):86–99, 2013.

[14] R.E. Kalman. Contributions to the theory of optimal control. *Boletín de la Sociedad Matemática Mexicana*, 5(2):102–119, 1960.

[15] R.E. Kalman. A new approach to linear filtering and prediction problems. *Journal of Basic Engineering*, 82(1):35–45, 1960.

[16] R.E. Kalman and B.L. Ho. Effective construction of linear state variable models from input output data. In *Proc. 3rd Allerton Conf*, pages 449–459, 1965.

[17] P. Van Overschee and B. De Moor. *Subspace Identification for Linear Systems: Theory, Implementation, Applications*. Springer Science & Business Media, 2012.

[18] K.J. Åström and T. Bohlin. Numerical identification of linear dynamic systems from normal operating records. In *Proc. IFAC Symposium on Self-Adaptive Systems*, pages 96–111, 1965.

[19] L. Ljung. *System Identification: Theory for the User*. Prentice-Hall, Englewood Cliffs, NJ, 1999.

[20] A. Fel'dbaum. Dual control theory. i. *Avtomatika i Telemekhanika*, 21(9): 1240–1249, 1960.

[21] K.J. Åström and B. Wittenmark. *Adaptive Control*. Reading, MA: Dover Publications, Inc, 2008.

[22] H. Hjalmarsson, M. Gevers, and F. De Bruyne. For model-based control design, closed-loop identification gives better performance. *Automatica*, 32(12):1659–1673, 1996.

[23] H. Hjalmarsson, S. Gunnarsson, and M. Gevers. Optimality and sub-optimality of iterative identification and control design schemes. In *Proc. of the American Control Conference*, volume 4, pages 2559–2563. IEEE, 1995.

[24] M. Gevers, X. Bombois, B. Codrons, G. Scorletti, and B.D.O. Anderson. Model validation for control and controller validation in a prediction error identification framework – Part I: Theory. *Automatica*, 39(3):403–415, 2003.

[25] A. Garulli, A. Vicino, and G. Zappa. Conditional central algorithms for worst case set-membership identification and filtering. *IEEE Transactions on Automatic Control*, 45(1):14–23, 2000.

[26] L. Ljung. Estimating linear time-invariant models of nonlinear time-varying systems. *European Journal of Control*, 7(2):203–219, 2001.

[27] M. Barenthin and H. Hjalmarsson. Identification and control: joint input design and H_∞ state feedback with ellipsoidal parametric uncertainty via LMIs. *Automatica*, 44(2):543–551, 2008.

[28] B. Wahlberg, H. Hjalmarsson, and M. Annergren. On optimal input design in system identification for control. In *49th IEEE Conference on Decision and Control*, pages 5548–5553. IEEE, 2010.

[29] K.S. Narendra and L.S. Valavani. Direct and indirect model reference adaptive control. *Automatica*, 15(6):653–664, 1979.

[30] M.G. Safonov and T.C. Tsao. The unfalsified control concept and learning. *IEEE Transactions on Automatic Control*, 42(6):843–847, 1997.

[31] H. Hjalmarsson. Iterative feedback tuning: an overview. *International Journal of Adaptive Control and Signal Processing*, 16(5):373–395, 2002.

[32] A. Karimi, L. Mišković, and D. Bonvin. Iterative correlation-based controller tuning. *International Journal of Adaptive Control and Signal Processing*, 18(8):645–664, 2004.

[33] H. Hjalmarsson. Efficient tuning of linear multivariable controllers using iterative feedback tuning. *International Journal of Adaptive Control and Signal Processing*, 13(7):553–572, 1999.

[34] M.C. Campi, A. Lecchini, and S.M. Savaresi. Virtual reference feedback tuning: a direct method for the design of feedback controllers. *Automatica*, 38(8):1337–1346, 2002.

[35] A. Karimi, K. van Heusden, and D. Bonvin. Noniterative data-driven controller tuning using the correlation approach. In *Proc. of the European Control Conference*, pages 5189–5195, 2007.

[36] K. van Heusden, A. Karimi, and D. Bonvin. Data-driven model reference control with asymptotically guaranteed stability. *International Journal of Adaptive Control and Signal Processing*, 25(4):331–351, 2011.

[37] H. Hjalmarsson. From experiment design to closed-loop control. *Automatica*, 41(3):393–438, 2005.

[38] S. Formentin, K. van Heusden, and A. Karimi. A comparison of model-based and data-driven controller tuning. *International Journal of Adaptive Control and Signal Processing*, 28(10):882–897, 2014.

[39] H. Hjalmarsson. Control of nonlinear systems using iterative feedback tuning. In *Proc. of the American Control Conference*, volume 4, pages 2083–2087. IEEE, 1998.

[40] M.C. Campi and S.M. Savaresi. Direct nonlinear control design: the virtual reference feedback tuning (VRFT) approach. *IEEE Transactions on Automatic Control*, 51(1):14–27, 2006.

[41] S. Formentin, P. De Filippi, M. Corno, M. Tanelli, and S.M. Savaresi. Data-driven design of braking control systems. *IEEE Transactions on Control Systems Technology*, 21(1):186–193, 2013.

[42] S. Formentin, D. Piga, R. Tóth, and S.M. Savaresi. Direct learning of LPV controllers from data. *Automatica*, 65:98–110, 2016.

[43] C. Novara, L. Fagiano, and M. Milanese. Direct feedback control design for nonlinear systems. *Automatica*, 49(4):849–860, 2013.

[44] T.E. Passenbrunner, S. Formentin, S.M. Savaresi, and L. Del Re. Direct multivariable controller tuning for internal combustion engine test benches. *Control Engineering Practice*, 29:115–122, 2014.

[45] J.D. Rojas, X. Flores-Alsina, U. Jeppsson, and R. Vilanova. Application of multivariate virtual reference feedback tuning for wastewater treatment plant control. *Control Engineering Practice*, 20(5):499–510, 2012.

[46] S. Formentin, A.L. Cologni, F. Previdi, and S.M. Savaresi. A data-driven approach to control of batch processes with an application to a gravimetric blender. *IEEE Transactions on Industrial Electronics*, 61(11):6383–6390, 2014.

[47] S. Formentin, M. Corno, S.M. Savaresi, and L. Del Re. Direct data-driven control of linear time-delay systems. *Asian Journal of Control*, 14(3):652–663, 2012.

[48] S. Formentin, S.M. Savaresi, and L. Del Re. Non-iterative direct data-driven controller tuning for multivariable systems: theory and application. *IET Control Theory & Applications*, 6(9):1250–1257, 2012.

[49] S. Yin, H. Gao, and O. Kaynak. Data-driven control and process monitoring for industrial applications: Part I. *IEEE Transactions on Industrial Electronics*, 61(11):6356–6359, 2014.

[50] I. Markovsky. A missing data approach to data-driven filtering and control. *IEEE Transactions on Automatic Control*, 62:1972–1978, 2017.

[51] M. Milanese, F. Ruiz, and M. Taragna. Direct data-driven filter design for uncertain LTI systems with bounded noise. *Automatica*, 46:1773–1784, 2010.

[52] M. Milanese, F. Ruiz, and M. Taragna. From model-based to data-driven filter design. In: d'Onofrio A. (ed.), *Bounded Noises in Physics, Biology, and Engineering. Modeling and Simulation in Science, Engineering and Technology.* New York: Birkhäuser, 2013.

[53] Y. Wang, G. Ma, S.X. Ding, and C. Li. Subspace aided data-driven design of robust fault detection and isolation systems. *Automatica*, 47:2474–2480, 2011.

[54] S. Yin, X. Yang, and H.R. Karimi. Data-driven adaptive observer for fault diagnosis. *Mathematical Problems in Engineering*, 2012:1–21, 2012.

[55] C. Novara, F. Ruiz, and M. Milanese. Direct identification of optimal SM-LPV filters and application to vehicle yaw rate estimation. *IEEE Transactions on Control Systems Technology*, 19(1):5–17, 2011.

[56] F. Ruiz, C. Novara, and M. Milanese. Direct design from data of optimal filters for LPV systems. *Systems and Control Letters*, 59/1:1–8, 2010.

[57] I. Arsie, A. Cricchio, M. De Cesare, F. Lazzarini, C. Pianese, and M. Sorrentino. Neural network models for virtual sensing of NO_x emissions in automotive diesel engines with least square-based adaptation. *Control Engineering Practice*, 61:11–20, 2017.

[58] S.B. Sun, Y.Y. He, S.D. Zhou, and Z.J. Yue. A data-driven response virtual sensor technique with partial vibration measurements using convolutional neural network. *Sensors (Switzerland)*, 17:1–22, 2017.

[59] W. Yan, D. Tang, and Y. Lin. A data-driven soft sensor modeling method based on deep learning and its application. *IEEE Transactions on Industrial Electronics*, 64:4237–4245, 2017.

[60] M. Milanese, C. Novara, K. Hsu, and K. Poolla. The filter design from data (FD2) problem: nonlinear set membership approach. *Automatica*, 45(10): 2350–2357, 2009.

[61] T. Poggi, M. Rubagotti, A. Bemporad, and M. Storace. High-speed piecewise affine virtual sensors. *IEEE Transactions on Industrial Electronics*, 59:1228–1237, 2012.

[62] S.A. Iliyas, M. Elshafei, M.A. Habib, and A.A. Adeniran. RBF neural network inferential sensor for process emission monitoring. *Control Engineering Practice*, 21:962–970, 2013.

[63] N. Cesario, C. Muscio, M. Farina, P. Amato, and M. Lavorgna. Modelling the rate of heat release in common rail diesel engines: a soft computing approach. In *SAE Paper*, 2004.

[64] S.X. Ding, S. Yin, Y. Wang, Y. Wang, Y. Yang, and B. Ni. Data-driven design of observers and its applications. In *IFAC Proceedings Volumes (IFAC-PapersOnline)*, 2011.

[65] L. Fortuna, S. Graziani, A. Rizzo, and M.G. Xibilia. *Soft Sensors for Monitoring and Control of Industrial Processes*. London: Springer, 2007.

[66] M. Milanese, F. Ruiz, and M. Taragna. Direct data-driven filter design for uncertain LTI systems with bounded noise. *Automatica*, 46(11):1773–1784, 2010.

[67] D. Slišković, R. Grbić, and Ž. Hocenski. Methods for plant data-based process modeling in soft-sensor development. *Automatika*, 52:306–318, 2011.

[68] V. Aran and M. Unel. Data driven disturbance observer design and control for diesel engine airpath. In *2017 Asian Control Conference, ASCC 2017*, 2018.

[69] P. Borodani. Virtual sensors: an original approach for dynamic variables estimation in automotive control systems. In *Proc. of the 9th International Symposium on Advanced Vehicle Control (AVEC)*, Kobe, Japan, 2008.

[70] I.V. de Bessa, R.M. Palhares, M.F.S.V. D'Angelo, and J.E. Chaves Filho. Data-driven fault detection and isolation scheme for a wind turbine benchmark. *Renewable Energy*, 87:634–645, 2016.

[71] C. Hametner, S. Jakubek, and W. Prochazka. Data-driven design of a cascaded observer for battery state of health estimation. *IEEE Transactions on Industry Applications*, 54:6258–6266, 2018.

[72] G. Heredia and A. Ollero. Virtual sensor for failure detection, identification and recovery in the transition phase of a morphing aircraft. *Sensors*, 10:2188–2201, 2010.

[73] P. Kadlec, B. Gabrys, and S. Strandt. Data-driven soft sensors in the process industry. *Computers & Chemical Engineering*, 33(4):795–814, 2009.

[74] M. Milanese, C. Novara, and I. Gerlero. Robust estimation of vehicle sideslip angle from variables measured by ESC system. In *15th Stuttgart International Symposium on Automotive and Engine Technology*, Stuttgart, Germany, 2015.

[75] E. Morgul, H. Yang, A. Kurkcu, *et al.* Virtual sensors. *Transportation Research Record: Journal of the Transportation Research Board*, 2014:27–37, 2014.

[76] P.F. Odgaard, B. Lin, and S.B. Jorgensen. Observer and data-driven-model-based fault detection in power plant coal mills. *IEEE Transactions on Energy Conversion*, 23:659–668, 2008.

[77] F. Ruiz, M. Taragna, and M. Milanese. Direct data-driven filter design for automotive controlled suspensions. In *European Control Conference ECC09*, Budapest, Hungary, 2009.

[78] P.A. Samara, J.S. Sakellariou, G.N. Fouskitakis, J.D. Hios, and S.D. Fassois. Aircraft virtual sensor design via a time-dependent functional pooling NARX methodology. *Aerospace Science and Technology*, 29:114–124, 2013.

[79] A. Torgashov and K. Zmeu. Identification of nonlinear soft sensor models of industrial distillation process under uncertainty. *IFAC-PapersOnLine*, 48: 45–50, 2015.

Part I

Data-driven modeling

Chapter 2

A kernel-based approach to supervised nonparametric identification of Wiener systems

Fei Xiong[1], Yongfang Cheng[2], Octavia Camps[3], Mario Sznaier[3], and Constantino Lagoa[4]

This chapter addresses the problem of nonparametric identification of Wiener systems using a Kernel-based approach. Salient features of the proposed framework are its ability to exploit both positive and negative samples, and the fact that it does not require prior knowledge of the dimension of the output of the linear subsystem. Thus, it can be considered as a generalization to dynamical systems of kernel-based nonlinear manifold embedding methods recently developed in the machine-learning field. The main result of the chapter shows that while in principle, the proposed approach results in a non-convex problem, a tractable convex relaxation can be obtained by using a combination of polynomial optimization and rank-minimization techniques. The main advantage of the proposed algorithm stems from the fact that, since it is based on kernel ideas, it uses scalar inner products of the observed data, rather than the data itself. Hence, it can comfortably handle cases involving systems with high dimensional outputs. A practical scenario where such situation arises is activity classification from video data, since here each data point is a frame in a video sequence, and hence its dimension is typically $\mathcal{O}(10^3)$ even when using low resolution videos.

2.1 Introduction and motivation

During the past decade, a large research effort has been directed towards the problem of identifying Wiener systems, consisting of the cascade of a linear time invariant (LTI) plant and a memoryless nonlinearity (see Figure 2.1), from experimental data, under various assumptions on the available a priori information. This effort was motivated in part by the fact that these models arise naturally in a large span of practical applications, ranging from communications to biology and computer vision [1–3]. In addition, Wiener systems can be considered "poor man's" nonlinear systems, providing tractable (local) approximations to more general models that, by

[1]Information Sciences Institute, USC, Arlington, VA, United States
[2]Bosch Research and Technology Center North America, Sunnyvale, CA, United States
[3]Department of Electrical and Computer Engineering, Northeastern University, Boston, MA, United States
[4]Department of Electrical Engineering, Penn State University, University Park, PA, United States

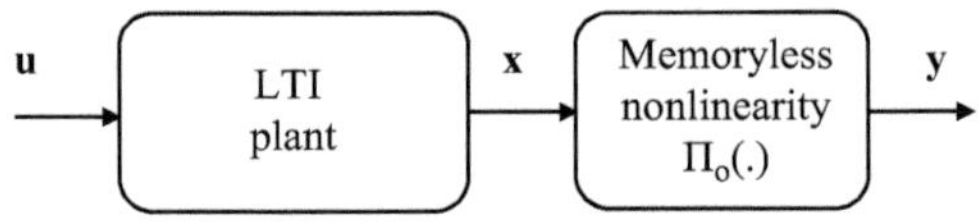

Figure 2.1 Wiener system model

virtue of incorporating a nonlinearity, are less conservative than piece-wise linear approximations.

Existing Wiener identification techniques can be roughly classified into statistical (e.g., [4–12]) and set membership ([13,14] and references therein), depending on whether the approach pursued is stochastic or worst case based and the corresponding assumptions on the available a priori information. A common feature of these techniques is that the dimension n_x of $\mathbf{x}$, the output of the linear portion of the system, is assumed to be known. While this assumption holds in many scenarios, it can be overly restrictive in others. For instance, in computer vision applications such as target tracking or activity recognition [15,16], pixels in an image (the vector $\mathbf{y}$) often exhibit a large degree of correlation and thus can be explained in terms of a smaller set of independent latent parameters $\mathbf{x}$. This leads naturally to Wiener models, where video sequences are represented by a vector $\mathbf{x(t)}$ evolving in a suitable manifold where the dynamics are linear, followed by a nonlinear transformation $\mathbf{x} \rightarrow \mathbf{y}$ to reconstruct the entire frame. In this scenario, the dimension of $\mathbf{x}$ must also be identified from the experimental data.

For the case of static data, the problem of finding low-dimensional nonlinear projections of high-dimensional correlated data (nonlinear manifold embedding) has be thoroughly studied in machine learning, leading to a number of computationally efficient methods such as locally linear embeddings [17], semidefinite embedding [18], global coordination of local linear models [19], and dynamic global coordinate model [20]. However, these methods generally do not perform well when dealing with dynamical data, e.g., data generated by an underlying dynamical system. This can be traced to the fact that in general, machine-learning motivated methods do not take advantage of the dynamical information, encapsulated in the temporal ordering of the data, leading to embeddings that overestimate the dimension of the embedded data $\mathbf{x}$, or of the dynamics underlying its time evolution [16]. These issues were partially addressed in [16], which proposed an embedding technique for dynamic data based upon the minimization of a combination of the dimension of the embedding manifold and the order of the underlying dynamics. However, since this paper was not intended as a tool for systems identification, no identifiability issues were taken into consideration when searching for the embedding manifold.

In this chapter, motivated by the work in [16], we seek a rapprochement between the systems identification and machine-learning frameworks described above. Our goal is, starting from experimental input output data and some minimal a priori assumptions about the nonlinearity, to identify a Wiener model that explains the observed data. The main idea is to, given input/output pairs $\{\mathbf{u}(t), \mathbf{y}(t)\}$, identify an embedded latent vector $\mathbf{x}(t)$ such that the dynamics relating $\{\mathbf{u}, \mathbf{x}\}$ are linear, low order, and consistent with the a priori information. Once the vector $\mathbf{x}(t)$ and associated dynamics are identified, the nonlinearity can be found by simply finding a mapping $\mathbf{x}(t) \rightarrow \mathbf{y}(t)$, for instance by solving a polynomial optimization problem. The main result of the chapter shows that the problem of jointly finding the embedding manifold and the linear dynamics can be recast as a constrained polynomial optimization and solved using recently introduced moments-based techniques. Further, this optimization problem has an inherent sparsity structure (it exhibits the so-called running intersection property), that can be exploited to substantially reduce the computational burden.

A key feature of the proposed identification framework, inspired by machine learning, is the ability to use both positive and negative samples, that is, experimental data generated both by the system to be identified and by other systems. This capability, to the best of our knowledge is not available in existing identification methods, improves robustness, specially in cases where the goal is to use the identified models to classify data, and the different classes are characterized by relatively similar models. Examples of this situation are activity classification, where models corresponding to running and skipping are close, but labeled sample clips of these activities are available for training, and in tracking where often a segmentation separating the target of interest from other targets and the background is known. Further, motivated by kernel-based methods and the papers [16,21], the proposed algorithm uses the inner products $\langle \mathbf{y}_i, \mathbf{y}_j \rangle$, rather than the raw data $\mathbf{y}$, resulting in substantial dimensionality reduction in the matrices involved. Thus, it can comfortably handle large-dimensional data, such as the one common in the motivating computer vision applications. The main result of the paper shows that in this context, the problem of jointly finding the embedding manifold and the linear dynamics can be recast into a convex optimization over a semi-algebraic set (a set defined by a collection of polynomial inequalities). In turn, the use of recent results from polynomial optimization allows for relaxing this problem to a tractable convex optimization.

The chapter is organized as follows: Section 2.2 introduces the notation used in the chapter, and recalls, for ease of reference, some results on polynomial optimization. Section 2.3 formally states the problem under consideration. Section 2.4 presents the main result of the chapter: a convex optimization-based algorithm for Wiener systems identification. This result is obtained by first formulating the problem as a constrained polynomial optimization and then using the results in Section 2.2 to obtain an equivalent convex optimization. Section 2.5 illustrates these results with two examples, one academic and one practical: activity classification from video data. Finally, Section 2.6 presents some concluding remarks and points out to directions for further research. A preliminary, shorter version of this chapter appeared in [22].

2.2 Preliminaries

For ease of reference, in this section, we summarize the notation used in the chapter and recall some results on sparse polynomial optimization that plays a key role in establishing the main result of this paper.

2.2.1 Notation and definitions

$\mathbb{R}, \mathbb{N}$	set of real number and nonnegative integers
$\mathbf{x}, \mathbf{M}$	a vector in $\mathbb{R}^n$ (matrix in $\mathbb{R}^{n\times m}$)
$\mathbf{M} \succeq \mathbf{N}$	the matrix $\mathbf{M} - \mathbf{N}$ is positive semidefinite.
$\mathbf{H}_y^{m,n}$	Hankel matrix associated with a vector sequence $\mathbf{y}_t$:

$$\mathbf{H}_y^{m,n} \doteq \begin{bmatrix} \mathbf{y}_0 & \mathbf{y}_1 \cdots & \mathbf{y}_m \\ \mathbf{y}_1 & \mathbf{y}_2 \cdots & \mathbf{y}_m \\ \vdots & \vdots \ddots & \vdots \\ \mathbf{y}_n & \mathbf{y}_{n+1} \cdots & \mathbf{y}_{m+n-1} \end{bmatrix}$$

In the sequel the indexes m, n may be omitted when clear from the context.

2.2.2 Solving polynomial optimization problems via convex optimization

A large class of problems arising in systems identification, including the one of interest in this chapter, leads to a constrained polynomial optimization of the form:

$$p^* = \min_{\mathbf{x}\in\mathscr{K}} p(\mathbf{x}) \doteq \sum_{\alpha} p_\alpha \mathbf{x}^\alpha \tag{2.1}$$

where $\mathbf{x}^\alpha = x_1^{\alpha_1} x_2^{\alpha_2} \cdots x_n^{\alpha_n}$. Here $\mathscr{K} \subset \mathbb{R}^n$ denotes a compact semi-algebraic set defined by polynomial inequalities of the form $g_k(\mathbf{x}) \geq 0$, $k = 1, \ldots, r$, where $g_k(\mathbf{x}) = \sum_\beta g_{k,\beta^{(l)}} \mathbf{x}^{\beta^{(l)}}$ are polynomials with coefficients $g_{k,\beta^{(l)}}$ and degree δ_k. Problem (2.1) is a very challenging, non-convex optimization. However, as shown in [23], (2.1) is equivalent to the following convex, infinite-dimensional optimization over $\mathscr{P}(\mathscr{K})$, the set of probability measures supported in $\mathscr{K}$:

$$p^*_{\mathscr{K}} = \min_{\mu\in\mathscr{P}(\mathscr{K})} \int p(\mathbf{x})\mu(d\mathbf{x}) = \min_{\mu} \sum_{\alpha} p_\alpha \mathbf{m}_\alpha \tag{2.2}$$

where $\mathbf{m}_\alpha \doteq \int_K \mathbf{x}^\alpha \mu(d\mathbf{x})$ denotes the αth moment with respect to μ. Problem (2.2) involves minimizing a linear function of the variables $\mathbf{m}_\alpha$, subject to the constraint that $\mathbf{m}_\alpha$ must be the moments of some probability measure $\mu \in \mathscr{P}(\mathscr{K})$. In turn, as shown in [24], this (convex) constraint is equivalent to a set of positive semidefinite constraints of the form $\mathbf{M}(\mathbf{m}_\alpha) \succeq 0$ and $\mathbf{L}(g_k \mathbf{m}_\alpha) \succeq 0$, where the entries of $\mathbf{M}$ (the moment matrix) and $\mathbf{L}$ (the localizing matrices) are given by

$$\begin{aligned} \mathbf{M}(\mathbf{m})(i,j) &= m_{\alpha^{(i)}+\alpha^{(j)}}, \\ \mathbf{L}(g_k\mathbf{m})(i,j) &= \sum_\beta g_{k,\beta^{(l)}} m_{\beta^{(l)}+\alpha^{(i)}+\alpha^{(j)}} \end{aligned} \tag{2.3}$$

In principle, these constraints involve infinite-dimensional matrices; however, finite-dimensional relaxations can be obtained considering truncated versions of these matrices, $\mathbf{M}_d$ and $\mathbf{L}_d$, containing moments of order up to $2d$. Since the LMIs involving these truncated matrices are necessary conditions for the existence of probability measure $\mu \in \mathcal{K}$, problem (2.1) can be reduced to a (monotonically convergent) sequence of convex semidefinite programs [23] of the form

$$\begin{aligned} p_d^* = \min_{\mathbf{m}} & \sum_{\alpha} p_\alpha m_\alpha \\ \text{s.t.} \quad & \mathbf{M}_d(\mathbf{m}) \succeq 0, \\ & \mathbf{L}_d(g_k \mathbf{m}) \succeq 0, k = 1, \ldots, r \end{aligned} \tag{2.4}$$

Note that the lowest value of d where $\mathbf{M}$ contains all the moments of the monomials appearing in (2.1) is $d_{\min} = \max\{\lceil n/2 \rceil, \delta_{\max}\}$, where n is the degree of p and $\delta_{\max} \doteq \max_k \lceil \delta_k/2 \rceil$. Thus, d above should satisfy $d \geq d_{\min}$. As d increases, $p_d^* \uparrow p_{\mathcal{K}}^*$ monotonically, and, if for some d, the *flat extension* property

$$\mathbf{rank}[\mathbf{M}_d(\mathbf{m})] = \mathbf{rank}[\mathbf{M}_{d-\delta_{\max}}(\mathbf{m})] \quad . \tag{2.5}$$

holds, then the corresponding finite-dimensional LMIs are sufficient for the existence of a measure μ supported on $\mathcal{K}$ (see Theorem 3.11 in [25]) and hence $p_d^* = p_{\mathcal{K}}^*$.

2.2.3 Exploiting sparsity in polynomial optimization

As we will show in the sequel, for the problem of interest here, both the objective and constraints in (2.1) are endowed with a sparse structure that can be used to reduce the computational complexity entailed in solving problem (2.4). Specifically, we will consider problems exhibiting the so-called running intersection property defined below.

Definition 2.1. *[26]: Suppose that the polynomial p can be partitioned into $p = p_1 + \cdots + p_r$ such that each p_k and that the constraints g_k that define the set $\mathcal{K}$ contain only variables indexed by elements of some subset $I_k \subset \{1, \ldots, n\}$. Then problem (2.1) is said to satisfy the running intersection property if there exists a reordering $I_{k'}$ of I_k such that for every $k' = 1, \ldots, r-1$:*

$$I_{k'+1} \cap \cup_{j=1}^{k'} I_j \subset I_s \quad \textit{for some } s \leq k' \tag{2.6}$$

As shown in [26], if (2.1) satisfies the *running intersection property*, the hierarchy (2.4) can be replaced by one where the semidefinite constraints have a smaller size. To this effect, partition the objective function $\{p_j\}_{j=1}^r$ according to the sets $\{I_k\}$ and consider the problem:

$$\begin{aligned} p_d^* = \min_{\mathbf{m}} & \sum_{j=1}^{l} \sum_{\alpha(j)} p_{j,\alpha(j)} m_{\alpha(j)} \\ \text{s.t.} \quad & \mathbf{M}_d(\mathbf{m}_{I_k}) \succeq 0, k = 1, \ldots, r \\ & \mathbf{L}_d(g_k \mathbf{m}_{I_k}) \succeq 0, k = 1, \ldots, r \end{aligned} \tag{2.7}$$

where $p_{j,\alpha(j)}$ is the coefficient of the $\boldsymbol{\alpha}(j)$th monomial in the polynomial p_j, $\mathbf{M}_d(\mathbf{m}_{I_k})$ denotes the moment matrix and $\mathbf{L}_d(g_k \mathbf{m}_{I_k})$ the localizing matrix for the subset of variables in I_k. Then, as shown in [26], $p_d^* \uparrow p^*$. Recall that for generic polynomials and constraints in n variables, a dth order relaxation requires considering moments and localization matrices containing $O(n^{2d})$ entries. On the other hand, the matrices in (2.7) contain variables only in I_k (i.e., each has at most $O(\kappa^{2d})$entries), where κ is the maximum cardinality of I_k. Thus, when $\kappa \ll n$ (as is the case in this chapter), exploiting the running intersection property structure substantially reduces the number of variables in the optimization (and hence the computational complexity, since the complexity of conventional LMI solvers scales as (number of variables)^3).

2.3 Problem statement

Consider the interconnection of an LTI system with a static nonlinearity shown in Figure 2.1, where $\mathbf{u} \in \mathbb{R}^{n_u}$ and $\mathbf{y} \in \mathbb{R}^{n_y}$ denote a known input/output pair and where the internal signal $\mathbf{x} \in \mathbb{R}^{n_x}$ is unmeasurable. In this chapter, we will consider a scenario where a known, fixed input $\mathbf{u}$ is repeatedly applied both to the system to be identified and to other systems (with dynamics that are in some sense close to the system of interest), generating output sequences $\{\mathbf{y}_t^{i,+}\}$, $i = 1, \ldots, n_+$ and $\{\mathbf{y}_t^{i,-}\}$, $i = 1, \ldots, n_-$ corresponding to different initial conditions and noise realizations. This situation is motivated by applications such as activity recognition from temporal data, where it is of interest to obtain models that can separate activities that could potentially be confused, such as running or skipping, and where the data can be assumed to be the impulse response of an unknown system [27]. In this context, the goal of this chapter is to solve the following problem:

Problem 2.1. *Given*

1. n_+ *output sequences* $\left\{\mathbf{y}_t^{(i,+)}\right\}$, $i = 1, \ldots, n_+$ *generated by applying the input* $\mathbf{u}$ *to the system under consideration for different initial conditions.*
2. n_- *output sequences* $\left\{\mathbf{y}_t^{(i,-)}\right\}$, $i = 1, \ldots, n_-$ *generated by applying* $\mathbf{u}$ *to systems known to be different from the one to be identified.*
3. *Some set-membership a priori information on the nonlinearity, that is, the nonlinearity is assumed to belong to some known family* $\mathcal{NL}$.[1]

find n_x *and the coefficients of an auto-regressive model of the form*

$$\mathbf{x}_t^{(i,+)} = \sum_{k=1}^{n_a} a_k \mathbf{x}_{t-k}^{(i,+)} + \sum_{k=1}^{n_b} \mathbf{B}_k \mathbf{u}_{t-k} \tag{2.8}$$

such that $\mathbf{y}_t^{i,+} = \Pi(\mathbf{x}_t^{i,+})$ *for some admissible nonlinearity* $\Pi(.) : \mathbb{R}^{n_x} \to \mathbb{R}^{n_y}$, $\Pi(.) \in \mathcal{NL}$.

The first step towards solving the problem above, is to recast it into a polynomial feasibility form that can be solved via convex optimization by exploiting the techniques outlined in Section 2.2.2. In order to obtain computationally tractable problems, we will assume (without loss of generality) that the input sequence $\mathbf{u}$ can be modeled as the impulse response of an n_u order exo-system $\mathcal{G}_u$ (see Figure 2.2), and find and embedding where the combined (plant/exo-system) dynamics are LTI.

[1]For example, the nonlinearity is known to be locally invertible almost everywhere, with known upper bounds on the Lipschitz constant of its local inverse.

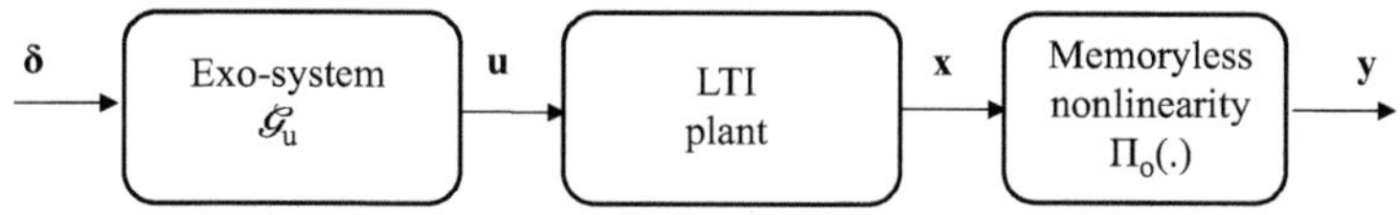

Figure 2.2 *Wiener system model incorporating the exo-system that generates the input used for identification*

Theorem 2.1. *Problem 2.1 is equivalent to finding scalars n_x and n, vector sequences $\mathbf{x}^{i,+} \in \mathbb{R}^{n_x}$ and a vector $\mathbf{w} \in \mathbb{R}^{n+1}$ such that*

$$\begin{aligned} &\|\mathbf{H}_{\mathbf{x}^{(i,+)}}\mathbf{w}\|_2^2 = 0 \; i = 1, \dots, n_+ \\ &\|\mathbf{H}_{\mathbf{x}^{(j,-)}}\mathbf{w}\|_2^2 > 0 \; j = 1, \dots, n_- \\ &\mathbf{y}^{(i,+)} = \Pi\left(\mathbf{x}^{(i,+)}\right); \mathbf{y}^{(j,-)} = \Pi\left(\mathbf{x}^{(j,-)}\right) \textit{ for some } \Pi \in \mathcal{NL} \\ &\qquad\qquad i = 1, \dots, n_+, j = 1, \dots, n_-. \end{aligned} \tag{2.9}$$

where $\mathbf{H}_{\mathbf{x}^{(i)}}$ denotes the Hankel matrix associated with an output sequence $\mathbf{x}^{(i)}$.

Proof. From Theorem 2 in [28], a given output sequence $\mathbf{x}^{(i,+)}$ of the combined system shown in Figure 2.2 satisfies a model of the form

$$\mathbf{x}_t^{(i,+)} = \sum_{k=1}^{n} \alpha_{n+1-k}\mathbf{x}_{t-k}^{(i,+)} \tag{2.10}$$

where $n \doteq n_a + n_u$ if and only if rank $(\mathbf{H}_{\mathbf{x}^{(i,+)}}) = n$. The first two equations in (2.9) force all of the matrices $\mathbf{H}_{\mathbf{x}^{(i,+)}}$ associated with the system to be identified (positive labels) to have the same right null space (spanned by $\mathbf{w}$) and hence satisfy a model of the form (2.10) with $\mathbf{w}_i = \alpha_i$, $i = 1, \dots, n$ and $\mathbf{w}_{n+1} = 1$, while, at the same time, guaranteeing that those associated with other systems (negative labels) do not share it. □

The inequality in the problem above guarantees separation between the system to be identified and others known to be different from it, thus avoiding trivial solutions. Once suitable sequences $\mathbf{x}^{(i,+)}$ have been found, the parameters of the linear model can be determined for instance by applying the multi-run identification algorithm proposed in [29] to the sequences $\left(\mathbf{u}, \mathbf{x}^{(i,+)}\right)$. Finally, the nonlinearity can be found for instance using nonlinear regression methods.

As mentioned earlier, Problem (2.9) is a polynomial feasibility problem that can be solved using the techniques described in Section 2.2.2. However, while this approach works well for cases where n_y and n_x are relatively small, due to the poor scaling of the underlying convex relaxations, it becomes unpractical for even moderately large-sized output vectors. To avoid this difficulty, in the next section, we present a computationally efficient algorithm, motivated in part by classification techniques used in machine learning.

2.4 Maximum margin Hankel classifiers

Next, we present a reformulation of (2.9) that can handle large data sets and is less sensitive to noise than the original formulation. The main idea is to replace (2.9) with

$$\begin{aligned}
&\|\mathbf{H}_{\mathbf{x}^{(i,+)}}\mathbf{w}\|_2^2 \le \xi_i^+,\ \xi_i^+ > 0,\ i = 1,\dots,n_+ \\
&\|\mathbf{H}_{\mathbf{x}^{(j,-)}}\mathbf{w}\|_2^2 + \xi_j^- \ge 1,\ \xi_j^- > 0,\ i = 1,\dots,n_- \\
&\mathbf{y}^{(i,+)} = \Pi\left(\mathbf{x}^{(i,+)}\right);\ \mathbf{y}^{(j,-)} = \Pi\left(\mathbf{x}^{(j,-)}\right) \text{ for some } \Pi \in \mathcal{NL} \\
&\qquad\qquad\qquad\qquad i = 1,\dots,n_+, j = 1,\dots,n_-.
\end{aligned} \tag{2.11}$$

When compared against (2.9), in the reformulation above, we seek a vector (model) that has small fitting error for the positive samples and a large one for the negative samples. Note in passing that the first inequality in (2.11) can be interpreted as allowing an error-in-model (or noise) term with magnitude ξ_i in the left hand side of (2.10). Motivated by maximum margin classification ideas, we will seek to minimize a combination of $\|\mathbf{w}\|_2$ and $\|\boldsymbol{\xi}\|_1$[2] leading to the following optimization problem:

Problem 2.2. *Let* $\mathbf{G}_i^+ \doteq \mathbf{H}^T_{\mathbf{x}^{(i,+)}}\mathbf{H}_{\mathbf{x}^{(i,+)}}$ *and* $\mathbf{G}_i^- \doteq \mathbf{H}^T_{\mathbf{x}^{(i,-)}}\mathbf{H}_{\mathbf{x}^{(i,-)}}$. *Solve*

$$\begin{aligned}
\min_{\mathbf{x},\mathbf{w},\xi^+,\xi^-} \quad & \tfrac{1}{2}\|\mathbf{w}\|_2^2 + C\left(\textstyle\sum \xi_i^+ + \sum \xi_i^-\right) \\
\textit{subject to} \quad & \mathbf{w}^T\mathbf{G}_i^+\mathbf{w} \le \xi_i^+, \quad i = 1,2,\dots,n_+ \\
& \mathbf{w}^T\mathbf{G}_i^-\mathbf{w} + \xi_i^- \ge 1, \quad i = 1,2,\dots,n_- \\
& \mathbf{y}^{(i,+)} = \Pi\left(\mathbf{x}^{(i,+)}\right);\ \mathbf{y}^{(j,-)} = \Pi\left(\mathbf{x}^{(j,-)}\right) \textit{ for some } \Pi \in \mathcal{NL} \\
& \qquad i = 1,\dots,n_+, j = 1,\dots,n_-.
\end{aligned} \tag{2.12}$$

Note that while the problem above is not convex, all of the constraints are third-degree multivariate polynomials in the unknowns $\mathbf{G}, \mathbf{w}, \xi$. This fact is key to obtaining computationally tractable relaxations, with optimality certificates.

Remark 2.1. *It is worth emphasizing that in the formulation above, the entries of* $\mathbf{G}_i$ *depend only on the inner products* $\left\langle \mathbf{x}_k^{(i)}, \mathbf{x}_j^{(i)} \right\rangle$ *rather than on the actual data* $\mathbf{x}^{(i)}$. *This fact leads to a substantial complexity reduction when compared against* (2.9) *for cases where* n_x *is not small, since the corresponding polynomials involve far fewer variables, and hence the associated moment matrices are substantially smaller.*

So far, we have not made use of any a priori information on the nonlinearity $\Pi(.)$. Typically, a weak assumption is to impose constraints on the (local) Lipschitz constant of its local inverse, that is, impose that

$$\|\mathbf{x}_r - \mathbf{x}_s\|_2^2 \le L_i\|\mathbf{y}_r - \mathbf{y}_s\|_2^2, \quad \forall (r,s) \in \mathcal{N}_i \tag{2.13}$$

[2] Recall that $\|w\|_2$ is a measure of the margin between classes. Minimizing the ℓ_1 norm of $\boldsymbol{\xi}$ allows for the existence of mislabeled samples, but attempts to minimize the number of misclassifications.

for some suitably defined neighborhoods $\mathcal{N}_i$. Consider now the Gram (or Kernel) matrix defined by its submatrices

$$\mathbf{K}_j^{i,l} = \begin{bmatrix} \mathbf{z}_j^{(i)T}\mathbf{z}_j^{(l)} & \mathbf{z}_j^{(i)T}\mathbf{z}_{j+1}^{(l)} & \cdots & \mathbf{z}_j^{(i)T}\mathbf{z}_{j+c}^{(l)} \\ \mathbf{z}_{j+1}^{(i)T}\mathbf{z}_j^{(l)} & \mathbf{z}_{j+1}^{(i)T}\mathbf{z}_{j+1}^{(l)} & \cdots & \mathbf{z}_{j+1}^{(i)T}\mathbf{z}_{j+c}^{(l)} \\ \vdots & \vdots & \ddots & \vdots \\ \mathbf{z}_{j+c}^{(i)T}\mathbf{z}_j^{(l)} & \mathbf{z}_{j+c}^{(i)T}\mathbf{z}_{j+2}^{(l)} & \cdots & \mathbf{z}_{j+c}^{(i)T}\mathbf{z}_{j+c}^{(l)} \end{bmatrix} \tag{2.14}$$

where

$$\mathbf{z}^{(i)} = \begin{cases} \mathbf{x}^{(i,+)} & \text{if } 1 \leq i \leq n_+ \\ \mathbf{x}^{(i-n_+,-)} & \text{if } n_+ + 1 \leq i \leq n_+ + n_- \end{cases}$$

and c is a design parameter. It can be easily shown that

$$\mathbf{G}_i^+ = \mathbf{H}_{\mathbf{x}^{(i,+)}}^T \mathbf{H}_{\mathbf{x}^{(i,+)}} = \sum_{j=1}^{T-c+1} \mathbf{K}_j^{i,i},$$

$$\mathbf{G}_i^- = \mathbf{H}_{\mathbf{x}^{(i,-)}}^T \mathbf{H}_{\mathbf{x}^{(i,-)}} = \sum_{j=1}^{T-c+1} \mathbf{K}_j^{i-n_+,i-n_+}.$$

Thus, (2.12) together with the additional constraint (2.13) can be rewritten in terms of **G** and **K** as

$$\begin{aligned} &\min_{\mathbf{K},\mathbf{w},\xi} \tfrac{1}{2}\|\mathbf{w}\|_2^2 + C\left(\sum \xi_i^+ + \sum \xi_i^-\right) - \lambda\,\mathrm{Tr}\,(\mathbf{K}) \\ &\text{subject to } \mathbf{w}^T\mathbf{G}_i^+\mathbf{w} \leq \xi_i^+ \quad, i = 1,\ldots,n_+ \\ &\mathbf{w}^T\mathbf{G}_i^-\mathbf{w} + \xi_i^- \geq 1, \quad i = 1,\ldots,n_- \\ &\mathbf{G}_i^+ = \sum_{j=1}^{T-c+1} \mathbf{K}_j^{i,i}, \quad i = 1,\ldots,n_+ \\ &\mathbf{G}_i^- = \sum_{j=1}^{T-c+1} \mathbf{K}_j^{i-n_+,i-n_+}, \quad i = 1,\ldots,n_-, \\ &k_{ii} + k_{jj} - 2k_{ij} \leq L_i\|\mathbf{y}_i - \mathbf{y}_j\|^2, \quad \forall (i,j) \in \mathcal{N}_i \\ &\qquad\qquad \text{for corresponding measurements } \mathbf{y}_i, \mathbf{y}_j \\ &\mathbf{K} \succeq 0, \quad \sum k_{ij} = 0 \end{aligned} \tag{2.15}$$

where the second term in the objective $(-\lambda\,\mathrm{Tr}(\mathbf{K}))$ favors lower dimensional embeddings (see [16,18]), while the constraint $\sum k_{ij} = 0$ enforces translational invariance of the embedding. As in (2.11), the problem above is a polynomial optimization problem involving at most third-order (multivariate) polynomials.

2.4.1 Further computational complexity reduction

When compared against (2.9), (2.11) entails a substantial complexity reduction due to the use of the Kernel matrix **K**. Still, its computational complexity grows quickly with the number of data points, since the number of free variables in **K** is equal to the square of the number of data points. To mitigate this growth, we will proceed along the lines of the approach proposed in [30] and parameterize $\mathbf{K}$ in terms of a much smaller matrix $\tilde{\mathbf{K}}$. Specifically, we will consider matrices of the form

$$\mathbf{K} = \mathbf{Q}\tilde{\mathbf{K}}\mathbf{Q}^T \tag{2.16}$$

where $\mathbf{Q}$ is a fixed matrix that encapsulates the local geometry of the problem. The reasoning behind (2.16) is to express each data point $\mathbf{y}$ as a linear combination of a small number of "landmarks" $\tilde{\mathbf{y}}$, that is, $\mathbf{y}_i \approx \sum_j \mathbf{Q}_{i,j}\tilde{\mathbf{y}}_j$ and to assume[3] that *locally* the same description applies to the manifold data $\mathbf{x}$. This leads to an expression of the form (2.16) for the products $\mathbf{x}_i^T\mathbf{x}_j$, where the elements of $\tilde{\mathbf{K}}$ correspond to the inner products of the embedded landmarks. As shown in [30], the value of $\mathbf{Q}$ that minimizes the ℓ_2 norm of the reconstruction error is obtained by selecting for each measurement $\mathbf{y}_i$ up to k nearest neighbors and solving the following optimization problem in $\mathbf{P}$:

$$\begin{aligned}
&\min_{\mathbf{P}} \sum_i \|\mathbf{y}_i - \sum_j \mathbf{P}_{ij}\mathbf{y}_j\|_2 \quad \text{subject to:}\\
&\sum_j \mathbf{P}_{ij} = \mathbf{1} \quad \text{and}\\
&\mathbf{P}_{ij} = 0 \quad \text{if } \mathbf{y}_j \text{ is not a } k \text{ nearest neighbor of } \mathbf{y}_i
\end{aligned} \tag{2.17}$$

The corresponding $\mathbf{Q}$ is given by

$$\mathbf{Q} = \begin{bmatrix} \mathbf{I} \\ -(\mathbf{\Phi}_{22})^{-1}\mathbf{\Phi}_{21} \end{bmatrix} \tag{2.18}$$

where $\mathbf{\Phi}_{ij}$ denotes the blocks of $\mathbf{\Phi} \doteq (\mathbf{I} - \mathbf{P})^T(\mathbf{I} - \mathbf{P})$, partitioned so that its top left $n_\ell \times n_\ell$ block corresponds to the landmarks, and where we have assumed, by reordering points if needed, that the landmarks correspond to the first n_ℓ elements of the set $\{\mathbf{y}_i\}$.

Rewriting the objective function and constraints in (2.15) in terms of $\tilde{\mathbf{K}}$ and $\mathbf{W} \doteq \mathbf{w}^T\mathbf{w}$ and dropping the constraint that $\text{rank}(\mathbf{W}) = 1$ leads to the following relaxation of (2.15):

$$\begin{aligned}
&\min_{\tilde{\mathbf{K}},\mathbf{W},\xi} \tfrac{1}{2}\,\text{Tr}(\mathbf{W}) + C\left(\sum \xi_i^+ + \sum \xi_i^-\right) - \lambda\,\text{Tr}(\mathbf{Q}\tilde{\mathbf{K}}\mathbf{Q}^T)\\
&\text{subject to } \text{Tr}(\mathbf{W}\mathbf{G}_i^+) \le \xi_i^+, \quad i = 1, \ldots, n_+\\
&\text{Tr}(\mathbf{W}\mathbf{G}_i^-) + \xi_i^- \ge 1, \quad i = 1, \ldots, n_-\\
&\mathbf{W} \succeq 0, \ \tilde{\mathbf{K}} \succeq 0\\
&\mathbf{K} = \mathbf{Q}\tilde{\mathbf{K}}\mathbf{Q}^T, \sum k_{ij} = 0\\
&\mathbf{G}_i^+ = \sum_{j=1}^{T-c+1} \mathbf{K}_j^{i,i}, \quad i = 1, \ldots, n_+\\
&\mathbf{G}_i^- = \sum_{j=1}^{T-c+1} \mathbf{K}_j^{i-n_+,i-n_+}, \quad i = 1, \ldots, n_-,\\
&k_{ii} + k_{jj} - 2k_{ij} \le (1+\varepsilon)\|\mathbf{y}_i - \mathbf{y}_j\|^2, \quad \forall \eta_{ij} = 1\\
&\qquad\qquad \text{for corresponding measurements } \mathbf{y}_i, \mathbf{y}_j
\end{aligned} \tag{2.19}$$

[3]This assumption simply corresponds to assuming that $\Pi^{-1}(.)$ is locally linearizable.

Algorithm 2.1: Wiener Id via Hankel classifiers

1. **Input:** C, λ, ε, $n_{neighbor}$ number of spatial neighbors and n_ℓ number of landmarks.
2. **Initialization:** Compute $\mathbf{Q}$, set $\mathbf{M}_N^0 = \mathbf{I}$, $\delta^{(0)} = 0$.
3. **Solve:**
   $\min_{\mathbf{M}_N^{(k+1)}} \ell(\mathbf{m} + \mathrm{Tr}((\mathbf{M}_N^{(k)} + \delta^{(k)}\mathbf{I})^{-1}\mathbf{M}_N^{(k+1)})$
   subject to
   $\mathbf{M}_N(\mathbf{m}) \succeq 0, \mathbf{L}_N(\mathbf{m}) \succeq 0, \mathbf{M}_\mathbf{W} \succeq 0$
   where $\ell(\mathbf{m})$ is the linear functional of the moments corresponding to the objective in (2.19), $\mathbf{M}_N(\mathbf{m})$ is the truncated moment matrix containing moments of order up to $2N$, $\mathbf{M}_\mathbf{W}$ is the submatrix of $\mathbf{M}_N$ associated with the elements of $\mathbf{W}$, and $\mathbf{L}_N(\mathbf{m})$ is the moments localizing matrix corresponding to the constraints.
4. Terminate if $\mathrm{rank}(\mathbf{M_N}) = 1$ or the maximum number of iterations has been reached. Otherwise, $\delta^{(k+1)} = \sigma_2(\mathbf{M_N})$ and repeat step 3).

Note that (2.19) involves only second-order polynomials, except for the constraint $\mathbf{W} \succeq 0$. In principle, this semidefinite constraint can be handled by enforcing non-negativity of polynomials of degree up to n_w^2 (e.g., the leading minors of $\mathbf{W}$). However, pursuing this approach requires considering high-order moment matrices, which is impractical even for moderately large problems. As an alternative, we will exploit the fact that if the moment matrix $\mathbf{M}$ associated with the polynomial optimization above is rank 1, then the constraint $\mathbf{W} \succeq 0$ is equivalent to $\mathbf{M}_\mathbf{W} \succeq 0$, where $\mathbf{M}_\mathbf{W}$ denotes the submatrix of $\mathbf{M}$ consisting of the moments of the elements of $\mathbf{W}$.[4] These observations motivate Algorithm 2.1, outlined below, where a reweighed heuristics is used to seek rank 1 solutions to the dual of (2.19).

2.4.2 Exploiting sparsity

As we briefly discuss next, computational complexity can be further reduced by exploiting the underlying sparse structure of the problem. Specifically, Problem (2.19) satisfies the running intersection property discussed in Section 2.2.3. To establish this, note that the objective in (2.19) can be partitioned as $\sum_{l=1}^{n_+ + n_-} p_l$, where $p_l = (1/(n_+ + n_-))\,\mathrm{Tr}(\mathbf{W}) + \xi_i - \lambda\,\mathrm{Tr}\,\mathbf{K}_l$ and $\mathbf{K}_l$ is the kernel matrix corresponding to the sequence $\mathbf{x}^{(l)}$. Similarly, the corresponding constraints g_l depend only on the variables $\{\mathbf{W}, \tilde{\mathbf{K}}, \xi_l\}$. Thus, the running intersection property holds. Hence, the semidefinite constraints $\mathbf{M}_N \succeq 0$ and $\mathbf{L}_N \succeq 0$ in Algorithm 2.1 can be replaced with c semidefinite constraints, each involving smaller matrices that depend only on $\{\mathbf{W}, \tilde{\mathbf{K}}, \xi_l\}$. Since the computational complexity of semidefinite program (SDP) solvers typically scales

[4]This follows from the fact that in this case the corresponding probability measure μ is atomic, with a single point mass and hence $\mathbf{W} = \mathscr{E}_\mu(\mathbf{W})$.

as n^6, where n denotes the size of the largest semidefinite constraint, it follows that exploiting the running intersection property leads to a significant reduction in computational complexity, specially in cases where n_y is large (i.e., video sequences).

2.5 Examples

In this section, we illustrate the advantages of the proposed method using both synthetic data and a computer-vision-motivated application: activity recognition from video data.

2.5.1 Synthetic data

The data for this example consist of 20 time sequences of length 9 (ten positive and ten negative) in $\mathbb{R}^2$ generated using two second-order linear models with random initial conditions and lifted to $\mathbb{R}^3$ by wrapping them on a cylinder surface. The positive samples were generated by a model with coefficients $\mathbf{r}_+ = [0.6418, -1]$ and the negative samples were generated by the model $\mathbf{r}_- = [1.9500, -1]$. Finally, these lifted sequences were corrupted with noise uniformly distributed in $[-\sigma\nu, \sigma\nu]$, where ν is the range of the signal and σ was varied between 0 and 0.3. Sample $\mathbb{R}^2$ sequences and their lifted counterparts are shown in Figure 2.3(a) and (b), respectively, where red and blue denote positive and negative classes. Two sets of experiments were performed using this data: one to assess the performance of the matrix $\mathbf{W}$ as a classifier and the second to explore the advantages of using negative examples in the training.

2.5.1.1 Performance of W as a classifier

To test the performance of $\mathbf{W}$ as a classifier, we divided the data into testing and training sets, each consisting of five positive and five negative examples, but only one labeled example from the negative class was included during training.[5] The values of $\mathrm{Tr}(\mathbf{WG}_i)$ obtained running Algorithm 2.1 for different noise levels, using hyper-parameter values $\lambda = 0.1$, $C = 10^5$, and 10^2 and Hankel matrices with five columns are shown in Table 2.1. As shown there, in most cases, the classifier incorrectly classifies seven of the negative examples as belonging to the positive class. This illustrates the fact that although the algorithm yields small values of $\mathrm{Tr}(\mathbf{WG}_i)$ for the positive samples (and hence can identify a Wiener system that explains the positive data), unless the condition $\mathrm{Tr}(\mathbf{WG}_i) > 1$ is explicitly enforced for several negative sequences, this model will also explain (within the noise level) some of the negative sequences. Thus, it will have relatively poor performance if used as a classification criteria.

2.5.1.2 Advantages of using negative sequences during training

In order to illustrate the advantages of using negative samples, in this second set of experiments, training was performed using five positive and five negative sequences.

[5]The other four negative training examples are only considered for the isometric constraints, i.e., they do not contribute with misclassification constraints.

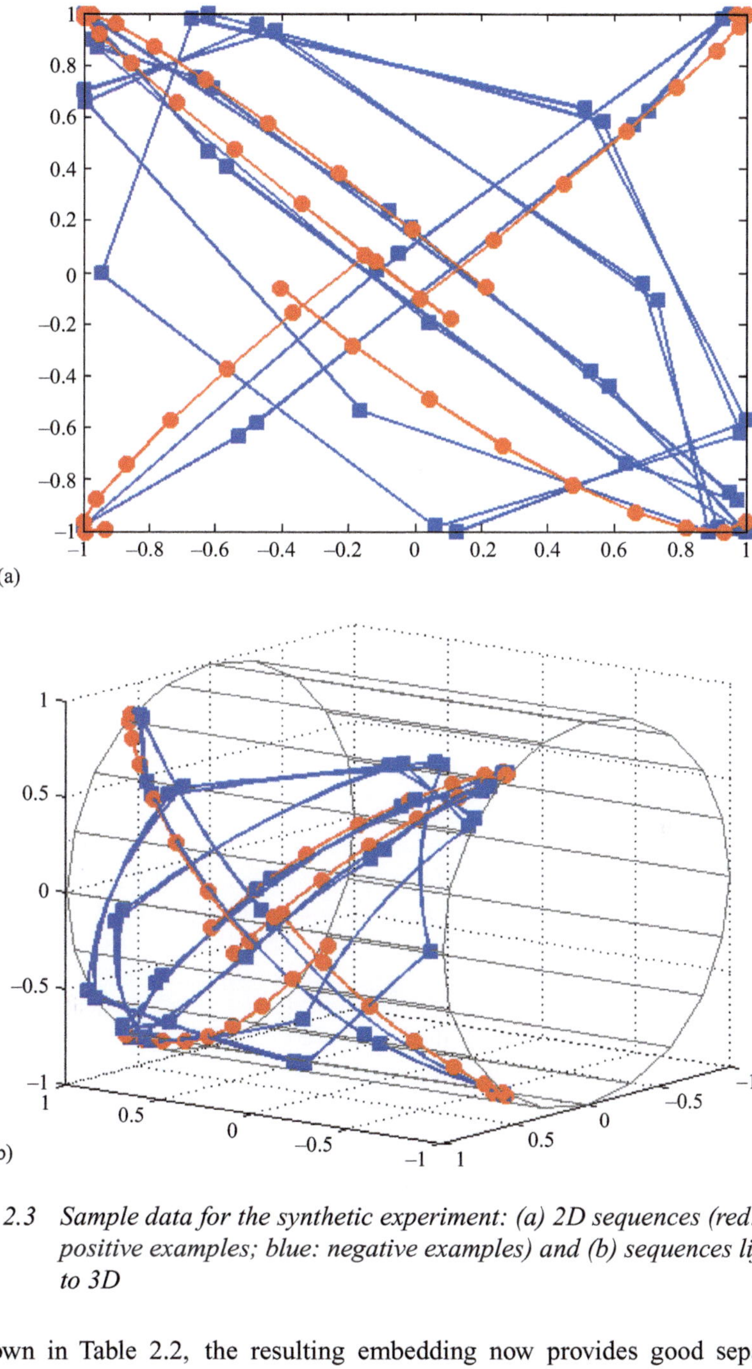

(a)

(b)

Figure 2.3 Sample data for the synthetic experiment: (a) 2D sequences (red: positive examples; blue: negative examples) and (b) sequences lifted to 3D

As shown in Table 2.2, the resulting embedding now provides good separation between classes both for the training and testing sets. In fact, using as criteria for separation, $\mathrm{sgn}(0.5 - \mathrm{Tr}(\mathbf{WG}_i))$ yields perfect classification with noise levels up to $\sigma = 0.3$.

*Table 2.1 Results of running Algorithm 2.1 with training data consisting of five sequences with positive labels, one sequence with negative label and four sequences with unknown labels (marked with ◇). Sequences that are incorrectly classified are marked with **

σ	$\lambda=0.1$, $C=1\times10^5$				$\lambda=0.1$, $C=1\times10^2$			
	Train Tr(WG$_i$)		Test Tr(WG$_i$)		Train Tr(WG$_i$)		Test Tr(WG$_i$)	
	+	−	+	−	+	−	+	−
0	0.0001	0.0905 ◇*	0.0105	0.0871 *	0.0039	0.0628 ◇*	0.0083	0.0617 *
	0.0005	0.0383 ◇*	0.0802	1.0680	0.0046	0.0718 ◇*	0.0471	0.7868
	0.0000	0.0894 ◇*	0.0711	0.0350 *	0.0034	0.0625 ◇*	0.0486	0.0905 *
	0.0001	1.1799 ◇	0.1071	0.0812 *	0.0033	1.0831 ◇	0.0495	0.0884 *
	0.0001	1.0000	0.0723	0.0822 *	0.0070	1.0000	0.0470	0.0898 *
	$\mu=0.0002$	$\mu=0.4796$	$\mu=0.0682$	$\mu=0.2707$	$\mu=0.0045$	$\mu=0.4560$	$\mu=0.0401$	$\mu=0.2235$
	$\sigma=0.0002$	$\sigma=0.5019$	$\sigma=0.0317$	$\sigma=0.3991$	$\sigma=0.0014$	$\sigma=0.4788$	$\sigma=0.0159$	$\sigma=0.2819$
0.1	0.0010	0.1432 ◇*	0.0415	0.1789 *	0.0064	0.2038 ◇*	0.0327	0.2377 *
	0.0003	0.1385 ◇*	0.0248	1.4449	0.0057	0.1393 ◇*	0.0253	1.3471
	0.0002	0.1307 ◇*	0.0625	0.0959 *	0.0038	0.1888 ◇*	0.0559	0.1379 *
	0.0004	1.2870 ◇	0.0726	0.1160 *	0.0020	1.1394 ◇	0.0617	0.1383 *
	0.0008	1.0000	0.0082	0.1119 *	0.0023	1.0000	0.0520	0.1393 *
	$\mu=0.0005$	$\mu=0.5399$	$\mu=0.0419$	$\mu=0.3895$	$\mu=0.0040$	$\mu=0.5343$	$\mu=0.0455$	$\mu=0.4000$
	$\sigma=0.0003$	$\sigma=0.5012$	$\sigma=0.0236$	$\sigma=0.5284$	$\sigma=0.0018$	$\sigma=0.4399$	$\sigma=0.0141$	$\sigma=0.4751$
0.2	0.0008	0.0866 ◇*	0.0026	0.0917 *	0.0208	0.0570 ◇*	0.1012	0.0657 *
	0.0001	0.1157 ◇*	0.0203	0.9033	0.0077	0.1221 ◇*	0.0496	0.8278
	0.0000	0.0932 ◇*	0.0340	0.1050 *	0.0065	0.0775 ◇*	0.0342	0.1125 *
	0.0000	1.3907 ◇	0.0370	0.0928 *	0.0086	0.2379 ◇*	0.0327	0.0795 *
	0.0000	1.0000	0.0141	0.0885 *	0.0097	1.0001	0.0061	0.0748 *
	$\mu=0.0002$	$\mu=0.5372$	$\mu=0.0216$	$\mu=0.2563$	$\mu=0.0107$	$\mu=0.2989$	$\mu=0.0448$	$\mu=0.2321$
	$\sigma=0.0003$	$\sigma=0.5515$	$\sigma=0.0127$	$\sigma=0.3236$	$\sigma=0.0052$	$\sigma=0.3561$	$\sigma=0.0315$	$\sigma=0.2983$
0.3	0.0006	0.1025 ◇*	0.0251	0.0823 *	0.0022	0.0699 ◇*	0.0253	0.1011 *
	0.0013	0.0601 ◇*	0.0541	0.9072	0.0101	0.1389 ◇*	0.0339	0.9705
	0.0017	0.1204 ◇*	0.0495	0.1405 *	0.0048	0.1435 ◇*	0.0573	0.2581 *
	0.0034	0.8828 ◇	0.0643	0.0923 *	0.0044	0.9476 ◇	0.0358	0.2125 *
	0.0023	1.0003	0.0091	0.0991 *	0.0059	1.0000	0.0463	0.1994 *
	$\mu=0.0019$	$\mu=0.4332$	$\mu=0.0405$	$\mu=0.2643$	$\mu=0.0055$	$\mu=0.4600$	$\mu=0.0397$	$\mu=0.3483$
	$\sigma=0.0009$	$\sigma=0.4172$	$\sigma=0.0203$	$\sigma=0.3221$	$\sigma=0.0026$	$\sigma=0.4207$	$\sigma=0.0111$	$\sigma=0.3153$

Table 2.2 Values of Tr(**WG**$_i$) *obtained running Algorithm 2.1 with training data consisting of five sequences with positive labels and five sequences with negative labels. Using as criteria for separation,* $sgn(0.5 - \mathrm{Tr}(\mathbf{WG}_i))$ *yields perfect classification*

	λ = 0.1, C = 1 × 10⁵				**λ = 0.1, C = 1 × 10²**			
	Train Tr(WG$_i$)		**Test Tr(WG$_i$)**		**Train Tr(WG$_i$)**		**Test Tr(WG$_i$)**	
σ	+	−	+	−	+	−	+	−
0	0.0023	1.0021	0.0392	0.9991	0.0015	1.0004	0.0337	1.0027
	0.0024	1.0000	0.1976	2.0338	0.0102	1.0005	0.1488	2.1446
	0.0037	1.0062	0.1435	0.6098	0.0006	1.0091	0.1044	0.5649
	0.0012	1.0120	0.1767	1.2579	0.0019	1.0062	0.1376	1.2247
	0.0011	2.4895	0.1393	1.2624	0.0006	2.9519	0.1201	1.2318
	$\mu = 0.0021$	$\mu = 1.3020$	$\mu = 0.1393$	$\mu = 1.2326$	$\mu = 0.0030$	$\mu = 1.3936$	$\mu = 0.1089$	$\mu = 1.2338$
	$\sigma = 0.0010$	$\sigma = 0.5938$	$\sigma = 0.0545$	$\sigma = 0.4660$	$\sigma = 0.0037$	$\sigma = 0.7791$	$\sigma = 0.0406$	$\sigma = 0.5159$
0.1	0.0024	1.0784	0.0698	1.1317	0.0031	1.1093	0.0860	1.1671
	0.0004	1.0027	0.0458	2.5370	0.0005	1.0067	0.0459	2.9623
	0.0009	1.0204	0.0613	0.9321	0.0005	1.0653	0.0574	0.9321
	0.0019	1.8523	0.1023	1.1913	0.0009	1.8465	0.1227	1.1849
	0.0012	1.3358	0.0332	1.1202	0.0017	1.3177	0.0373	1.1137
	$\mu = 0.0013$	$\mu = 1.2579$	$\mu = 0.0625$	$\mu = 1.3825$	$\mu = 0.0013$	$\mu = 1.2691$	$\mu = 0.0699$	$\mu = 1.4720$
	$\sigma = 0.0007$	$\sigma = 0.3204$	$\sigma = 0.0236$	$\sigma = 0.5838$	$\sigma = 0.0010$	$\sigma = 0.3072$	$\sigma = 0.0311$	$\sigma = 0.7505$
0.2	0.0036	1.0000	0.0078	1.1013	0.0055	1.0001	0.0159	1.1416
	0.0008	1.0475	0.0247	1.0054	0.0005	1.0389	0.0136	0.9845
	0.0018	1.0307	0.0470	1.0760	0.0001	1.0467	0.0283	1.0661
	0.0025	1.7412	0.0786	1.0107	0.0003	1.6865	0.0497	1.0179
	0.0008	1.0000	0.0317	0.9852	0.0001	1.0057	0.0199	0.9802
	$\mu = 0.0019$	$\mu = 1.1639$	$\mu = 0.0380$	$\mu = 1.0357$	$\mu = 0.0013$	$\mu = 1.1556$	$\mu = 0.0255$	$\mu = 1.0381$
	$\sigma = 0.0011$	$\sigma = 0.2892$	$\sigma = 0.0239$	$\sigma = 0.0447$	$\sigma = 0.0021$	$\sigma = 0.2661$	$\sigma = 0.0131$	$\sigma = 0.0602$
0.3	0.0008	1.0905	0.1367	1.2337	0.0007	1.0739	0.1311	1.2250
	0.0014	1.0038	0.0835	0.9304	0.0010	1.0031	0.0794	0.9505
	0.0023	1.1720	0.1082	1.6196	0.0024	1.1675	0.1057	1.5548
	0.0006	1.2602	0.1311	1.5305	0.0005	1.2798	0.1198	1.4697
	0.0015	1.0700	0.0418	1.9144	0.0014	1.0775	0.0451	1.8125
	$\mu = 0.0013$	$\mu = 1.1193$	$\mu = 0.1003$	$\mu = 1.4457$	$\mu = 0.0012$	$\mu = 1.1204$	$\mu = 0.0962$	$\mu = 1.4025$
	$\sigma = 0.0006$	$\sigma = 0.0886$	$\sigma = 0.0348$	$\sigma = 0.3369$	$\sigma = 0.0007$	$\sigma = 0.0953$	$\sigma = 0.0308$	$\sigma = 0.2938$

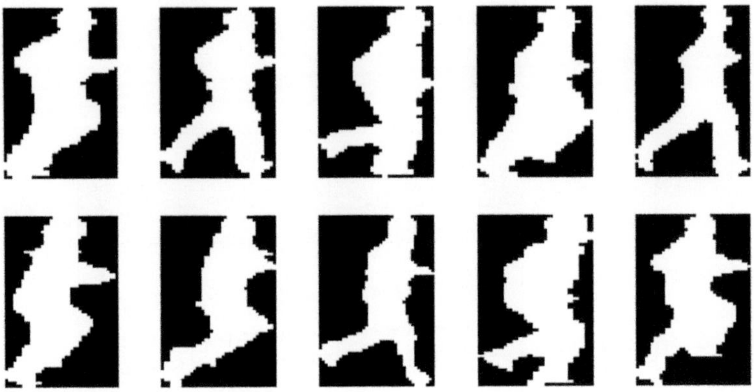

Figure 2.4 Aligned actor silhouette frames of Weizmann running activity (top row) and skipping (bottom row) recognition dataset

2.5.2 Application: activity recognition from video data

In this experiment, we used the proposed algorithm to classify human activities using video data from the Weizmann dataset [31]. The data consisted of 13 frames with 2,400 pixel/frame from 18 aligned silhouette videos with an equal number of "running" and "skipping" activities, performed by different actors. Figure 2.4 shows sample frames from the data. The data were divided into ten sequences (five of each type) for training and the remaining eight sequences for testing. The singular values of the moment matrix $\mathbf{M}$ found by the algorithm are 2.2615 and 0—i.e., the algorithm converged to a rank 1 solution. Table 2.3 shows the outputs of the algorithm achieving 100% classification rate.

2.6 Conclusions

This chapter proposes a new algorithm, based on a combination of ideas from systems theory and machine learning, for nonparametric identification of Wiener systems in scenarios where there is no a priori available information on the dimension of the output of the linear dynamics. A salient feature of this algorithm is its ability to exploit both positive and negative samples and the use of Kernel ideas to avoid high computational complexity, even when dealing with high-dimensional data. While in principle, the proposed approach requires solving a nontrivial polynomial optimization problem, tractable convex relaxations with optimality certificates can be obtained through a combination of moments-based polynomial optimization and matrix rank minimization ideas. Moreover, the resulting problem exhibits a sparse structure (the running intersection property) that can be exploited to further reduce the computational complexity. The advantages of the proposed algorithm were illustrated with experiments using synthetic and video data in the context of activity recognition.

Table 2.3 Classification results for the Weizmann dataset, with $\lambda = 0.1$ *and* $C = 1 \times 10^2$*: 100% accuracy*

Train Tr(WG$_i$)		**Test Tr(WG$_i$)**	
+	−	+	−
0.0116	1.9365	0.1912	2.6805
0.0104	17.9540	0.1025	0.6422
0.0169	1.0007	0.0521	0.5266
0.0063	1.8692	0.0985	0.7232
0.0239	1.0279		
$\mu = 0.0138$	$\mu = 4.7577$	$\mu = 0.1111$	$\mu = 1.1431$
$\sigma = 0.0061$	$\sigma = 6.6102$	$\sigma = 0.0503$	$\sigma = 0.8903$

In particular, these experiments highlighted the fact that including negative examples in the training data significantly improves classification accuracy.

Acknowledgments

This work was supported in part by NSF grants IIS-1318145, ECCS-1404163, CMMI-1638234 and CNS-1646121 and AFOSR grant FA9550-15-1-0392.

References

[1] Cripps SC. RF Power Amplifiers for Wireless Communications. Norwood, MA: Artech House; 1999.

[2] Celka P, Colditz P. Nonlinear nonstationary Wiener model of infant EEG seizures. IEEE Transactions on Biomedical Engineering. 2002;49:556–564.

[3] Matveev A, Hu X, Frezza R, *et al.* Observers for systems with implicit output. IEEE Transactions on Automatic Control. 2000;45(1):168 –173.

[4] Westwick D, Verhaegen M. Identifying MIMO wiener systems using subspace model identification methods. Signal Processing. 1996;52:235–258.

[5] Bruls J, Chou CT, Haverkamp BRJ, Verhaegen M. Linear and non-linear system identification using separable least squares, European Journal of Control. 1999;5(1):116–128.

[6] Greblicki W. Nonparametric identification of Wiener systems. IEEE Transactions on Information Theory. 1992;38:1487–1493.

[7] Greblicki W. Nonparametric approach to Wiener system identification. IEEE Transactions on Circuits and Systems. 1997;44:538–545.

[8] Bai EW. An optimal two-stage identification algorithm for Hammerstein–Wiener nonlinear systems. Automatica. 1998;34:333–338.

[9] Bai EW. A blind approach to the Hammerstein–Wiener model identification. Automatica. 2002;38:967–979.

[10] Bai EW. Frequency domain identification of Wiener models. Automatica. 2003;39:1521–1530.

[11] Raich R, Zhou T, Viberg M. Subspace based approaches for Wiener system identification. IEEE Transactions on Automatic Control. 2005;50:1629–1634.

[12] Lia K, Pengan JX, Bai EW. A two-stage algorithm for identification of nonlinear dynamic systems. Automatica. 2006;42(7):1189–1197.

[13] Cerone V, Regruto D. Parameter bounds evaluation of Wiener models with noninvertible polynomial nonlinearities. Automatica. 2006;42:1775–1781.

[14] Yilmaz B, Ayazoglu M, Sznaier M, *et al.* Convex relaxations for robust identification of Wiener systems and applications. In: Proc. 2011 IEEE Conf. Dec. Control; 2011. p. 2812–2818.

[15] Lim H, Camps OI, Sznaier M, *et al.* Dynamic appearance modelling for human tracking. In: IEEE Computer Vision and Pattern Recognition; 2006. p. 751–757.

[16] Xiong F, Camps O, Sznaier M. Low order dynamics embedding for high dimensional time series. In: 2012 IEEE ICCV; 2012. p. 2368–2374.

[17] Saul LK, Roweis ST. Think globally, fit locally: unsupervised learning of low dimensional manifolds. Journal on Machine Learning Research. 2003;4: 119–155.

[18] Weinberger KQ, Saul LK. Unsupervised learning of image manifolds by semidefinite programming. In: CVPR. vol. 2; 2004. p. 988–995.

[19] Roweis S, Saul L, Hinton GE. Global coordination of local linear models. Advances in Neural Information Processing Systems. 2001;14:889–896.

[20] Lin RS, Liu CB, Yang MH, *et al.* Learning nonlinear manifolds from time series. In: ECCV. vol. LNCS 3952. Springer-Verlag; 2006. p. 245–256.

[21] Toth R, Laurain V, Zheng WX, *et al.* Model structure learning: a support vector machine approach for LPV linear-regression models. In: Proc. 2011 IEEE Conf. Dec. Control; 2011. p. 3192–3197.

[22] Xiong F, Cheng Y, Camps O, *et al.* Hankel based maximum margin classifiers: a connection between machine learning and Wiener systems identification. In: 52nd IEEE Conference on Decision and Control; 2013. p. 6005–6010.

[23] Lasserre JB. Global optimization with polynomials and the problem of moments. SIAM Journal on Optimization. 2001;11:796–817.

[24] Curto RE, Fialkow LA. Truncated K-moment problems in several variables. Journal of Operator Theory. 2005;54(1):189–226.

[25] Lasserre JB. Moments, Positive Polynomials and Their Applications. vol. 1. London: Imperial College Press; 2010.

[26] Lasserre J. Convergent SDP-relaxations in polynomial optimization with sparsity. SIAM Journal on Optimization. 2006;17(3):822–843.

[27] Li B, Ayazoglu M, Mao T, *et al.* Activity recognition using dynamic subspace angles. In: CVPR 2011; 2011. p. 3193–3200.

[28] Moonen M, Moor BD, Vandenberghe L, *et al.* On- and off-line identification of linear state-space models. International Journal of Control. 1989;49(1): 219–232.

[29] Yilmaz B, Bekiroglu K, Lagoa C, *et al.* A randomized algorithm for parsimonious model identification. IEEE Transactions on Automatic Control. 2018;63(2):532–539.

[30] Weinberger KQ, Packer BD, Saul LK. Nonlinear dimensionality reduction by semidefinite programming and kernel matrix factorization. In: Proceedings of the Tenth International Workshop on Artificial Intelligence and Statistics; 2005. p. 381–388.

[31] Gorelick L, Blank M, Shechtman E, *et al.* Actions as space-time shapes. Transactions on Pattern Analysis and Machine Intelligence. 2007;29(12): 2247–2253.

Chapter 3

Identification of a quasi-LPV model for wing-flutter analysis using machine-learning techniques

Rodrigo Alvite Romano[1], Marcelo Mendes Lafetá Lima[1], Paulo Lopes dos Santos[2], and Teresa Paula Azevedo Perdicoúlis[3,4]

Aerospace structures are often submitted to air-load tests to check possible unstable structural modes that lead to failure. These tests induce structural oscillations stimulating the system with different wind velocities, known as flutter test. An alternative is assessing critical operating regimes through simulations. Although cheaper, model-based flutter tests rely on an accurate simulation model of the structure under investigation. This chapter addresses the data-driven flutter modeling using state-space linear parameter varying (LPV) models. The estimation algorithm employs support vector machines to represent the functional dependence between the model coefficients and the scheduling signal, which values can be used to account for different operating conditions. Besides versatile, that model structure allows the formalization of the estimation task as a linear least-squares problem. The proposed method also exploits the ensemble concept, which consists of estimating multiple models from different data partitions. These models are merged into a final one, according to their ability to reproduce a validation data segment. A case study based on real data shows that this approach resulted in a more accurate model for the available data. The local stability of the identified LPV model is also investigated to provide insights about critical operating ranges as a function of the magnitude of the input and output signals.

3.1 Introduction

Aerospace structures are subject to oscillations that can lead to their collapse. Possible structural vulnerabilities and critical operating regimes can be investigated from

[1]Escola de Engenharia Mauá, Instituto Mauá de Tecnologia, São Caetano do Sul, SP, Brazil
[2]INESC TEC—Institute for Systems and Computer Engineering, Technology and Science and FEUP—Faculty of Engineering, University of Porto, Porto, Portugal, funded by FCT – Fundação para a Ciência e a Tecnologia within project: UID/EEA/50014/2019.
[3]Universidade de Trás-os-Montes e Alto Douro, Vila Real, Portugal
[4]Instituto de Sistemas e Robótica (ISR), Coimbra, Portugal

simulations. Obviously, the validity of these analyses depends on the accuracy of the model used in the simulations. In this chapter, the wing-flutter modeling from test data is addressed. Flutter is a complex aeroelastic phenomenon where structural modes are simultaneously coupled and excited by aerodynamic loads [1–3]. More specifically, a data-driven modeling method based on LPV state-space models is considered to capture the flutter dynamics. In this model class, the relationship between input and output signal is linear, while the model coefficients have a functional dependence on a parameter-varying signal denoted as scheduling variable. LPV models have proven to be effective tools for describing nonlinear and time-varying systems [4,5]. As many linear analysis and control design techniques can be adapted to the LPV framework, in the last years, they have been widely used with successful applications.

The standard approach to handle the LPV model estimation is by using parametric representations, which require the parameterization of such functional dependences, which may vary from simple polynomials to complex rational or even discontinuous functions. This choice is critical to obtaining accurate models; however, it depends on prior knowledge which is not always available beforehand. To circumvent this issue, which means to be able to "learn" from the data the (possible nonlinear) dependence of the model coefficients on the scheduling signal, we employ a recently developed nonparametric model estimation algorithm [4], formulated in a LS-SVM setting. This feature makes it possible to depict a wide range of nonlinear dependencies as linear combinations of infinitely many functions defined through the choice of a particular inner product and a relative low-dimension parameter vector [5,6].

The flutter identification problem here considered also poses an additional challenge which is also found in other practical situations: the lack of information for estimation and validation purposes. Motivated by ensemble methods from the machine-learning literature [7], which consists in generating multiple models from partitioned data, we adapted the LPV model estimation algorithm to handle disjoint time-series segments so that the available data is used more efficiently. Then, a final model is derived by merging the obtained models. The LPV model estimated with our algorithm fits the validation dataset substantially better than the ones produced by classical LTI methods, often used to characterize flutter parameters. In addition, the local stability of the identified LPV model is also investigated in order to provide an indication of critical operating ranges with respect to the magnitude of the input and output signals.

This chapter has the following structure: the motivation for the application of LPV model estimation to the wing-flutter identification problem is presented in Section 3.1. In Section 3.2, an LPV model parametrization that leads to a linear predictor with respect to the model parameters is presented. The estimation procedure using the LS-SVM setting is formulated in Section 3.3. Ensemble model estimation, where the algorithm is reformulated to deal with partitioned data, so that multiple models are obtained using different data partitions, is developed in Section 3.4. These models are merged to provide a single final one. A wing-flutter identification case study using real data is discussed in Section 3.5. At last, concluding remarks are drawn in Section 3.6.

3.2 LPV state-space model parameterization

In this work, LPV models of the form

$$x(k+1) = (A + L(p_k)C)\,x(k) + B(p_k)u(k) \tag{3.1}$$

$$y(k) = Cx(k), \tag{3.2}$$

where $x_k \in \Re^{n_x}$ is the state vector, $u_k \in \Re$ and $y_k \in \Re$ are the model input and output, respectively, are considered. The matrices A and C can be arbitrarily chosen, provided the pair (A, C) is observable, and the eigenvalues of A are located inside the unity circle. The time-varying parameter vectors $L(p_k) \in \Re^{n_x}$ and $B(p_k) \in \Re^{n_x}$

$$L(p_k) = \sum_{j=1}^{n_f} L_j f_j(p_k) \tag{3.3}$$

$$B(p_k) = \sum_{j=1}^{n_f} B_j f_j(p_k), \tag{3.4}$$

are linear combinations of arbitrary basis functions $f_j(p_k)$, for $j \in \{1, \ldots, n_f\}$. The scheduling signal p_k is assumed to be known (or measured) at each time instant k, and it takes values in a scheduling space $\mathbb{P} \subset \Re^{n_p}$. In fact, a quasi-LPV approach is considered in Section 3.5, which means that p_k is composed of input–output data.

The previous state-space LPV model structure is based on the matchable-observable LTI parameterization proposed initially in the context of adaptive control theory [8] and applied more recently to multivariable system identification problems [9]. For the sake of simplicity, we shall restrict the discussion to single-output model descriptions. However, the proposed contribution can be generalized quite straightforwardly to the multi-output case, provided the model structural indices (i.e., the set of observability indices) are assumed to be time invariant.

It is important to mention that, due to the observability of the pair (C, A), the frozen (local) models achieved by fixing p_k to a constant value $\bar{p}$ in (3.1) and (3.2) can match any linear time-invariant transfer function of order n_x. In addition, such representation is particularly useful for model estimation as it enables to derive a predictor whose model is a linear function of the basis functions weights L_j and B_j. Such predictor is built as follows:

First, substitute (3.3) and (3.4) into (3.1) to yield

$$x(k+1) = Ax(k) + \sum_{j=1}^{n_f} L_j f_j(p_k)y(k) + \sum_{j=1}^{n_f} B_j f_j(p_k)u(k). \tag{3.5}$$

Next, introduce the auxiliary variables

$$\breve{y}^{(j)}(k) = C(qI - A)^{-1} f_j(p_k)y(k) \tag{3.6}$$

$$\breve{u}^{(j)}(k) = C(qI - A)^{-1} f_j(p_k)u(k), \tag{3.7}$$

where q denotes the time shift operator. It is emphasized that in a parameter-varying context $q\left[f_j(p_k)u(k)\right] = f_j(p_{k+1})u(k+1)$. Then, from (3.2) and (3.5), the output predictor can be written as

$$\hat{y}(k) = CA^{k-1}x(1) + \sum_{j=1}^{n_f} \breve{y}^{(j)}(k)L_j + \sum_{j=1}^{n_f} \breve{u}^{(j)}(k)B_j,$$

where $x(1)$ denotes the initial state condition. The notation $\hat{y}$ is introduced to distinguish between the observed output y and the predicted one.

Notice that denoting the parameter vector as

$$\theta = \begin{bmatrix} x(1) & L_1 & \cdots & L_{n_f} & B_1 & \cdots & B_{n_f} \end{bmatrix}^\top \tag{3.8}$$

and the regressor/information vector as

$$\varphi(k) = \begin{bmatrix} CA^{k-1} & \breve{y}^{(1)}(k) & \cdots & \breve{y}^{(n_f)}(k) & \breve{u}^{(1)}(k) & \cdots & \breve{u}^{(n_f)}(k) \end{bmatrix}^\top, \tag{3.9}$$

the output predictor can be written as $\hat{y}(k) = \varphi^\top(k)\theta$. Thus, for a data window of length N, we get

$$\hat{Y} = \Phi\theta \tag{3.10}$$

where

$$\hat{Y} = \left[\hat{y}(1) \cdots \hat{y}(N)\right]^\top$$

and

$$\Phi = \left[\varphi(1) \cdots \varphi(N)\right]^\top.$$

The parameter vector (3.8) can be estimated using least squares, provided the basis functions $f_j(p_k)$ are selected beforehand. But these functions need to be adequately chosen to capture the unknown dependencies on the scheduling variable. An inappropriate choice of the basis functions can lead to over-parameterized models. Hence, the estimated parameters tend to have a large variance [5]. In what follows, a nonparametric approach using LS-SVM is presented. This approach allows that the dependencies of the model coefficients on the scheduling signal are "learned" from the plant data.

3.3 Model estimation

The model estimation is formulated as the optimization problem

$$\hat{\theta} = \arg\min_{\theta} \frac{1}{2}\theta^\top\theta + \frac{\gamma}{2}E^\top E, \tag{3.11}$$

subject to $E = Y - \Phi\theta$, where

$$Y = \begin{bmatrix} y(1) & \cdots & y(N) \end{bmatrix}^\top$$
$$E = \begin{bmatrix} e(1) & \cdots & e(N) \end{bmatrix}^\top$$

and γ is a positive scalar used to adjust the bias-variance trade-off. To solve this constrained optimization problem, it is introduced the Lagrangian

$$\mathfrak{L}(E,\theta,\lambda) = \frac{1}{2}\theta^\top\theta + \frac{\gamma}{2}E^\top E - \lambda^\top(E - Y + \Phi\theta).$$

The Karush–Kuhn–Tucker conditions of the above problem are

$$\frac{\partial \mathfrak{L}}{\partial E} = \gamma E^\top - \lambda^\top = 0 \Rightarrow E = \frac{\lambda}{\gamma} \tag{3.12}$$

$$\frac{\partial \mathfrak{L}}{\partial \theta} = \theta^\top - \lambda^\top \Phi = 0 \Rightarrow \theta = \Phi^\top \lambda \tag{3.13}$$

$$\frac{\partial \mathfrak{L}}{\partial \lambda} = E - Y + \Phi\theta = 0 \Rightarrow E = Y - \Phi\theta \tag{3.14}$$

Substituting (3.12) and (3.13) into (3.14) yields

$$Y = \frac{\lambda}{\gamma} + \Phi\Phi^\top\lambda \Rightarrow \lambda = \left(\frac{I_N}{\gamma} + \Phi\Phi^\top\right)^{-1} Y. \tag{3.15}$$

The Lagrange multiplier $\lambda \in \Re^N$ is responsible to capture the information contained in the data, which relies on the computation of $\Phi\Phi^\top$. Although the matrix Φ cannot be computed without the specification of $f_j(p_k)$, it is possible to exploit the structure of $\Omega \triangleq \Phi\Phi^\top \in \Re^{N\times N}$ to circumvent such a necessity. First, notice that the (k_1, k_2)th entry of Ω is given by

$$\begin{aligned}\Omega(k_1,k_2) &= CA^{k_1-1}\left(CA^{k_2-1}\right)^\top \\ &\quad + \sum_{j=1}^{n_f}\left(\breve{y}^{(j)}(k_1)\breve{y}^{(j)}(k_2)^\top + \breve{u}^{(j)}(k_1)\breve{u}^{(j)}(k_2)^\top\right).\end{aligned} \tag{3.16}$$

From (3.6), the inner product of the output signal is

$$\begin{aligned}\sum_{j=1}^{n_f}\breve{y}^{(j)}(k_1)\breve{y}^{(j)}(k_2)^\top &= \sum_{j=1}^{n_f} C(q_1 I - A)^{-1} f_j(p_{k_1})y(k_1)\left(C(q_2 I - A)^{-1} f_j(p_{k_2})y(k_2)\right)^\top \\ &= \sum_{j=1}^{n_f} C(q_1 I - A)^{-1} f_j(p_{k_1})y(k_1) f_j(p_{k_2})y(k_2)(q_2 I - A^\top)^{-1}C^\top \\ &= C(q_1 I - A)^{-1}\psi\left(p_{k_1},p_{k_2}\right)y(k_1)y(k_2)(q_2 I - A^\top)^{-1}C^\top,\end{aligned}$$

where

$$\psi\left(p_{k_1}, p_{k_2}\right) = \sum_{j=1}^{n_f} f_j(p_{k_1}) f_j(p_{k_2})$$

is a positive definite kernel function. The symbols q_1 and q_2 are introduced to discriminate between the row and column shift operator, respectively. Analogously, from (3.7), we can write

$$\sum_{j=1}^{n_f} \breve{u}^{(j)}(k_1)\breve{u}^{(j)}(k_2)^\top = C(q_1 I - A)^{-1}\psi\left(p_{k_1}, p_{k_2}\right) u(k_1)u(k_2)(q_2 I - A^\top)^{-1}C^\top.$$

Hence, (3.16) becomes

$$\Omega(k_1, k_2) = CA^{k_1-1}\left(CA^{k_2-1}\right)^\top + C(q_1 I - A)^{-1}\mathscr{U}(k_1, k_2)(q_2 I - A^\top)^{-1}C^\top,$$

where $\mathscr{U}(k_1, k_2) = \psi\left(p_{k_1}, p_{k_2}\right)\left[y(k_1)y(k_2) + u(k_1)u(k_2)\right]$.

It is worth to note that the computation of Ω is a 2D (two-dimensional) filtering operation on $\mathscr{U}(k_1, k_2)$. This can be represented in the state-space form as

$$\begin{aligned}
\mathscr{X}_1(k_1 + 1, k_2) &= A^\top \mathscr{X}_1(k_1, k_2) + C^\top \mathscr{U}(k_1, k_2) \\
\mathscr{X}_2(k_1, k_2 + 1) &= \mathscr{X}_1(k_1, k_2) + A\mathscr{X}_2(k_1, k_2) \\
\Omega\,(k_1, k_2) &= C\mathscr{X}_2(k_1, k_2) + CA^{k_1-1}(CA^{k_2-1})^\top.
\end{aligned} \tag{3.17}$$

As mentioned previously, the model estimation is addressed using a nonparametric approach, that is, instead of explicitly parameterizing the basis functions $f_j(p_k)$, the kernel function $\psi(p_{k_1}, p_{k_2})$ is used to represent the inner products $\sum_{j=1}^{n_f} f_j(p_{k_1})f_j(p_{k_2})$. This strategy is called the *kernel trick* [10,11] and enables the reconstruction of $L(p_k)$ and $B(p_k)$ from λ, the vector of Lagrange multipliers, and the kernel functions. For the wing-flutter identification problem described in Section 3.5, Gaussian radial-basis functions with a shape parameter of $1/\sigma$ are used to characterize the inner products $\sum_{\ell=1}^{n_f} f_\ell(p_i)f_\ell(p_j)$, i.e.,

$$\psi\left(p_i, p_j\right) = \exp\left(-\frac{\|p_i - p_j\|_2^2}{\sigma^2}\right),$$

where $\|\cdot\|_2$ denotes the Euclidean norm of the argument.

Next, the equations used to retrieve the model parameters are developed.

3.3.1 Parameter reconstruction

From (3.13), it is straightforward to see that

$$\theta = \Phi^\top \lambda \rightarrow \begin{bmatrix} x(1) \\ L_1 \\ L_2 \\ \vdots \\ L_{n_f} \\ B_1 \\ B_2 \\ \vdots \\ B_{n_f} \end{bmatrix} = \begin{bmatrix} (CA^0)^\top & (CA^1)^\top & \cdots & (CA^{N-1})^\top \\ \breve{y}^{(1)}(1)^\top & \breve{y}^{(1)}(2)^\top & \cdots & \breve{y}^{(1)}(N)^\top \\ \breve{y}^{(2)}(1)^\top & \breve{y}^{(2)}(2)^\top & \cdots & \breve{y}^{(2)}(N)^\top \\ \vdots & \vdots & & \vdots \\ \breve{y}^{(n_f)}(1)^\top & \breve{y}^{(n_f)}(2)^\top & \cdots & \breve{y}^{(n_f)}(N)^\top \\ \breve{u}^{(1)}(1)^\top & \breve{u}^{(1)}(2)^\top & \cdots & \breve{u}^{(1)}(N)^\top \\ \breve{u}^{(2)}(1)^\top & \breve{u}^{(2)}(2)^\top & \cdots & \breve{u}^{(2)}(N)^\top \\ \vdots & \vdots & & \vdots \\ \breve{u}^{(n_f)}(1)^\top & \breve{u}^{(n_f)}(2)^\top & \cdots & \breve{u}^{(n_f)}(N)^\top \end{bmatrix} \begin{bmatrix} \lambda(1) \\ \vdots \\ \lambda(N) \end{bmatrix}$$

The initial-state estimate is directly obtained from

$$x(1) = \sum_{k=1}^{N} \left(CA^{k-1}\right)^\top \lambda(k).$$

The basis functions weights become

$$L_j = \sum_{k=1}^{N} \left(\breve{y}^{(j)}(k)\right)^\top \lambda(k)$$

$$B_j = \sum_{k=1}^{N} \left(\breve{u}^{(j)}(k)\right)^\top \lambda(k).$$

Then, evaluating (3.3) for a particular scheduling signal value $\bar{p} \in \mathbb{P}$ leads to

$$L(\bar{p}) = \sum_{j=1}^{n_f} L_j f_j(\bar{p}) = \sum_{j=1}^{n_f} \left[\sum_{k=1}^{N} \left(\breve{y}^{(j)}(k)\right)^\top \lambda(k) \right] f_j(\bar{p}),$$

and using (3.6), it follows that

$$\begin{aligned} L(\bar{p}) &= \sum_{j=1}^{n_f} \sum_{k=1}^{N} \left(C(qI - A)^{-1} f_j(p_k) y(k)\right)^\top \lambda(k) f_j(\bar{p}) \\ &= \sum_{k=1}^{N} \left(qI - A^\top\right)^{-1} C^\top y(k) \sum_{j=1}^{n_f} f_j(p_k) f_j(\bar{p}) \lambda(k). \end{aligned}$$

Lastly, performing analogously for (3.4) and (3.7) one obtains the equations

$$L(\bar{p}) = \sum_{k=1}^{N} \left[\left(qI - A^\top\right)^{-1} C^\top y(k) \psi(p_k, \bar{p}) \right] \lambda(k)$$

$$B(\bar{p}) = \sum_{k=1}^{N} \left[\left(qI - A^\top\right)^{-1} C^\top u(k) \psi(p_k, \bar{p}) \right] \lambda(k)$$

that give the parameters for a particular value $\bar{p}$ of the scheduling signal.

3.4 Ensemble estimation approach

Ensemble techniques are often used in the big-data literature to improve regression estimates by using multiple models obtained from different data segments. The motivation for this strategy is that multiple models can focus on different data partitions [7,12]. Then, those models are combined into a final one. The variance error of the combined model is smaller than the one of the model estimated directly from the entire dataset [13], Section 7.2 and [14].

We shall adopt the so-called K-fold approach in which the data is split into $n_w + 1$ segments of the same length. A model is then estimated using a partition composed of n_w segments. The remaining segment is deployed to assess this model and is referred here as the *tuning window* because it is also used to adjust the algorithm hyperparameters, namely, γ and σ. This is repeated for all possible *tuning windows* so that $n_w + 1$ models are obtained.

In contrast to static (memoryless) models, it is not straightforward to estimate dynamic models from partitions composed of segmented data. This issue arises from the influence of past data on the present system behavior. In other words, one cannot simply concatenate disjoint segments because it may cause transients that degrade the estimates. For this reason, it is proposed to introduce the initial state of each segment to capture the influence of the removed preceding data window. In what follows, the estimation algorithm proposed in the previous section is recast to handle partitioned data.

We begin by writing the predictor equation for an arbitrary segment

$$\hat{y}_m(k) = CA^{k-1}x_m(1) + \sum_{j=1}^{n_f} \breve{y}_m^{(j)}(k)L_j + \sum_{j=1}^{n_f} \breve{u}_m^{(j)}(k)B_j. \tag{3.18}$$

The subscript m denotes the signals of a specific segment, e.g., $\breve{y}_3^{(j)}(1)$ refers to the first sample of the filtered signal defined by (3.6) in the third segment. The same applies to $\hat{y}_3(1)$, $\hat{x}_3(1)$, and $\breve{u}_3(1)$. According to (3.18), for n_w (possibly disjoint) segments, the partitioned regression model becomes

$$\underbrace{\begin{bmatrix} \hat{Y}_1 \\ \vdots \\ \hat{Y}_{n_w} \end{bmatrix}}_{\hat{Y}} = \underbrace{\left[\begin{array}{ccc|c} \mathcal{O}(N_1) & & \mathbf{O} & \eta_1 \\ & \ddots & & \vdots \\ \mathbf{O} & & \mathcal{O}(N_{n_w}) & \eta_{n_w} \end{array}\right]}_{\Upsilon} \vartheta, \tag{3.19}$$

where $\mathcal{O}(N_m)$ is an observability matrix of dimension N_m, the number of data points in the mth segment,

$$\hat{Y}_m = \begin{bmatrix} \hat{y}_m(1) & \cdots & \hat{y}_m(N_m) \end{bmatrix}^\top$$

$$\eta_m = \begin{bmatrix} \breve{y}_m^{(1)}(1) & \cdots & \breve{y}_m^{(n_f)}(1) & \breve{u}_m^{(1)}(1) & \cdots & \breve{u}_m^{(n_f)}(1) \\ \vdots & & \vdots & \vdots & & \vdots \\ \breve{y}_m^{(1)}(N_m) & \cdots & \breve{y}_m^{(n_f)}(N_m) & \breve{u}_m^{(1)}(N_m) & \cdots & \breve{u}_m^{(n_f)}(N_m) \end{bmatrix}$$

for $m \in \{1, \ldots, n_w\}$, and

$$\vartheta = \left[x_1(1) \quad \cdots \quad x_{n_w}(1) \middle| L_1 \quad \cdots \quad L_{n_f} \quad B_1 \quad \cdots \quad B_{n_f} \right]^\top.$$

Notice that ϑ is an augmented version of the parameter vector (3.8), as it also includes the initial state of each segment.

Then, solving the optimization problem (3.11) using the partitioned regression model (3.19) yields

$$\boldsymbol{Y} = \frac{\boldsymbol{\lambda}}{\gamma} + \Upsilon\Upsilon^\top\boldsymbol{\lambda} \;\Rightarrow\; \boldsymbol{\lambda} = \left(\frac{I_N}{\gamma} + \Upsilon\Upsilon^\top\right)^{-1} \boldsymbol{Y} \tag{3.20}$$

where $\boldsymbol{Y} = \begin{bmatrix} Y_1^\top & \cdots & Y_{n_w}^\top \end{bmatrix}^\top$ is the segmented output observation vector and the ijth block of $\Upsilon\Upsilon^\top$ is

$$\left[\Upsilon\Upsilon^\top\right]_{ij} = \begin{cases} \mathcal{O}(N_i)\mathcal{O}^\top(N_i) + \eta_i\eta_i^\top & \text{, if } i = j \\ \eta_i\eta_j^\top & \text{, if } i \neq j, \end{cases} \tag{3.21}$$

for $i, j \in \{1, \ldots, n_w\}$. Each block $\left[\Upsilon\Upsilon^\top\right]_{ij}$ is computed using the 2D filter (3.17), such that for the blocks outside the diagonal (i.e., $i \neq j$), the last element on the right-hand side of (3.17) is not included. The step-by-step estimation procedure is summarized in Algorithm 3.1.

Once the estimation algorithm is adapted to handle segmented data, the estimation data can be divided into $n_w + 1$ sets. Figure 3.1 shows these estimation datasets. In each set, the estimation segments are the white blocks while the tuning window is the gray block. The hyper-parameters γ and σ are adjusted using a grid search for the pair that best fits the tuning segment. The process is repeated for every set, yielding

Algorithm 3.1: Estimate model for specific windowed scenario

Data: Input, output and scheduling signal of an estimation dataset with n_w possibly disjoint segments.

Initialize
$N = \sum_{m=1}^{n_w} N_m$.
Compute the elements of $\Upsilon\Upsilon^\top$*:*
for $i \leftarrow 1$ **to** n_w **do**
  for $j \leftarrow 1$ **to** n_w **do**
    Determine the terms $\left[\Upsilon\Upsilon^\top\right]_{i,j}$ *using* (3.21).
  end
end
Compute the Lagrange multipliers from (3.20).
Reconstruct the time-varying parameters at specific points $p_k = \bar{p}$:
$L(\bar{p}) = \sum_{k=1}^{N} \left[\left(qI - A^\top\right)^{-1} C^\top y(k) \psi(p_k, \bar{p}) \right] \boldsymbol{\lambda}(k)$,
$B(\bar{p}) = \sum_{k=1}^{N} \left[\left(qI - A^\top\right)^{-1} C^\top u(k) \psi(p_k, \bar{p}) \right] \boldsymbol{\lambda}(k)$.

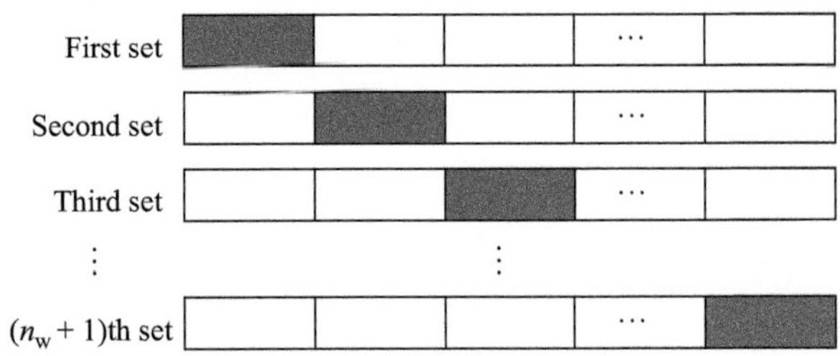

Figure 3.1 K-fold segmentation methodology. In each set, the white blocks are the estimation segments and the gray block is the tuning window.

$n_w + 1$ models. It is proposed a simple weighted average procedure to merge these models, so that

$$\boldsymbol{L}(\bar{p}) = \sum_{m=1}^{n_w} w_m L_m(\bar{p}) \tag{3.22}$$

$$\boldsymbol{B}(\bar{p}) = \sum_{m=1}^{n_w} w_m B_m(\bar{p}), \tag{3.23}$$

where w_m are weighting terms, while $L_m(\bar{p})$ and $B_m(\bar{p})$ are the parameter vectors obtained for each segment $m \in \{1, \ldots, n_w\}$.

3.5 Wing-flutter model identification

This section presents a case study with the application of the LPV model estimation algorithm to experimental wing-flutter data. The wing-flutter identification problem had already been approached using LPV techniques in [15,16]. In both works, subspace techniques based on affine[1] LPV model structures were applied to simulated flutter data.

It is known that flutter model parameters are significantly influenced by velocity and dynamic pressure [2,15]. Although it is reasonable to use such information as scheduling signal, we adopted a quasi-LPV framework, with the scheduling parameter

$$p_k = \left[y(k)\ u(k)\right]^\top,$$

due to the lack of other signals to better describe the operating conditions.

The data is available from the DaISy database [17]. Due to industrial secrecy agreements, no prior information is available, besides that the plant is excited using a highly colored input. Figure 3.2 illustrates the input–output data consisting of 720 samples. The first 360 points were partitioned into 5 data windows ($n_w = 4$). Thus, four were used for estimation and the other for tuning the hyper-parameters according to the proposed algorithm. The second half of the dataset is reserved to evaluate the model capability to reproduce the flutter behavior. We emphasize that the ensemble

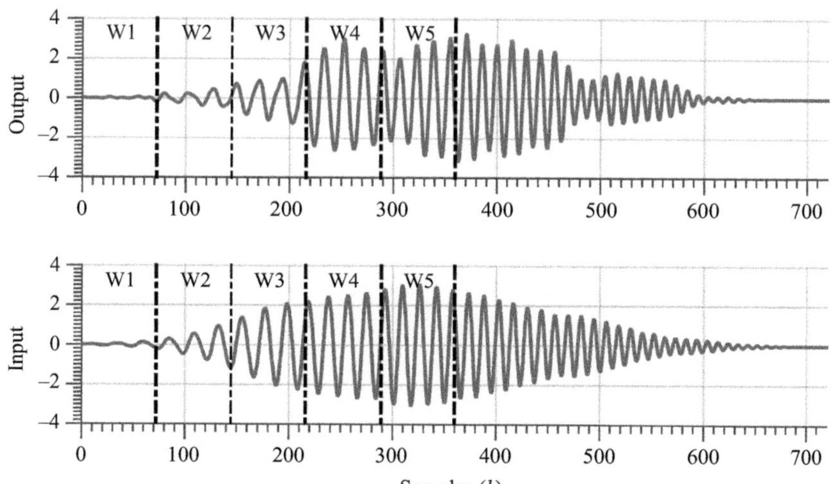

Figure 3.2 Wing-flutter input–output signals

[1]The dependence of the model parameters on the scheduling signal is represented using linear affine functions.

technique is specially compelling in this case because of the limited data with different excitation regimes encompassed in the dataset. The estimated models were weighted by terms proportional to γ, given by

$$w_m = \frac{\gamma_m}{\sum_{m=1}^{n_w} \gamma_m},$$

where γ_m is the regularization hyper-parameter of the mth dataset. We follow the rationale that model fit is favored when γ becomes higher (one should recall that the cross-data nature of the tuning procedure avoids over-modeling). Therefore, during the merging process, more weight is given to the models that are more accurate in reproducing the flutter behavior in the tuning segment.

The observability of the user defined pair (C, A) (3.1) and (3.2) is assured by choosing $C = \begin{bmatrix} 0 & \cdots & 1 \end{bmatrix}$ and A in the companion form, i.e.,

$$A = \begin{bmatrix} 0 & \cdots & 0 & -\alpha_{n_x} \\ 1 & & & -\alpha_{n_x-1} \\ & \ddots & & \vdots \\ & & 1 & -\alpha_1 \end{bmatrix}.$$

The characteristic polynomial of A provides a low-pass filtering mechanism. Although the poles of the filter can be tuned using data-driven optimization strategies, see, for example, [4,18], preliminary tests do not indicated significant improvement of model accuracy, possibly because the amount of high-frequency noise in the flutter data is negligible. Therefore, the filtering mechanism is disregarded by setting the eigenvalues of A to zero.

A grid composed of 15 candidate values of γ, logarithmically spaced in the range [0.01, 100] and 19 values of σ, linearly spaced between [0.1, 3.7], was used to search for the hyper-parameters. The pair (γ, σ) providing the best fit with respect to the tuning data window was chosen. The strategy of using a fresh data segment to tune these hyper-parameters avoids over-modeling issues. The accuracy of each pair was quantified using the *best fit rate* criterion

$$\text{BFR}(\%) = 100 \cdot \max\left(1 - \frac{\|Y - Y_s\|_2}{\|Y - \bar{Y}\|_2}, 0\right),$$

where Y_s is the output simulated by the estimated model and $\bar{Y}$ is the mean of the observed output sequence Y.

Figure 3.3 depicts the contour maps of the BFR for the range of the candidate values of (γ, σ) in the five datasets for the second-order model ($n_x = 2$). The whiter the fill, the better the model fit in the tuning window. Each subplot refers to one of the five datasets. For example, Set 1 employs data segments W2–W5 for model estimation and W1 as tuning window. The best hyper-parameter pair is indicated by a red cross, and the respective BFR is reported in the top of each subplot. The search for a suitable model order is discussed below. The model from Set 1 has the poorest fit (3.39%) due to the poor excitation in the tuning window (W1). However, this model has negligible weight in the merging process (3.22) and (3.23) as the obtained regularization factor γ is small compared to the ones in Sets 2, 3, and 4, that provided

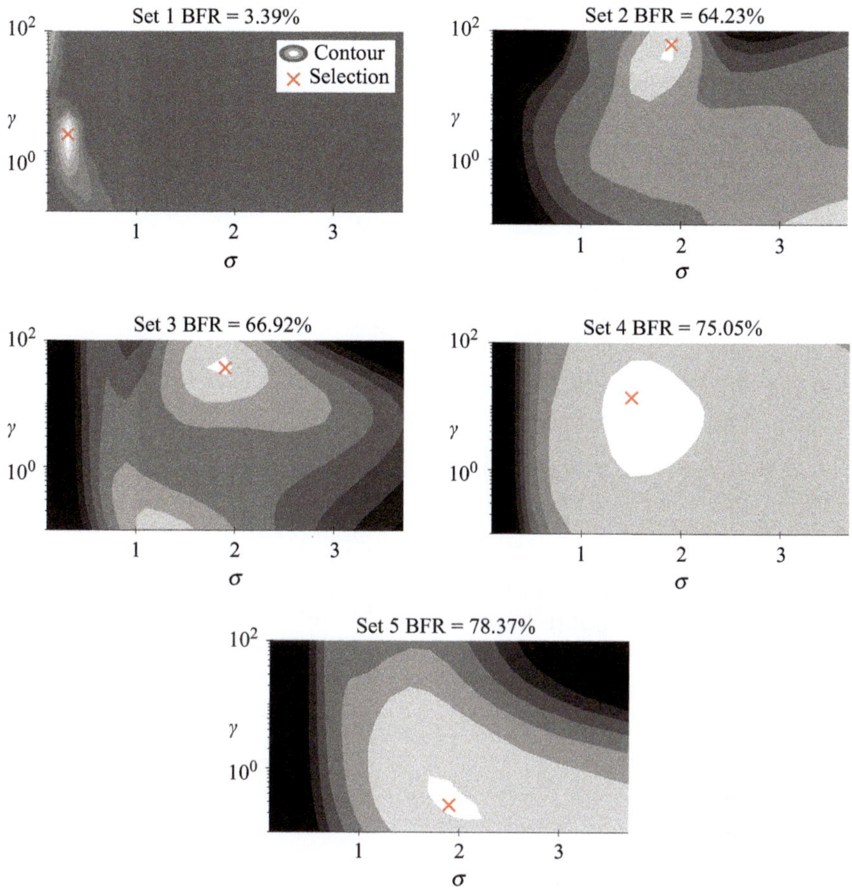

Figure 3.3 Hyper-parameters (γ, σ) contour maps for the second-order models

more accurate models. Besides, indicating that best values of the hyper-parameters are within the range of the candidate values, the contour plots also illustrate that the sensibility of the estimated model fit with respect to these hyper-parameters is different in the partitions considered. While such choice is critical for Sets 2 and 3, the range of better fits is wider in the other ones.

Preliminary tests yielded models with unexpected non-minimal phase zeros, indicating the existence of a transport delay. Even in a flexible setting provided by the support vector machines, if not addressed adequately, a transport delay may degrade substantially the estimation process [19]. A known delay of τ samples is easily incorporated by shifting the output data backward by the same amount of samples. Table 3.1 reports the achieved BFR calculated over the entire dataset of 720 samples depicted in Figure 3.2 of several models with distinct delays (up to $\tau = 5$) and distinct orders

Table 3.1 Overall BFRs of the estimated models with different delays (τ) and orders (n_x)

τ	1	2	3	4	5
$n_x = 1$	50.52%	57.96%	67.00%	70.73%	61.42%
$n_x = 2$	66.11%	72.05%	**80.55%**	68.17%	29.21%
$n_x = 3$	73.13%	74.61%	79.97%	61.52%	67.82%
$n_x = 4$	71.92%	74.93%	76.99%	61.92%	73.45%

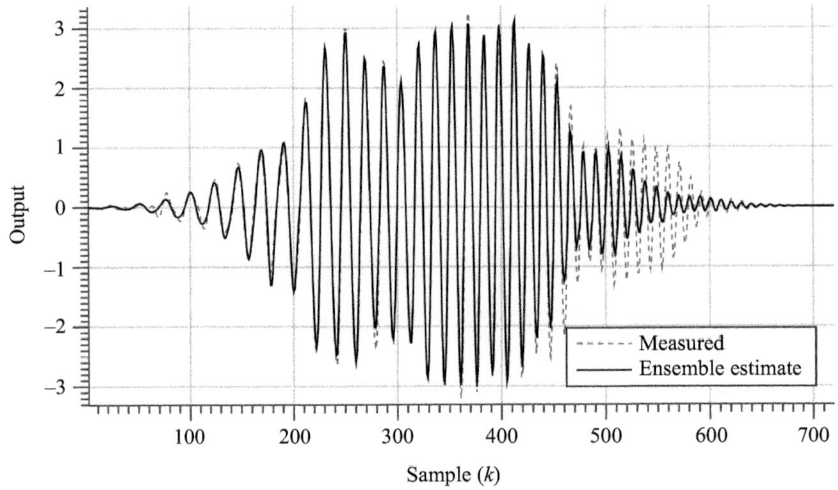

Figure 3.4 Simulated and actual system output

(up to $n_x = 4$). From this table, the model structure that best fits the entire dataset was chosen. The best choice was a second-order model with a delay of $\tau = 3$, providing an overall BFR of 80.55%, as highlighted in Table 3.1. A comparison between the actual output and the one simulated by the estimated LPV model is shown in Figure 3.4. The model fits almost perfectly to the estimation dataset (up to instant $k = 360$) and also reproduces the flutter behavior of the validation data, except for the output oscillation magnitude around $k = 550$. The BFR over the validation data was 73.38%. For the same flutter data, a previous work[2] [20] reported LTI models estimated by the `pem` function of the MATLAB™System Identification toolbox and a structured low-rank approximation method, achieving fit indexes (equivalent to the BFR) of

[2]In that work, the flutter data is split into identification (70%) and validation (30%) segments.

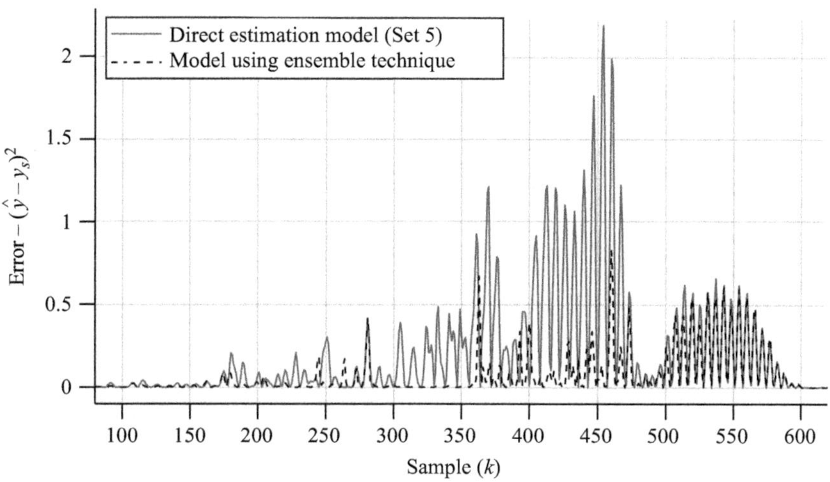

Figure 3.5 Squared residuals (quadratic error) of the most accurate model directly estimated from a dataset (Set 5) and the model estimated by the ensemble approach

approximately 40% and 35%, respectively. The modest fit achieved by LTI System Identification methods suggests the significant presence of nonlinear behaviors in the data. Another work [21] employs the LS-SVM setting to estimate a nonlinear flutter model which provides a root-mean-square error for the estimation dataset of 0.471, while our approach obtains 0.232. Although a nonlinear model structure is also considered in that work, the better result achieved by our LPV method can be attributed, in large part, to the ensemble technique associated with the merging process proposed in this work.

Therefore, it is also worth to assess into what extent did the ensemble techniques improved the model accuracy in this case. Figure 3.5 shows the squared residuals of two models: one directly estimated from Set 5 (W1–W4 for estimation and W5 for tuning), which provided the best BFR in the training data and the other obtained using the ensemble technique presented Section 3.4. The ensemble strategy clearly led to the most accurate model both over the estimation data ($k \leq 360$) and the validation data ($k > 360$). In the range $500 < k < 570$, both models had some difficulty in reproducing the amplitude of the oscillation, possibly because this is a characteristic that is not present in the first half of the data.

A complementary manner to evaluate the models accuracy is through the spectrum of the residuals depicted in Figure 3.6. For the sake of comparison, the spectrum of the output signal is also shown. This representation shows that the residuals of the model estimated with the ensemble approach have a spectrum concentrated in a frequency band where the output signal has low power. On the other hand, the residuals of the model directly estimated from Set 5 have a spectrum concentrated in a frequency

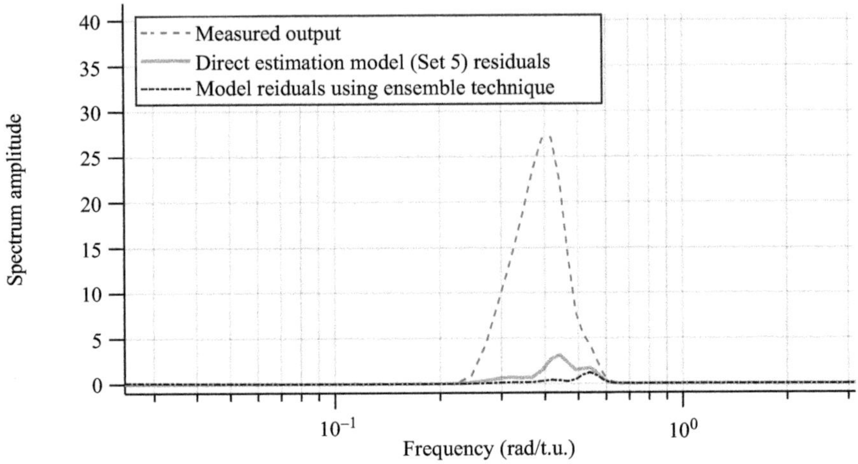

Figure 3.6 Spectrum of the residuals and the output signal

range where the magnitude of the output spectrum is relevant. In addition, the power of the residuals generated by the model estimated with the ensemble techniques is smaller in the whole frequency range.

An interesting feature of the parameterization used in this work is to allow an LTI interpretation of the estimated LPV model. For example, one can evaluate the local dynamic behavior using a set of LTI models obtained by fixing ("freezing") the scheduling signal at specific values,

$$p_k = \bar{p} = \begin{bmatrix} \bar{y} & \bar{u} \end{bmatrix}^\top,$$

such that each pair $(\bar{y}, \bar{u})$ denotes a particular operating point. An important analysis of flutter problems is the definition of an operating envelope in which the system is globally stable [3]. Although the stability of local LTI models does not necessarily imply global stability, we shall use the frozen models to assess which are the most critical operating regions to the flutter phenomenon. The operating points are the (assumed) fixed values of the input and output signals (the components of the scheduling vector). The analyzed region is the square where these signal values are in the interval $[-2.2, 2.2]$. This is almost the entire operating envelope delimited by the input and output signals, so it is reasonable to admit that the plant is excited in this region. Stable[3] local models are indicated in Figure 3.7 as "□", while unstable local models are shown using "*". An asymmetric behavior is observed on the stability map of the frozen models. More specifically, in the region around $\bar{y} = -2$ and $\bar{u} = 2$ (lower right corner) some unstable local models arise.

[3]The eigenvalues of matrix $A + L(\bar{y}, \bar{u})C$ are inside the unity circle.

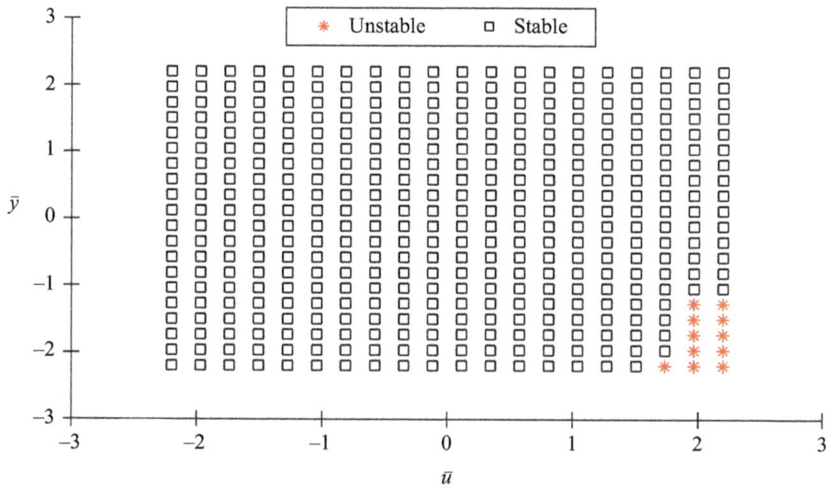

Figure 3.7 Stability map of local (frozen) LTI models

3.6 Concluding remarks

The problem of flutter identification was addressed using state-space LPV models. The proposed method contains two ingredients of the machine-learning area. The first is to use support vector machines to characterize the dependence of the model coefficients as a function of the scheduling signal. From support vectors, one can reconstruct the parameters of the model at certain operating points. The model structure adopted allows the formalization of the estimation task through an optimization problem with linear constraints, whose solution is obtained analytically.

The second ingredient is the application of the ensemble technique in which the estimation data are subdivided into segments and then arranged in different partitions so that one of these segments is always reserved for tuning design variables of the algorithm, the so-called hyper-parameters. Then, the different models obtained from each partition are fused into a final model. The application of this technique required the reformulation of the algorithm (in relation to the original proposal [4]), in order to allow the use of segmented data. This approach resulted in a more accurate model for the available data. It is worth mentioning that the segmentation strategy was very attractive in the present study because it provides an effective way of using the available data. Therefore, other identification problems subject to data scarcity may benefit from such strategy that, to the best of the authors' knowledge, have been overlooked by data-driven modeling community, particularly in the context of LPV systems.

References

[1] Kehoe MW. A Historical Overview of Flight Flutter Testing. NASA–TM–4720; 1995.

[2] Lind R, Brenner MJ, NASA Dryden Flight Research Center. Robust Flutter Margin Analysis That Incorporates Flight Data. Edwards, USA: National Aeronautics and Space Administration, Dryden Flight Research Center; National Technical Information Service; 1998.
[3] Feron E, Brenner M, Paduano J, *et al.* Time-Frequency Analysis for Transfer Function Estimation and Application to Flutter Clearance. Journal of Guidance, Control, and Dynamics. 1998;21(3):375–382.
[4] Romano RA, dos Santos PL, Pait F, Perdicoúlis TP, Ramos JA. Machine Learning Barycenter Approach to Identifying LPV State-Space Models. In: Proceedings of the 2016 American Control Conference (ACC). 2016, p. 6351–6356.
[5] dos Santos PL, Azevedo-Perdicoulis TP, Ramos JA, *et al.* LPV System Identification Using a Separable Least Squares Support Vector Machines Approach. In: Proceedings of the 53rd IEEE Conference on Decision and Control; 2014. p. 2548–2554.
[6] Tóth R, Laurain V, Zheng WX, *et al.* Model Structure Learning: A Support Vector Machine Approach for LPV Linear–Regression Models. In: Proceedings of the 50th IEEE Conference on Decision and Control. Orlando, USA; 2011. p. 3192–3197.
[7] Dietterich TG. Ensemble Methods in Machine Learning. In: Multiple Classifier Systems. MCS 2000. Lecture Notes in Computer Science. vol. 1857. Berlin, Heidelberg: Springer Berlin Heidelberg; 2000. p. 1–15.
[8] Pait F, Morse AS. A Smoothly Parameterized Family of Stabilizable, Observable Linear Systems Containing Realizations of All Transfer Functions of McMillan Degree Not Exceeding N. In: Proceedings of the 29th IEEE Conference on Decision and Control. vol. 5; 1990. p. 2722–2723.
[9] Romano RA, Pait F. Matchable–Observable Linear Models and Direct Filter Tuning: An Approach to Multivariable Identification. IEEE Transactions on Automatic Control. 2017;62(5):2180–2193.
[10] Cortes C, Vapnik V. Support-Vector Networks. Machine Learning. 1995; 20(3):273–297.
[11] Vapnik V. The Nature of Statistical Learning Theory. New York, USA: Springer New York; 2000.
[12] Baesens B. Analytics in a Big Data World: The Essential Guide to Data Science and Its Applications. Wiley & SAS Business Series. Hoboken, USA: John Wiley & Sons, Inc; 2014.
[13] Haykin SS. Neural Networks: A Comprehensive Foundation. 2nd ed. Upper Saddle River, USA: Prentice Hall; 1999.
[14] Pham TD, Wells C, Crane DI. Analysis of Microarray Gene Expression Data. Current Bioinformatics. 2006;1(1):37–53.
[15] van Wingerden JW, Gebraad P, Verhaegen M. LPV Identification of an Aeroelastic Flutter Model. In: Proceedings of the 49th IEEE Conference on Decision and Control (CDC); 2010. p. 6839–6844.

[16] Buchholz M, Larimore WE. Subspace Identification of an Aircraft Linear Parameter-Varying Flutter Model. In: Proceedings of the 2013 American Control Conference (ACC). 2013, p. 2263–2267.
[17] De Moor B, De Gersem P, De Schutter B, *et al.* DAISY: A Database for Identification of Systems. Journal A. 1997;38:4–5.
[18] Pait FM. The Barycenter Method for Direct Optimization. ArXiv e-prints. 2018; Available from: https://arxiv.org/abs/1801.10533.
[19] Zhu H, Shao H. A Novel Method of Process Dead-Time Identification: Support Vector Machine Approach. In: Proceedings of the 2004 American Control Conference. vol. 1; 2004. p. 880–884.
[20] Markovsky I. A Software Package for System Identification in the Behavioral Setting. Control Engineering Practice. 2013;21(10):1422–1436.
[21] Santos JDA, Mattos CLC, Barreto GA. Performance Evaluation of Least Squares SVR in Robust Dynamical System Identification. In: Rojas I, Joya G, Catala A, editors. Advances in Computational Intelligence. vol. 9095. Cham: Springer International Publishing; 2015. p. 422–435.

Chapter 4

Experimental modeling of a web-winding machine: LPV approaches

Jose Vuelvas[1], Fredy Ruiz[1], and Carlo Novara[2]

This chapter presents the identification of a web-winding system as a linear parameter-varying (LPV) system with the reel radius as the time-varying parameter. This system is nonlinear, time-varying and input–output unstable. Two identification methods are considered: in the first one, an LPV model is estimated in a single step using a novel approach based on sparse identification and set membership optimality evaluation. In the second one, several local linear time-invariant (LTI) models are identified using classical identification algorithms, and the overall LPV model is constructed as a weighted sum of the local models. The two methods are applied to experimental data measured on a real web-winding machine.

4.1 Introduction

Web-winding systems are very common in industry, for example, in the production of materials such as paper, cloth, plastic, wire, sheets and steel. In order to guarantee integrity and quality standards, automatic control systems are implemented to regulate the tension and speed of the material as it is transported from one reel to another, so that the process is run at the desired time, ensuring that the final product does not suffer deformations or breaks that may block the process line. One characteristic of these systems is their time-varying behavior, due to the accumulation of material that causes variations in the reel radius, resulting in changes in the moment of inertia of the motor load. That is, these systems have the reel radius as a time-varying parameter.

Several approaches to modeling of web-winding systems can be found in the literature. White-box models, based on first principles, are presented in [1–4]. These models are nonlinear, time-varying and input–output unstable. Black-box approaches have also been proposed, based on different identification techniques. In [5,6], MIMO (multi-input–multi-output) LTI models are estimated by subspaces techniques. In [7],

[1]Departamento de Ingeniería Electrónica, Pontificia Universidad Javeriana, Bogotá, Colombia
[2]Dipartimento di Elettronica e Telecomunicazioni, Politecnico di Torino, Turin, Italy

a set of MISO (multiple-input–single-output) models are estimated by instrumental variables methods. Gray-box models, which are a combination of the two previous types, are developed in [8,9], where the friction in each motor is estimated using least squares with data acquired with constant speed. The moment of inertia of the motors and the gain of the torques are identified using model-fitting techniques.

Due to the nonlinear and time-varying nature of the web-winding system, LTI control systems do not give in general adequate performances in terms of disturbance rejection, while gain-scheduling controllers are expected to provide high performances, since these controllers can effectively deal with nonlinearities and time-varying parameters. This motivates the need of accurate LPV models of the web-winding system, which may allow the design of efficient gain-scheduling controllers. Note that, because of the intrinsic system instability, model identification and validation are quite challenging tasks.

In the literature, several LPV-identification techniques have been proposed, e.g., [10,11]. In parametric approaches, fixed basis of transfer functions are weighted by nonlinear functions of the time-varying parameter. The works [12,13] claim that the critical step is how to select a priori the basis functions to describe the system dynamics. An improper selection causes a structural bias. LPV state space-model identification methods have been proposed in [14,15]. They allow the estimation of low-order models but maintain the inconvenience of basis selection. On the other hand, nonparametric techniques have been proposed in [16–18]. They yield an alternative way to face the bias problem by obtaining nonparametric models directly from data. For instance, in [19,20], kernel-based LPV state-space models are identified. Nevertheless, this method requires an appropriate choice of a kernel in order to provide a suitable model that explains the nonlinearity of the physical system.

In this chapter, two techniques are employed to estimate LPV models for a laboratory scale web-winding machine, extending the results presented in [21]. First, a unique LPV model is estimated for the complete range of variation of the reel radius. This model is estimated by means of the set membership state-space method recently proposed in [22], where the matrix coefficients are described by a linear combination of basis functions. Relevant features of this method are that the obtained model is almost optimal (in a worst case sense) and sparse (i.e., the vector of linear combination coefficients has only a few nonzero elements). Then, a set of local LTI models are estimated applying predictor-based subspace methods that allow us to handle closed-loop data, see [23]. Then, an LPV model is obtained as a weighted sum of local models employing radial basis functions with fixed centers. The weighting functions depend on the reel radius, and the models are selected minimizing a quadratic criterion solving a convex optimization problem.

The chapter is organized as follows. First, Section 4.2 describes the sparse set membership (SSM) identification method of state-space LPV systems. Then, the interpolated identification of state-space LPV systems is described in Section 4.3. The experimental setup and a comparison of the results obtained with the two identification methods are presented in Section 4.4. Finally, the conclusions of the work are presented in Section 4.5.

4.2 Sparse set membership identification of state-space LPV systems

In this section, a variant of the SSM approach proposed in [22] for state-space LPV system identification is described.

Consider a discrete-time LPV system in the following state-space observability form:

$$\begin{aligned} \mathbf{x}(k+1) &= \mathbf{A}(\mathbf{p}(k))\mathbf{x}(k) + \mathbf{B}(\mathbf{p}(k))\mathbf{u}(k) + \mathbf{B}_w w(k) \\ y(k) &= \mathbf{C}\mathbf{x}(k) + v(k) \end{aligned} \tag{4.1}$$

$$\mathbf{A}(\mathbf{p}(k)) = \begin{bmatrix} 0 & 1 & \cdots & 0 \\ \vdots & \vdots & \ddots & \vdots \\ 0 & 0 & \cdots & 1 \\ f_{n_x}(\mathbf{p}(k)) & f_{n_x-1}(\mathbf{p}(k)) & \cdots & f_1(\mathbf{p}(k)) \end{bmatrix}$$

$$\mathbf{B}(\mathbf{p}(k)) = \begin{bmatrix} f_{n_x+1}(\mathbf{p}(k)) & \cdots & f_{2n_x}(\mathbf{p}(k)) \end{bmatrix}^T$$

$$\mathbf{B}_w = \begin{bmatrix} 0 & \cdots & 0 & 1 \end{bmatrix}^T$$

$$\mathbf{C} = \begin{bmatrix} 1 & 0 & \cdots & 0 \end{bmatrix}$$

where $\mathbf{x}(k) \in \mathbb{R}^{n_x}$ is the state, $\mathbf{u}(k) \in \mathbb{R}^{n_u}$ is the input, $y(k) \in \mathbb{R}$ is the output, $\mathbf{p}(k) \in P \subset \mathbb{R}^{n_p}$ is the time-varying parameter, P is a compact set, $w(k) \in \mathbb{R}$ accounts for all the process noises acting on the system, $v(k) \in \mathbb{R}$ is a measurement noise, $\mathbf{A}(\mathbf{p}(k))$, $\mathbf{B}(\mathbf{p}(k))$, $\mathbf{B}_w$ and $\mathbf{C}$ are matrices/vectors of suitable dimensions, and

$$f_i : \mathbb{R}^{n_p} \to \begin{cases} \mathbb{R}, & 1 \le i \le n_x, \\ \mathbb{R}^{1\times n_u}, & n_x < i \le 2n_x. \end{cases}$$

We consider here the MISO case for simplicity of notation. The generalization to the MIMO case is straightforward, see [22].

Suppose that this system is not known, but a set of input–output data

$$DS = \{\mathbf{u}(k), y(k), \mathbf{p}(k)\}_{k=1-n_x}^{L} \tag{4.2}$$

is available. Then, assume that the noise

$$d(k) \doteq w(k-n_x) + v(k) - \sum_{i=1}^{n_x} f_i(\mathbf{p}(k-n_x))v(k-i), \tag{4.3}$$

accounting for both the noises $w(k)$ and $v(k)$, is unknown but bounded as

$$\|d\|_h \le \mu \tag{4.4}$$

where $\|\cdot\|_h$ is a vector ℓ_h norm.

In this section, the problem of identifying the system (4.1) from the data set (4.2) is considered. Note that bound (4.4) on the noise affecting the measurements of $y(t)$

does not assume any statistical property on $d(t)$, then the method presented here is valid for data generated in open loop or in closed loop.

Since system (4.1) is univocally defined by the vector-valued function $f = [f_1 \cdots f_n]^T$, $n \doteq (1 + n_u)n_x$, its identification can be performed by finding an estimate $\widehat{f}$ of f giving a "small" identification error

$$E(f,\widehat{f}) \doteq \left\| \xi\left(f,\widehat{f}\right) \right\|_q \tag{4.5}$$

where

$$\begin{aligned} \xi\left(f,\widehat{f}\right) &\doteq \left[\xi\left(f_1,\widehat{f}_1\right) \ \cdots \ \xi\left(f_n,\widehat{f}_n\right)\right]^T \\ \xi\left(f_i,\widehat{f}_i\right) &\doteq \left\| f_i - \widehat{f}_i \right\|_s, \end{aligned}$$

$\|\cdot\|_q$ is a vector ℓ_q norm and $\|\cdot\|_s$ is a functional L_s norm evaluated over the set $P \subset \mathbb{R}^{n_p}$.

An algorithm for deriving such an estimate is now presented. For simplicity, we consider here the case where the first $n_x - 1$ elements $f_{n_x+1}, \ldots, f_{2n_x-1}$ of the matrix $B(p(k))$ are null. The general case can be treated in a similar way, at the expense of a more complicated notation. It can be noted that, even supposing null these $n_x - 1$ elements, the LPV model structure (4.1) is general enough to capture quite complex systems, as demonstrated in Section 4.4.1, where the identification of a real-web winding machine is carried out.

Let $BF \doteq \{\beta_j : \mathbb{R} \to \mathbb{R}, j = 1, \ldots, m\}$ be a set of basis functions (guidelines for the choice of these functions are given in [22]). Define the following quantities:

$$\begin{aligned} \mathbf{Y} &\doteq [y(1) \ \cdots \ y(L)]^T \\ \Lambda_i &\doteq \begin{bmatrix} \beta_1(\rho_i(1)) & \cdots & \beta_m(\rho_i(1)) \\ \vdots & \ddots & \vdots \\ \beta_1(\rho_i(L)) & \cdots & \beta_m(\rho_i(L)) \end{bmatrix} \in \mathbb{R}^{L\times m} \\ \Psi &\doteq \left[\phi_1\Lambda_1 \ \cdots \ \phi_n\Lambda_n\right] \in \mathbb{R}^{L\times N} \end{aligned} \tag{4.6}$$

where $i = 1, \ldots, n$, $N \doteq nm$,

$$\begin{aligned} \rho_i(k) &\doteq \begin{cases} \mathbf{p}(k - n_x), & 1 \le i \le n_x \\ \mathbf{p}(k + n_x + 1 - i), & n_x < i \le n \end{cases} \\ \varphi(k) &\doteq [y(k-1) \ \cdots \ y(k-n_x) \ \cdots \\ &\qquad \mathbf{u}(k-1) \ \cdots \ \mathbf{u}(k-n_x)]^T \\ \phi_i &\doteq \mathrm{diag}([\varphi_i(1) \cdots \varphi_i(L)]), \end{aligned} \tag{4.7}$$

and φ_i denotes the ith component of the regressor $\varphi(k)$. Note that, while the set BF consists of m functions, the total number of basis functions actually used for identification is $N \doteq nm$. Indeed, in the matrix Ψ, each function of BF is multiplied by each component of the regressor $\varphi(k)$, and the total number of regressor components is $n = (1 + n_u)n_x$.

SSM algorithm

1. Solve the optimization problem

$$\begin{aligned} \theta^1 &= \arg\min_{\theta\in\mathbb{R}^N} \|\theta\|_1 \\ \text{subject to} &\quad \|\mathbf{Y} - \Psi`\theta\|_h \leq \mu. \end{aligned} \tag{4.8}$$

2. Consider the set of indices

$$\overline{\zeta}(\theta^1) \doteq \{i \in \{1,2,\ldots,N\} : \theta_i^1 = 0\}$$

where θ_i^1 are the components of the vector θ^1.
Obtain the coefficient vector θ^* as the solution of the following optimization problem:

$$\begin{aligned} \theta^* &= \arg\min_{\theta\in\mathbb{R}^N} \|\mathbf{Y} - \Psi\theta\|_h \\ \text{subject to} &\quad \theta_i = 0,\ i \in \overline{\zeta}(\theta^1). \end{aligned} \tag{4.9}$$

3. Define the coefficients $\alpha_{i,j}^* \doteq \theta_{(i-1)m+j}^*$. An estimate of the unknown function f is given by $f^* = [f_1^* \ \cdots \ f_n^*]^T$, where

$$f_i^*(\rho) = \sum_{j=1}^{m} \alpha_{i,j}^* \beta_j(\rho). \tag{4.10}$$

4. The resulting LPV model is a system of the form (4.1) with $f = f^*$. ■

Remark 4.1. *In [22], it is shown that, assuming a Lipschitz continuous* residue functions $\Delta_i \doteq f_i - f_i^*$, $i = 1, \ldots, n$, *the estimate* f^* *is almost optimal (in a worst case sense) for any* ℓ_q *and* ℓ_h *vector norms, and any* L_s *functional norm. That is,*

$$EW(f^*) \leq 2 \inf_{\hat{f}} EW(\hat{f})$$

where EW *is the* worst case identification error, *defined as*

$$EW(\hat{f}) \doteq \sup_{g\in FFS} E(g,\hat{f})$$

and FFS is the feasible function set, *i.e., the set of all functions consistent with prior assumptions and measured data. This shows that the worst case identification error of* f^* *is at most twice the lowest achievable one. See [22] for more details and for a rigorous proof of this result.* ■

In the reminder of this section, the sparsity properties of the SSM algorithm and of the resulting estimate f^* are briefly discussed. A *sparse* function is a linear combination of many basis functions, where the vector of linear combination coefficients is sparse, i.e., it has only a few nonzero elements. The *sparsity* of a vector θ is typically measured by the ℓ_0 quasi-norm, defined as the number of its nonzero elements. Sparse identification can thus be performed by looking for a coefficient vector of the basis function linear combination with a small ℓ_0 quasi-norm. However, the ℓ_0 quasi-norm is a non-convex function, and its minimization is in general an nondeterministic polynomial time (NP-hard) problem. Two main approaches are commonly adopted to deal with this issue: convex relaxation and greedy algorithms. In convex relaxation, a suitable convex function, e.g., the ℓ_1 norm, is minimized instead of the ℓ_0 quasi-norm [24–26]. Indeed, the ℓ_1 norm is the convex envelope of the ℓ_0 quasi-norm, and its minimization yields a sparse vector. In greedy algorithms, the sparse solution is obtained iteratively [27].

The above SSM algorithm is essentially an improved ℓ_1 algorithm: In step 1, an optimization problem is solved, where the ℓ_0 quasi-norm is replaced by the ℓ_1 norm. The vector θ^1 derived in step 1 is thus sparse and guarantees a bounded prediction error $\|\mathbf{Y} - \Psi\theta^1\|_h$. However, θ^1 is not ensured to give the minimum prediction error among all vectors with the same sparsity. Then, a vector θ^* is obtained in step 2, having the same sparsity of θ^1 and giving minimum prediction error. In [28], a condition is provided, under which θ^* is maximally sparse (i.e., with minimum ℓ_0 quasi-norm).

Note that the sparsity property is of high practical interest. Indeed, if θ^* is sparse, the evaluation of $f_i^*(\rho)$ is very "fast" and can be performed in online applications where "small" sampling periods are utilized (e.g., model predictive control, gain scheduling).

4.3 Interpolated identification of state-space LPV systems

A classic approach to the identification of systems of the form (4.1) is to perform local experiments guaranteeing constant values of the parameter $\mathbf{p}(k) = \bar{\mathbf{p}} \in \mathbb{R}$ and then to build an ILM. The combination of local models to build an LPV system is not a trivial task. If common basis are employed in the local identification step, the structure proposed in [29] can be employed. When state-space models are estimated, similarity transformations that allow the combination of local models must be pursuit, see, e.g., [30]. The complexity of the aforementioned methods stems from the fact that a common state vector is required for the local models to build the interpolation. To avoid this limitation, in this paper, the interpolated model is constructed as the sum (*parallel connection*) of local models.

For a given constant value $\bar{p}_i$, the system (4.1) can be expressed as

$$\begin{aligned} \mathbf{x}(k+1) &= \mathbf{A}(\bar{p}_i)\mathbf{x}(k) + \mathbf{B}(\bar{p}_i)\mathbf{u}(k) + \mathbf{B}_w w(k) \\ y_i(k) &= \mathbf{C}(\bar{p}_i)\mathbf{x}(k) + v(k) \end{aligned}$$

and thus it can be identified using standard estimation techniques for LTI systems. When it is not possible to maintain a constant parameter p during the experiment, but f_i are smooth functions of p, it can be assumed that the system dynamics remain almost constant when p belongs to an interval $[\underline{p}, \overline{p}]$, where the width of the interval depends on the smoothness of the functions f_i.

The interpolated model generates an output formed by a weighted sum of the outputs predicted by the local models,

$$\hat{y}_{LPV}(k) = \sum_{i=1}^{l} \alpha_i \sigma_i(p(k)) y_i(k), \tag{4.11}$$

where l is the number of local models, α_i is a fixed weighting parameter and $\sigma_i(p(k))$ is an activation function of the time-varying parameter $p(k)$.

The concept behind (4.11) is a gain-scheduling structure. The activation functions $\sigma_i(p(k))$ select as global output $\hat{y}_{LPV}(k)$, the output of the local system i, whose dynamics behave as system (4.1) for the current parameter $p(k)$.

Given a set of input–output data

$$DS = \{\mathbf{u}(k), y(k), p(k)\}_{k=1-n_x}^{L},$$

the aim of the procedure presented here is to estimate several local LTI models for different intervals of the parameter $p(k)$, and then to combine them in order to construct an interpolated LPV model (ILM). The algorithm consists of two main parts, as shown in the following:

ILM algorithm

1. *Local model estimation algorithm*
   i. Group the samples in l identification data set as

   $$DS_i = \left\{\mathbf{u}(k), y(k), p(k) : \underline{p}_i \leq p(k) < \overline{p}_i\right\}, \quad i = 1, 2, \ldots l.$$

   where l is the number of local models.
   ii. Estimate an appropriate order of the local model for each data set DS_i and obtain the model matrices $\hat{\mathbf{A}}_i, \hat{\mathbf{B}}_i, \hat{\mathbf{C}}_i$ and noise variances $var(w)$ and $var(v)$. Note that different orders can be selected for local models as no state combination is performed in the interpolation. Diverse LTI identification methods can be employed to estimate local models. For example, when the system is operated in open loop, robust subspace algorithms, such as `N4SID` [31], can be applied. When closed-loop data are employed, recursive identification techniques can be exploited [32,33].
   iii. Obtain the simulated output $\hat{y}_i(k)$ for the estimated model as

   $$\hat{y}_i(k) = \sum_{r=0}^{k-1} \hat{\mathbf{C}}_i \hat{\mathbf{A}}_i^{k-r-1} \hat{\mathbf{B}}_i u(r), \; k = 1, \ldots, L.$$

2. *Interpolation function estimation*
   i. Compute the vector $\alpha = [\alpha_1 \; \cdots \; \alpha_l]^T$ by means of the following quadratic optimization problem:

   $$\begin{aligned} \alpha^* &= \arg\min_{\alpha \in \mathbb{R}^l} \|\mathbf{Y} - \hat{\mathbf{Y}}_{LPV}\|_2 \\ &\quad \text{subject to } \; 0 \leq \alpha_i \leq 1. \end{aligned}$$

   where $\hat{\mathbf{Y}}_{LPV} \doteq [\hat{y}_{LPV}(1) \; \cdots \; \hat{y}_{LPV}(L)]^T$ and

   $$\hat{y}_{LPV}(k) = \sum_{i=1}^{l} \alpha_i \sigma_i(p(k)) \hat{y}_i(k)$$

   with

   $$\sigma_i(p) = e^{-\beta_i (p - p_i^c)^2}$$

   where $p_i^c = 0.5(\underline{p}_i + \overline{p}_i)$ and β_i is a fixed parameter. Radial basis functions with Gaussian form are selected to interpolate the models output; however, any radial basis set can be employed.

ii. The LPV model is built as a weighted sum of the local models output:

$$\hat{y}_{LPV}(k) = \sum_{r=1}^{l} \alpha_i^* \sigma_i(p(k))\hat{y}_i(k).$$ ■

Remark 4.2. *The order of the complete system is*

$$n_{ILM} = \sum_{r=1}^{l} n_r.$$

That is, the sum of the orders (state dimension) of the local LTI models. ■

4.4 Web-winding system identification

4.4.1 The web-winding system

The considered web-winding system is a laboratory scale plant (see Figure 4.1), where it is required to transport material at a desired tension and speed, ensuring that the web (in this case magnetic tape) is not broken or deteriorated. This system includes the following components:

- Two DC motors with optical encoder (and with angular speeds ω_1, ω_2) drive the material, each one connected to a reel.
- An optical encoder coupled to one of the pulleys that guides the web measures the web transport speed (angular speed ω_3).
- A dancer arm system measures the web tension. Its rotation axis is connected to a linear potentiometer.
- A radius meter for one of the reels (the radius is denoted by r_2).

Figure 4.1 A picture of the web-winding system

Figure 4.2 shows a scheme of the plant operation when the web is rolled around the reel connected to motor 2. The difference between tensions T_1 and T_2 causes the transport of the material with a web speed ω_3 and a variation of the height h.

The voltages applied to motors 1 (V_1) and 2 (V_2) are the system input, the material transport speed (ω_3) is the output of interest, and the radius r_2 is the time-varying parameter.

Assuming that the motors 1 and 2 have equal electromechanical characteristics, that the slip between the tape and bearings is negligible, that the dancer arm moves vertically only, and that all the fixed axes of the plant have null moment of inertia, the system dynamics can be described by the following set of equations:

$$J_i(t)\dot{\omega}_i(t) = -B_i\omega_i(t) + \frac{k_T}{R_a}(V_i(t) - k_V\omega_i(t)) + r_i(t)T_i(t);\ \ i = 1,2 \tag{4.12}$$

$$\dot{r}_i(t) = \frac{s}{2\pi}\omega_i(t);\ \ i = 1,2 \tag{4.13}$$

$$\dot{\omega}_3(t) = \frac{r_3}{J_3}\left(T_2(t) - T_1(t)\right) \tag{4.14}$$

$$\dot{T}_1(t) = k_E(r_3\omega_3(t) - r_1(t)\omega_1(t)) + D(r_3\dot{\omega}_3(t) - r_1(t)\dot{\omega}_1(t) - \dot{r}_1(t)\omega_1(t)) \tag{4.15}$$

$$\dot{T}_2(t) = k_E(r_2(t)\omega_2(t) - r_3\omega_3(t)) + D(r_2(t)\dot{\omega}_2(t) + \dot{r}_2(t)\omega_2(t) - r_3(t)\dot{\omega}_3(t)) \tag{4.16}$$

where ω_i are the motors speed, J_i are the time-varying moments of inertia of each motor axis, r_i are time-varying radius of the motor loads. The model constant parameters are the following: B_i are the axis friction coefficients, k_T is the motor torque constant, k_V is the motor voltage constant, R_a is the armature resistance, s is the tape thickness,

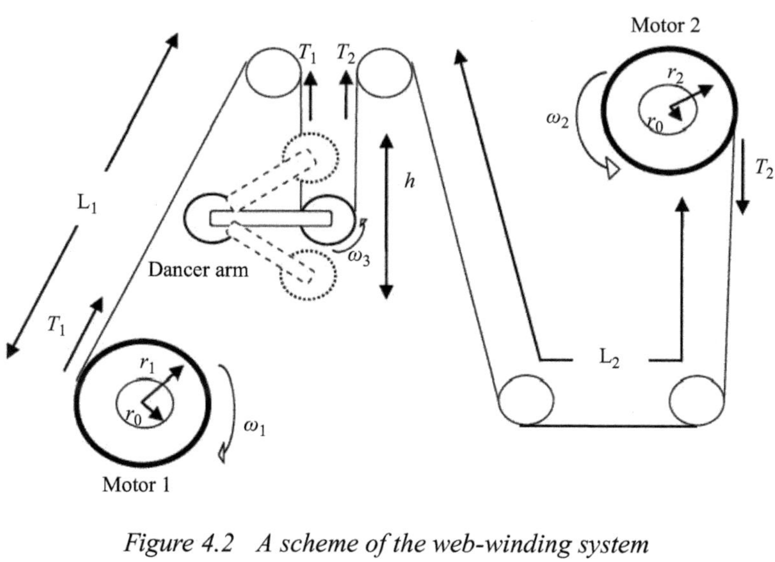

Figure 4.2 A scheme of the web-winding system

r_3 is the dancer arm radius, J_3 is the dancer arm moment of inertia, k_E is the tape elastic constant, and D is the tape damping constant.

From (4.12) to (4.16), it can be seen that the web-winding system is a quite complex plant, approximately described by a set of seven nonlinear and time-varying differential equations. Moreover, the web-winding system, by nature, is input–output unstable [1–4]. Due to these features, linear control systems do not give in general adequate performances in terms of disturbance rejection. On the contrary, gain-scheduling controllers are expected to provide high performances, since these controllers can effectively deal with nonlinearities and time-varying parameters. This motivates the need of accurate LPV models, which can be used to design efficient gain-scheduling controllers. Note that obtaining an LPV model from (4.12) to (4.16) is not practicable, since these equations involve several uncertain parameters, which are in general difficult to measure or estimate. Note also that, because of the intrinsic instability of the system, model identification and validation are in general quite difficult tasks.

4.4.2 Experiment description

The web-winding system previously described is input–output unstable in open loop; therefore the identification experiments have been performed in closed loop. Two decoupled proportional-integral-derivative (PID) controllers have been used to regulate the web transport speed ω_3 and web tension T_1 by controlling the dancer arm position θ, see Figure 4.3. These controllers have been tuned experimentally in order to stabilize the plant in the complete operation range.

For the system-identification process, the signals V_1, V_2, and ω_3 were recorded with a sample time of 40 ms. It has been observed that the system bandwidth does not exceed 10 Hz. Each experiment started with no material on the reel attached to motor 2 and finished when it reached its maximum capacity. The duration of each experiment was about 90 s (2,250 samples).

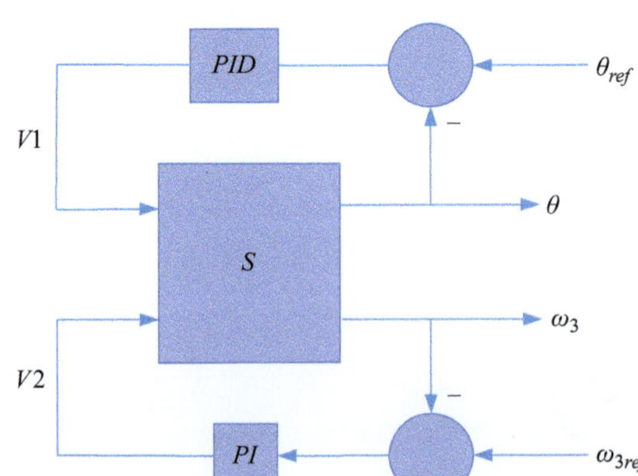

Figure 4.3 Control system employed for generating the data sets

The input signals for each experiment is formed by a sequence of set-point commands for the two PID loops. Five experiments have been carried out, where different set-point signals have been used:

- Pseudorandom binary sequence (PRBS). This signal has a peak amplitude of 3.5 degrees around the set point for the dancer arm control loop and 21 rad/s for the web speed control loop. Two experiments, called PRBS1 and PRBS2, have been performed using a PRBS input signal generated as a 4,095-step random sequence, the only difference between these two experiments is the realization of measurement noise and uncontrollable disturbances.
- Random binary sequence (RBS). This signal has a peak amplitude of 5 degrees around the set-point for the dancer arm control loop and 20 rad/s for the web speed control loop. The switching time is ten sample instants, with switching probability 0.2.
- Sum of sinusoids (SINE). This signal is formed by 20 sinusoids and has a maximum amplitude of 8 degrees to the dancer arm control loop and 48 rad/s to the speed control loop.
- Random Gaussian sequence (RGS): This signal has a peak amplitude of 10 degrees around the set-point for the dancer arm control loop and 40 rad/s for the web speed control loop. It is generated as a white Gaussian sequence filtered by a low-pass filter with cut-off frequency 1 Hz.

The data set has been partitioned as follows:

- *Estimation data set*: Data generated in the PRBS1 and RBS experiments.
- *Validation data set*: Data generated in all the five experiments.

4.4.3 Sparse set membership LPV model

Identification of the web-winding system has been performed by means of the SSM approach summarized in Section 4.2. The optimization problems in the SSM algorithm have been solved using the MATLAB®-based modeling system for convex optimization (CVX) package [34].

The following set of $m = 8$ basis functions has been used:

$$BF \doteq \{1, \rho, \rho^2, \rho^3, \rho^4, \rho^{-1}, \rho^{-2}, \rho^{-3}\}.$$

Model orders n_x in the range $[1, 15]$ have been considered. For each order, a model has been identified on the estimation data set by means of the SSM algorithm. The value of μ has been taken slightly larger than the minimum value for which the optimization problem (4.8) was feasible (see [22] for more details). The chosen values of μ resulted to range in the interval $[30, 40]$.

The identified models have been compared in simulation on the estimation data set, and the one with order $n_x = 13$ provided the best results. Note that, while the set BF consists of $m = 8$ functions, the total number of basis functions actually used for the identification of this model is $N \doteq nm = (1 + n_u)n_x m = 312$. Indeed, each function of BF has been multiplied by each regressor to build the matrix Ψ in (4.6), and the total number of regressors is $n = (1 + n_u)n_x = 39$. Among these 312 basis functions, the SSM algorithm selected 37 functions. A model of order $n_x = 13$ has thus been chosen and called SSM.

4.4.4 Interpolated LPV model

Identification of the web-winding system has been performed by means of the interpolation-based approach described in Section 4.3. In the case of the web-winding machine, it is not possible to maintain a constant reel radius, but it is known that the web moves always from motor 1 to 2, then r_1 and r_2 are monotonically decreasing and increasing functions of time, respectively.

The estimation set DS has been split into l subsets DS_i. Each one containing input and output data corresponding to a radius interval $r_2 \in [\underline{p}_i, \overline{p}_i]$, forming equally spaced intervals of the reel radius range. The ILM algorithm has been applied for several amounts of subsystems in the range $[2, \ldots, 15]$. Note that for $l = 15$, each local data set has 150 samples on average.

The predictor-based subspace identification toolbox [35] has been employed for local models estimation. For each data set DS_i, an order 4 has been suggested by the predictor-based subspace identification toolbox. For the construction of the weighting functions, the value of β has been adjusted according to the number of local models. Centers p_i^c of the radial basis functions have been set to the interval centers $(\underline{p}_i + \overline{p}_i)/2$. This quadratic programming (QP) problem has been solved using the CVX package [34].

The identified models have been compared in simulation on the estimation data set, and the one with $l = 3$ sub-models provided the best results. Then, an interpolated model of order $n_{ILM} = 12$ has been selected and called ILM.

Finally, a single LTI model has been identified from the estimation set employing the predictor-based subspace identification toolbox [35]. Several orders have been tried and the model that showed better performance on the estimation set was a model with $n = 6$. This model is named LTI in the following.

4.4.5 Model validation and results

The three models identified in the previous subsections have been simulated in open-loop, using as inputs the measured voltages of the validation data sets. Then, the outputs provided by the models have been compared to the measured output signals. The goodness of fit between test and reference data (*FIT*) index and the relative maximum (*RM*) error have been used to evaluate the model quality. These indexes have been computed as

$$FIT = 100\left(1 - \frac{\|\widetilde{\mathbf{Y}} - \mathbf{Y}_s\|_2}{\|\widetilde{\mathbf{Y}} - \text{mean}(\widetilde{\mathbf{Y}})\|_2}\right) \tag{4.17}$$

$$RM = 100\frac{\|\widetilde{\mathbf{Y}} - \mathbf{Y}_s\|_\infty}{\|\widetilde{\mathbf{Y}} - \text{mean}(\widetilde{\mathbf{Y}})\|_\infty} \tag{4.18}$$

where $\widetilde{\mathbf{Y}} = [y(1) \cdots y(L_e)]^T$ and $\mathbf{Y}_s = [y_s(1) \cdots y_s(L_e)]^T$ are the measured and simulated output signals, respectively, and L_e is the experiment length.

The obtained results are shown in Tables 4.1 and 4.2 for all the five experiments in the validation set. A comparison between the measured and simulated outputs is shown in Figures 4.4–4.7 for four experiments of the validation set. From these tables and figures, it can be seen that both the LPV models are able to explain quite accurately data not employed for identification.

Note that the ILM model is formed by three LTI models of the third order, yielding an overall model of order 12, whereas the SSM model is of order 13. It is important to remark that some of the 37 basis functions selected by the SSM algorithm during the

Table 4.1 Model FIT indexes on the validation data set

Model	PRBS1	PRBS2	RBS1	SINE1	RGS1
SSM	95	93	92	90	90
ILM	86	85	89	83	90
LTI	86	85	89	83	90

Table 4.2 Model RM errors on the validation data set

Model	PRBS1	PRBS2	RBS1	SINE1	RGS1
SSM	5	5	6	8	6
ILM	15	15	18	17	8
LTI	86	85	89	83	90

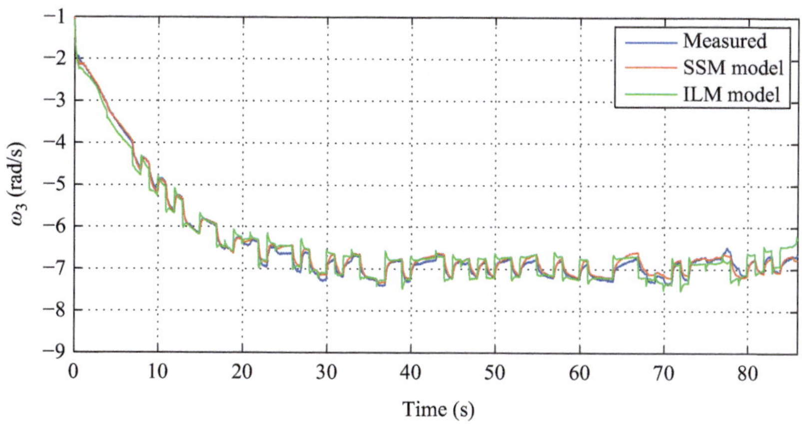

Figure 4.4 Experiment PRBS2: comparison between measured and simulated outputs

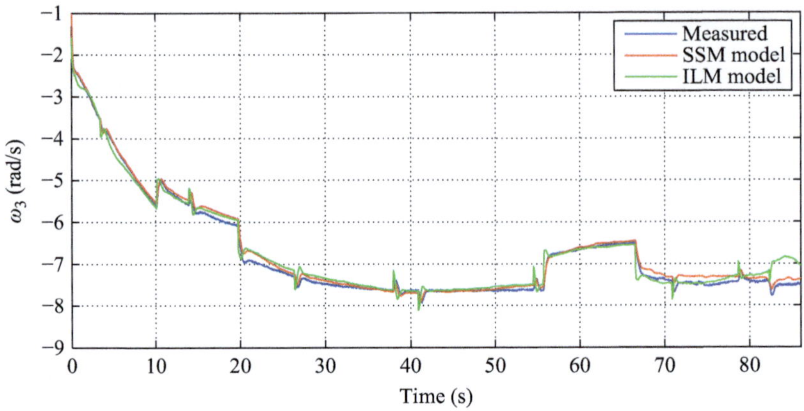

Figure 4.5 Experiment RBS1: comparison between measured and simulated outputs

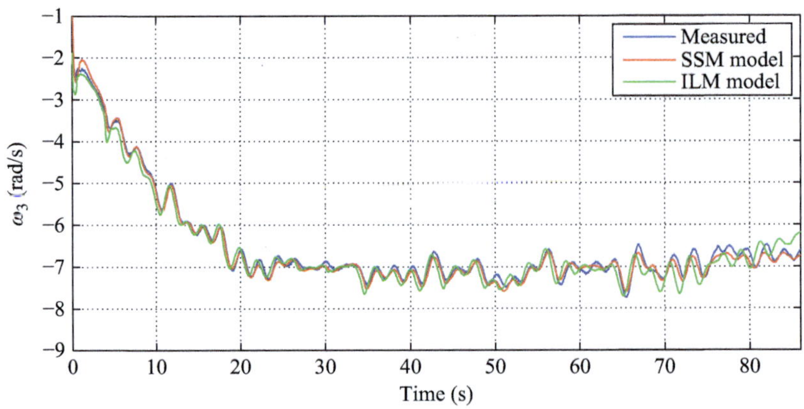

Figure 4.6 Experiment SINE1: comparison between measured and simulated outputs

identification of the SSM model involve regressors of orders 12 and 13, confirming that such orders are necessary to obtain accurate results.

4.5 Conclusion

Two methods for the identification of LPV models have been applied to a laboratory scale web-winding machine, describing the web-speed dynamics and employing the reel radius as the time-varying parameter. A unique LPV model in state-space form has been estimated, where the matrix coefficients are described by a linear combination of basis functions. The coefficients of the basis expansion have been selected in order to maximize the model sparsity and to guarantee at the same time its almost

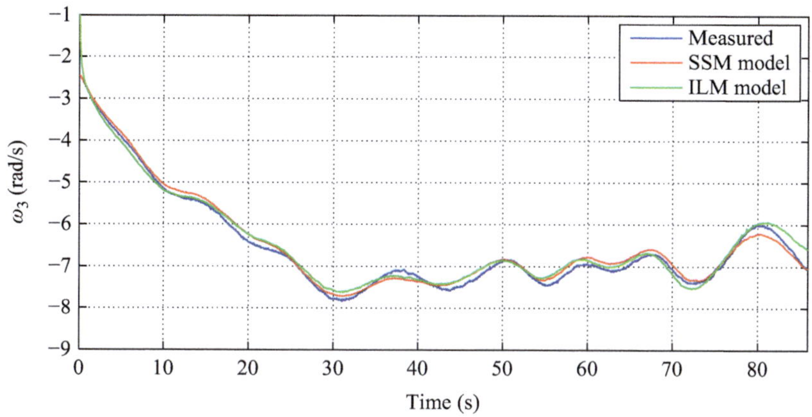

Figure 4.7 Experiment RGS1: comparison between measured and simulated outputs

optimality (in a set membership framework). A second model was constructed as an interpolation of local LTI models, estimated by subspace methods. Three time-invariant models have been identified for equally spaced reel radius intervals, and a weighted sum of the LTI models outputs has been formed minimizing a quadratic criterion. Both the models are able to explain fresh data, not used in the identification phase. Better results have been obtained with the unique LPV model in terms of FIT coefficient and maximum error, while the interpolated model required a simpler identification process.

References

[1] Liu Z. Dynamic analysis of center-driven web winder controls. In: Industry Applications Conference, 1999. Thirty-Fourth IAS Annual Meeting. Conference Record of the 1999 IEEE. vol. 2. IEEE; 1999. p. 1388–1396.

[2] Benlatreche A, Knittel D, Ostertag E. Robust decentralised control strategies for large-scale web handling systems. Control Engineering Practice. 2008;16(6):736–750.

[3] Lynch AF, Bortoff SA. Nonlinear tension observers for web machines. Automatica. 2004;40(9):1517–1524.

[4] Chen CL, Chang KM, Chang CM. Modeling and control of a web-fed machine. Applied Mathematical Modelling. 2004;28(10):863–876.

[5] Noura H, Bastogne T. Tension optimal control of a multivariable winding process. In: American Control Conference, 1997. Proceedings of the 1997. vol. 4. IEEE; 1997. p. 2499–2503.

[6] Bastogne T, Noura H, Sibille P, *et al.* Multivariable identification of a winding process by subspace methods for tension control. Control Engineering Practice. 1998;6(9):1077–1088.

[7] Garnier H, Gilson M, Young P, *et al.* An optimal IV technique for identifying continuous-time transfer function model of multiple input systems. Control Engineering Practice. 2007;15(4):471–486.

[8] Xu Y, Wang D, Zhang Q. Modeling and robust control of web winding system with sinusoidal tension disturbance. In: Mechatronics and Automation, Proceedings of the 2006 IEEE International Conference on. IEEE; 2006. p. 1958–1963.

[9] Xu Y. Modeling and LPV control of web winding system with sinusoidal tension disturbance. In: Control and Decision Conference, 2009. CCDC'09. Chinese. IEEE; 2009. p. 3815–3820.

[10] Tóth R. Modeling and Identification of Linear Parameter-Varying Systems. The Netherlands: Springer; 2010.

[11] dos Santos PL. Linear Parameter-Varying System Identification: New Developments and Trends. Advanced Series in Electrical and Computer Engineering. World Scientific Publishing Company; 2012. Available from: https://books.google.com.co/books?id=4AjU84zbhiMC.

[12] van Wingerden JW, Verhaegen M. Subspace identification of multivariable LPV systems: A novel approach. In: Proceedings of the 2008 IEEE International Symposium on Computer-Aided Control System Design. vol. 2; 2008. p. 840–845.

[13] Toth R, Abbas HS, Werner H. On the state-space realization of LPV input-output models: Practical approaches. IEEE Transactions on Control Systems Technology. 2012;20(1):139–153.

[14] Larimore WE, Cox PB, Toth R. CVA identification of nonlinear systems with LPV state-space models of affine dependence. In: Proceedings of the American Control Conference. 2015;2015 July. pp. 831–837.

[15] Tanelli M, Ardagna D, Lovera M. Identification of LPV state space models for autonomic web service systems. IEEE Transactions on Control Systems Technology. 2011;19(1):93–103.

[16] Hsu K, Vincent TL, Poolla K. Nonparametric methods for the identification of linear parameter varying systems. In: IEEE International Conference on Computer-Aided Control Systems; 2008. p. 846–851.

[17] Goos J, Pintelon R. Continuous time frequency domain LPV state space identification via periodic time-varying input-output modeling. In: 53rd IEEE Conference on Decision and Control; 2014. p. 2567–2572.

[18] Rizvi SZ, Mohammadpour J, Roland T. An IV-SVM-based approach for identification of state-space LPV models under generic noise conditions. In: Conference on Decision and Control (CDC); 2015. p. 7380–7385.

[19] Proimadis I, Bijil HJ, Van Wingerden JW. A kernel based approach for LPV subspace identification. IFAC Proceedings Volumes (IFAC-PapersOnline). 2015;48(26):97–102.

[20] Rizvi SZ, Mohammadpour J, Tóth R, Meskin N. A kernel-based approach to MIMO LPV state-space identification and application to a nonlinear process system. IFAC Proceedings Volumes (IFAC-PapersOnline). 2015;48(26): 85–90.

[21] Ruiz F, Vuelvas J, Novara C. LPV model identification for a web winding system. IFAC Proceedings Volumes (IFAC-PapersOnline). 2012;16(part I): 1779–1784.

[22] Novara C. Set membership identification of state-space LPV systems. In: dos Santos PL, Perdicoúlis TPA, Novara C, *et al.*, editors. Linear Parameter-Varying System Identification – New Developments and Trends, Advanced Series in Electrical and Computer Engineering Vol. 14. Singapore: World Scientific Publishing Co; 2012. p. 65–93.

[23] Houtzager I, van Wingerden JW, Verhaegen M. VARMAX-based closed-loop subspace model identification. In: Decision and Control, 2009 held jointly with the 2009 28th Chinese Control Conference. CDC/CCC 2009. Proceedings of the 48th IEEE Conference on; 2009. p. 3370–3375.

[24] Fuchs JJ. Recovery of exact sparse representations in the presence of bounded noise. IEEE Transactions on Information Theory. 2005;51(10):3601–3608.

[25] Tropp JA. Just relax: Convex programming methods for identifying sparse signals in noise. IEEE Transactions on Information Theory. 2006;52(3): 1030–1051.

[26] Donoho DL, Elad M, Temlyakov VN. Stable recovery of sparse overcomplete representations in the presence of noise. IEEE Transactions on Information Theory. 2006;52(1):6–18.

[27] Tropp JA. Greed is good: Algorithmic results for sparse approximation. IEEE Transactions on Information Theory. 2004;50(10):2231–2242.

[28] Novara C. Sparse identification of nonlinear functions and parametric Set Membership optimality analysis. In: American Control Conference. San Francisco, USA; 2011.

[29] Tóth R, Heuberger P, Van den Hof P. Asymptotically optimal orthonormal basis functions for LPV system identification. Automatica. 2009;45(6):1359–1370.

[30] De Caigny J, Camino JF, Swevers J. Interpolation-based modeling of MIMO LPV systems. IEEE Transactions on Control Systems Technology. 2011;19(1):46–63.

[31] Overchee V, De Moor B. Subspace Identification for Linear Systems. Dordrecht, Holanda: Kluwer Academic Publishers; 1996.

[32] Chiuso A, Picci G. Consistency analysis of some closed-loop subspace identification methods. Automatica. 2005;41(3):377–391.

[33] Houtzager I, van Wingerden JW, Verhaegen M. Fast-array recursive closed-loop subspace model identification. In: Proceedings of the 15th IFAC Conference on System Identification (SYSID 2009). Saint Malo, France; 2009.

[34] Grant M, Boyd S. CVX: Matlab Software for Disciplined Convex Programming, version 1.21; 2010. http://cvxr.com/cvx.

[35] Houtzager I, Gebraad PMO, van Wingerden JW, *et al.* Predictor-based Subspace Identification Toolbox Version 0.5; 2012. http://www.dcsc.tudelft.nl/~datadriven/pbsid/.

Chapter 5

In situ identification of electrochemical impedance spectra for Li-ion batteries

Tyrone Vincent[1], Peter J. Weddle[2], Aleksei La Rue[1], and Robert J. Kee[2]

The monitoring and control of battery systems can be enhanced by data collection and analysis that provide insight into the internal behavior of the battery. A well-known example is electrochemical impedance spectroscopy (EIS), which is equivalent to estimating the frequency response of the battery impedance at a particular operating condition. System identification provides a method for implementing EIS using hardware commonly found in advanced battery-management systems. In this chapter, a possible implementation of online system identification is discussed and illustrated using both simulation and experimental data.

5.1 Introduction

EIS is a standard process that identifies the frequency response of an electrochemical system by observing voltage/current responses due to sinusoidal current/voltage perturbations. EIS techniques have been used to study a variety of electrochemical phenomena, which include ionic/electronic conduction, charge-transfer reactions, diffusion through porous media and others [1]. As electrochemical systems themselves, lithium-ion (Li-ion) batteries are the subject of many studies using EIS [2–5]. In this chapter, a method is developed for efficiently extracting the frequency response of electrochemical systems using system-identification techniques. Because this method is compatible with typical battery-management hardware, this in situ technique opens up the possibility of identifying a battery's EIS during operation.

5.1.1 Motivation: understanding battery dynamics

The frequency response of a Li-ion battery can be used to understand many internal physical characteristics and operations. Figure 5.1 illustrates a typical notional view of a Li-ion battery. On the left most side, an 18650 cell is shown with internal

[1]Department of Electrical Engineering, Colorado School of Mines, Golden, CO, USA
[2]Department of Mechanical Engineering, Colorado School of Mines, Golden, CO, USA

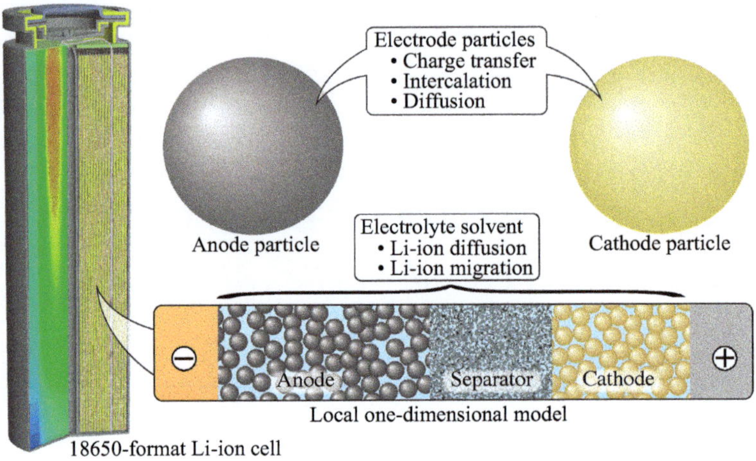

Figure 5.1 Notional image of 18650 battery cell with call outs indicating the pseudo-two-dimensional modeling framework

temperature gradients and mesoscale architecture including tabbing, safety features, and the common jelly-roll configuration. These cells are simulated using a "pseudo-two-dimensional" (P2D) model (bottom of Figure 5.1). These P2D models resolve the lithium concentrations and phase electrostatic potentials in the anode (left), separator (middle), cathode (right), and percolating electrolyte [6–11]. The electrodes (anode and cathode) comprise percolating active material, electrolyte, and a conductivity aid. Electrochemical reactions occur at the electrode/electrolyte interface where lithium is absorbed/desorbed into the active material (illustrated as spheres in Figure 5.1). Because the active material is typically a poor electron conductor, a conductivity aid is added to reduce resistance in transporting electrons to the electrochemically active interface from the current collectors at the far ends of the electrodes.

A variety of materials are used in Li-ion batteries. The most common active materials used in the electrodes are lithium metal oxide for the cathode and lithiated graphite for the anode. A lithium-ion conducting (and ideally electron blocking) electrolyte percolates through both electrodes and the intermediate separator. The electrolyte shuttles lithium between composite anode and cathode and provides the conduction pathway for Li-ions. A microporous separator allows lithium ions to pass between the cathode and anode and prevents direct contact between cathode and anode active particles.

During charging, an external potential is applied to the current collectors. This external potential moves electrons from the cathode to the anode through the external circuit. In response, lithium cations in the electrolyte electrochemically react on the anode/electrolyte surface. The incorporated lithium on the anode-active material surfaces diffuse interior to the particles due to electrochemical potentials gradients. To maintain electroneutrality, Li-ions are added to the electrolyte through electrochemical reactions on the cathode side. The electrochemical reactions on the cathode active particle surface in turn reduce the lithium concentration at the surface. Lithium then diffuses from the interior of the cathode particles to the particle surface due to electrochemical gradients. The electrolyte shuttles lithium ions between the electrodes through the separator by diffusion and migration under the influence of electric potential and concentration gradients. During discharge, the process is reversed. Lithium leaves the surface of the anode active particles to the electrolyte and lithium enters the surface of the cathode particles from the electrolyte. The electrons, transported through the external circuit, release energy and can drive an external load.

The transport, charge-transfer, and storage mechanisms occur at different timescales. This gives the voltage/current frequency response of the battery distinctive features that can be ascribed to specific physics. Changes in the frequency response during the operation of the battery can thus be evaluated to determine important characteristics [10,12,13]. Interpreting the EIS can provide quantitative information about battery performance, state-of-charge, and state-of-health [14–19]. For example, the battery electrolyte is known to decompose on the electrode surface, forming a surface-electrolyte-interface (SEI) layer. Increasing this passivization layer decreases the available lithium for charge/discharge cycles and can increase the capacitance and resistance measured in EIS testing. Zhang *et al.* thoroughly investigate the influence of SEI growth on EIS spectra in lithium battery systems [20].

5.1.2 Traditional methods for measuring EIS

Electrochemical impedance is typically determined by measuring a battery's voltage response to a low-amplitude sinusoidal current perturbation [21]. Suppose a harmonic current is imposed on the battery with frequency ω and magnitude I. The battery voltage, in response to the current perturbation, is sinusoidal at the same frequency. The voltage response will have magnitude E and a phase shift relative to the imposed current ϕ. By definition, the complex electrochemical impedance at frequency ω is represented as

$$Z(\omega) = \frac{E}{I} e^{j\phi}, \tag{5.1}$$

where $j \equiv \sqrt{-1}$. Widely available laboratory-scale electrochemical equipment is designed to impose such harmonic perturbations and evaluate the responses.

Although sinusoidal perturbations follow intuitively from the EIS definition (Equation (5.1)), there are alternatives. Especially for real-time in-operando applications, sinusoidal perturbations suffer from the need for relatively complicated instrumentation [22]. Alternative perturbation sequences can be more easily created and interpreted. A particularly attractive approach is to generate a binary (on/off) current-perturbation sequence, which can be produced using a single solid-state switch.

Processing non-sinusoidal data to obtain the frequency response was recognized fairly early in the electrochemistry community [1], where the data processing has focused on the use of Fourier transforms to obtain an estimate of the frequency response [23]. Alternatively, galvanostatic intermittent titration techniques (GITT) and potentiostatic intermittent titration techniques (PITT) have also been used to understand battery physics. In these tests, a square pulse of either voltage or current is applied, and the response recorded both during and after the pulse (similar to a step response test). GITT is the case of imposed current pulses and PITT is the case of imposed voltage pulses. The GITT and PITT tests can be interpreted in different ways but is most often used to determine the overall chemical diffusion of the electrode particles by plotting the data w.r.t. the square root of time.

A disadvantage of using Fourier transforms for data processing is the need for windowing or other regularization methods to obtain a reliable estimate when sensor noise exists [24]. An alternative data-processing technique is to recognize that since the data represents the impedance of a dynamic system, a system identification method can be utilized to determine the transfer function that represents the system. Once the transfer function is identified, the impedance is evaluated via the frequency response. The advantage of using a system-identification technique is that the user can regularize the estimate in terms of the number of poles and zeros. This approach enables a more structured method to manage estimation errors and window the frequency domain appropriately.

5.1.3 Related work

Because low-order, computationally efficient, dynamical models are directly useful for online battery-management systems, there are numerous published examples of system identification for batteries [25–30], but the focus is on prediction, rather than estimating the frequency response. The perturbations applied include pulses [27–29], random switching to a load [26], and the natural load cycling during operation [30]. Model structures include lumped-impedance models with unknown parameters [27–30] and general discrete-time transfer functions with unknown coefficients [26]. One of the challenges of identifying battery dynamics is the wide range of timescales. One method to obtain models valid over a wide bandwidth is via the use of multiple experiments [31–33], which is also the approach taken here. We will focus on identifying the battery system using multiple prescribed pseudorandom binary sequence perturbations to efficiently identify the wide bandwidth frequency response. A significant difference from prior work is here the frequency response estimates of multiple models are combined, rather than compiled to create a single model.

5.1.4 Outline of approach

The method presented here assumes that a current perturbation is applied to the battery using hardware typically found in battery management systems. Using measurements of the voltage and estimates of the current, system-identification techniques are used to identify a dynamic model for the battery. The frequency response of the identified model is used as an estimate of the frequency response of the physical battery. However, because the estimate is derived from sampled data, the identified model validity is restricted to a particular range of frequencies. To extend the accurate frequency range of the estimate, perturbations and measurements are applied at a range of sampling rates, and the results are combined into a single frequency response estimate. When combining different identified frequency responses, special attention is given to appropriately weight the response of the individual models to minimize error.

5.2 Method

In this section, the EIS estimation procedure is outlined in several steps. To help illustrate the steps, data from a simulated battery will be used. Figure 5.2 shows the simulated battery circuit diagram. The example battery is chosen here to illustrate the methodology and accuracy of the identification method. After the method is outlined, the same procedure is presented on a physical battery where the method's expected frequency response will be compared to standard EIS-identification techniques.

5.2.1 Data collection

Large-capacity storage systems consist of multiple cells within a module that are in a series and parallel arrangement. The modules themselves may also be grouped in a series or parallel arrangement or interconnected via power electronics. The number of batteries in series determines the working voltage of the module, and larger capacity is obtained by placing multiple series cells in parallel. Battery-management systems are added to monitor and manage the system.

Advanced battery-management systems will include stack monitors that measure the voltage of each individual cell, as well as monitor the temperature via thermocouples placed at various points within the cell pack. Individual voltage monitors are important when multiple cells are placed in series to ensure the cells remain balanced, that is, that the cells contain the same amount of charge [34,35]. To prevent overcharging/overdischarging, the charging capacity of a cell string will be limited by the cell with the highest state-of-charge, while the discharge capacity of a cell string is limited by the cell with the lowest state-of-charge. Cell balancing can be accomplished by charge shuttling (a.k.a. active cell balancing), which use power electronics and inductive coupling to draw current from one cell and apply that current to other cells, or to the overall series stack. An alternate method, called passive cell balancing, uses resistors to draw current out of the cell and dissipate it as heat. In either case, this mechanism, along with voltage monitoring, can used to provide a perturbation sequence that can be processed via system identification to estimate the impedance of a single battery.

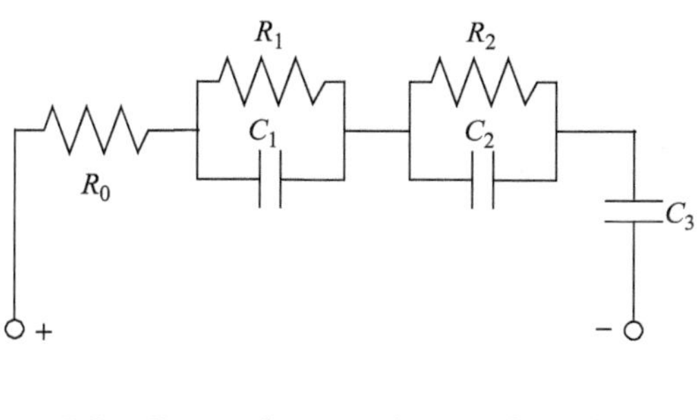

Component	Value
C_1	1 F
C_2	100 F
C_3	500 F
R_0	0.005 Ω
R_1	0.08 Ω
R_2	0.01 Ω

Figure 5.2 Circuit diagram for simulated battery, along with component values

An example stack monitor is the Analog Devices LTC6811, which can measure 12 battery cells in series with a 1.2 mV accuracy. This stack monitor is used in the experimental section to validate the system-identification method. In this system, passive cell balancing is used with a 33 Ω balancing resistor.

It is assumed that a passive balancing method is implemented; however, active cell balancing can also be used to create the necessary perturbation sequence. Figure 5.3 illustrates the general setup for the passive-balancing method for a single cell. A balancing resistor is connected and disconnected across the desired cell string. The voltage across the cell string $v(t)$ is measured in discrete samples at specified times t_s. Assuming that filtering and other stray capacitance can be ignored, and that the string is not being charged or discharged, this perturbation creates the current perturbation sequence

$$i(t) = \begin{cases} -v(t)/R_b & R_b \text{ connected} \\ 0 & \text{otherwise} \end{cases}. \tag{5.2}$$

Since the change in voltage $v(t)$ will be small, the resulting current sequence is close to (although not exactly) a series of pulses.

For the physical battery testing, the experiment is implemented using the embedded processor. At a regular sampling rate of t_s seconds, the state of the switch connecting the balancing resistor and the voltage across the battery is recorded (i.e., the voltage is discretely sampled). Using (5.2), the current at the sampling time is calculated. For use in the identification process discussed below, the discretely measured current is indicated as $u_k = i(kt_s)$, and the discretely measured voltage is indicated as $y_k = v(kt_s)$, where k is the sample number.

To accurately estimate the frequency response over a wide bandwidth, the input perturbation needs to have significant energy over as wide a range as possible. To this aim, psuedo-random-binary-sequences (PRBS) are used. PRBS signals are binary signals that have an autocorrelation function that is similar to white noise [36]. To be more explicit, let x_k be a full length ($N = 2^k - 1$ for some k) PRBS that is converted

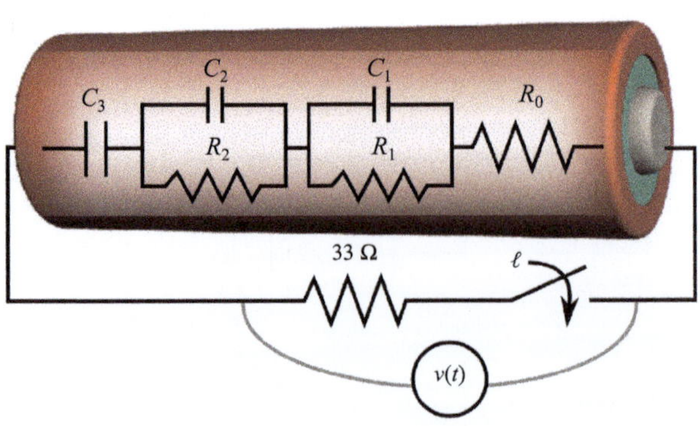

Figure 5.3 General passive cell balancing setup with an 33 Ω balancing resistor

to a sequence of 1 and −1. (Note: For the method described for system identification −1 will be replaced by zeros.) The circular, scaled autocorrelation of x_k is

$$C(\tau) = \frac{1}{N}\sum_{k=1}^{N} x_{k+\tau}x_k = \begin{cases} 1 & \tau = 0 \pmod N \\ \dfrac{-1}{N} & \text{otherwise} \end{cases} \tag{5.3}$$

When N is large, the autocorrelation for the PRBS signal is close to the expected autocorrelation for white noise with variance $\sigma^2 = 1$. This gives pseudorandom-binary-sequences a wide power spectrum. When implemented using a zero-order sample and hold, the PRBS sequence becomes a series of square pulses.

To mitigate the effects of sampling jitter, anti-aliasing, and other effects close to the sampling frequency, the PRBS signal will be held for several sampling intervals, so the clock period of the PRBS signal is slower than the sampling rate t_s. The number of samples held is denoted as h, so R_b is connected at sample k depending on the value of $\ell = x_{\text{ceil}(k/h)}$, where ceil($x$) is smallest integer greater than or equal to x.

Figure 5.4 illustrates example data collected from the simulation of the circuit illustrated in Figure 5.2. In this data, simulated independent Gaussian measurement noise with mean zero and variance 2×10^{-4} has been added to the voltage. The sample time is $t_s = 2.5$ ms and $h = 4$. The current is estimated using (5.2). Note that the current well approximates a binary sequence despite the noise variation added to the voltage. Current is defined as positive entering the battery, so the current is negative when discharging.

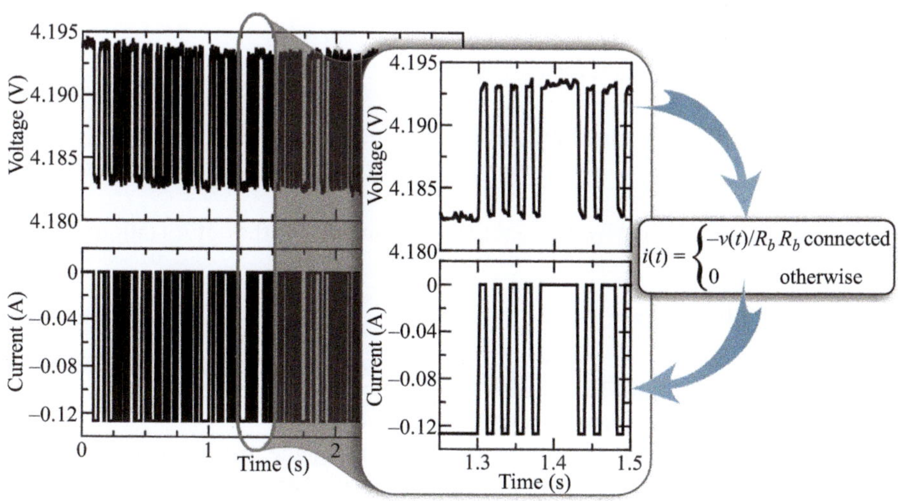

Figure 5.4 Simulated voltage collection from the equivalent circuit model with added Gaussian noise using a switch with a prescribed PRBS signal. The current is computed from the voltage data using (5.2)

5.2.2 Identification

Each experiment is processed to identify a discrete-time dynamic system that best represents the current/voltage behavior. A classical prediction error method will be used, so the description will be fairly brief, with more details available in standard textbooks, e.g., [37].

We will utilize the autoregressive with exogenous input (ARX) and output error (OE) model structures, each of which can be expressed via

$$A(q)y_k = \frac{B(q)}{F(q)}u_k + e_k, \tag{5.4}$$

where u_k is the input sequence, y_k is the output sequence, and e_k is an i.i.d. random sequence representing measurement noise and disturbances. The sample index of each sequence is k. $A(q)$, $B(q)$, and $F(q)$ are order n polynominals in q^{-1}, which is the shift operator, i.e. $q^{-1}u_k = u_{k-1}$. When $F(q) = 1$, this is called an ARX model structure, and when $A(q) = 1$, this is called an output error model structure. The coefficients of the polynomials define a specific model and are the parameters to be selected using the simulated/experimental data. The parameters to be adjusted are collected into the vector θ. In the ARX case, this contains the coefficients of $A(q)$ and $B(q)$, and in the OE case, this contains the coefficients of $F(q)$ and $B(q)$. Note that the zeroth-order term for $A(q)$ and $F(q)$ are set to one, and thus not included in θ.

The prediction error method finds the parameters θ such that the one-step-ahead prediction of the output based on past measurements of the output and the (assumed known) input sequence best matches the observed output using an appropriate fit function. For the model structure (5.4), under the assumption that e_k is an zero mean i.i.d sequence, the minimum mean squared error one-step ahead predictor is given by

$$\hat{y}_k = \frac{B(q)}{F(q)}u_k + (1 - A(q))y_k \tag{5.5}$$

where $\hat{y}_k$ is the one-step ahead prediction sequence. As discussed above, the polynomial coefficients are determined by θ, but this notation is not included for simplicity.

The model is identified by estimating the parameters via the optimization problem

$$\hat{\theta} = \arg\min_{\theta} \sum_{k=1}^{N} (\hat{y}_k - y_k)^2. \tag{5.6}$$

That is, the parameters are found such that the squared error between the one-step ahead predictor and the measured output is minimized.

Because the current, $i(t)$, is estimated from the voltage, the input signal u_k contains measurement noise. This would usually indicate the use of error-in-variables identification methods such as instrument variables [36]. However, because the noise level is quite low, we propose to use a simple preprocessing of the current data instead. Using an ARX model structure with $n = 6$, a preliminary model is found. A filtered voltage estimate $\bar{v}_k$ is obtained by using a one-step ahead predictor based on the ARX model, and the smoothed current estimate $\bar{u}_k$ is calculated using (5.2) based on the smoothed voltage estimate.

After the preprocessing to estimate current, the data is divided into two sections, with two-thirds used for identification and one-third used for validation. Using the OE model structure, parameters are identified for $n = 2$ through $n = 10$. The mean squared one-step ahead prediction error for each model is calculated using the validation data, with initial conditions estimated. The model with the lowest prediction error is selected.

As an example, the simulated data shown in Figure 5.4 is processed as described in this section. The number of samples in the experiment is $N = 1{,}845$, so 1,230 samples are used for identification and the remainder for validation. Based on the validation fit, $n = 7$ is chosen for the model order. The one-step ahead prediction for this model is compared to the measured data in Figure 5.5.

5.2.3 Frequency response and uncertainty estimation

Each model provides an estimate of the frequency response of the battery. However, the accuracy of this estimate varies dramatically by frequency. In addition, the identified models are discrete time, while the system of interest is continuous time, and it is the continuous-time frequency response that is to be estimated.

When a linear-time-invariant continuous-time system has an input applied using a zero-order hold, with the output sampled at rate synchronized with the zero-order hold (as shown in Figure 5.6), it is well known that the resulting discrete time system (from the input of the zero-order hold to the output of the sampler) is also linear time invariant, and there is a mapping between the continuous system and the discrete-time system representation (see, e.g., [38]). This mapping can be calculated using either transfer functions, or using state-space representation. For our purposes, given

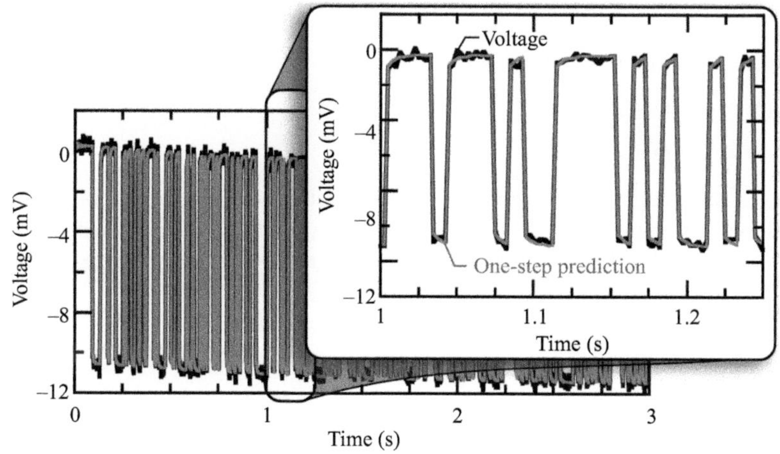

Figure 5.5 Comparison of one-step ahead prediction of voltage using identified system with voltage measurement

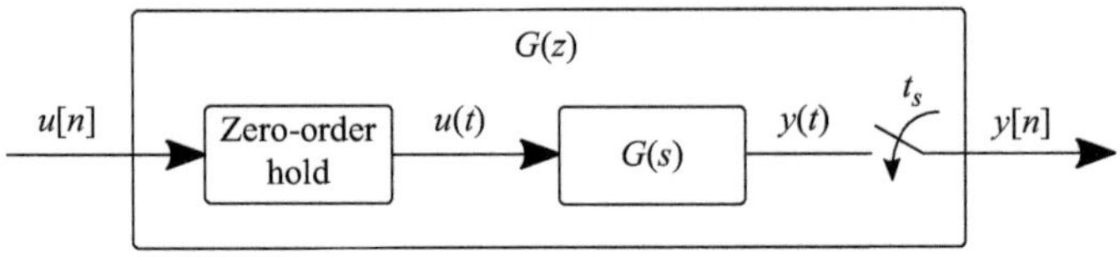

Figure 5.6 Relationship between a continuous system (G(s)) and the discrete system (G(z)) obtained by connecting a zero-order hold digital to analog converter and sampling analog to digital converter

an identified discrete-time system, when this system is in the range space of this mapping, the equivalent continuous-time system can be obtained, and the frequency response of this continuous time system will be the EIS estimate.

When the transfer function of the continuous time system is $G(s)$, (where s is the Laplace transform variable) the discrete time transfer function $G(z)$ (where z is the z-transform variable) can be calculated via

$$G(z) = \mathscr{Z}\left\{ \mathscr{L}^{-1}\left\{ \frac{G(s)}{s} \right\}_{t=t_s k} \right\}(1 - z^{-1}) \tag{5.7}$$

where $\mathscr{L}$ is the one-sided Laplace transform operator and $\mathscr{Z}$ is the one-sided z-transform operator, with $\mathscr{L}^{-1}$ indicating the inverse operator.

The inverse mapping can be obtained via the inverse of each of the elements in (5.7), resulting in

$$G(s) = \mathscr{L}\left\{ \mathscr{Z}^{-1}\left\{ \frac{G(z)}{1 - z^{-1}} \right\}_{k=t/t_s} \right\} s. \tag{5.8}$$

When $G(s)$ is a proper, rational transfer function, these calculations are most easily made by utilizing a state-space representation. With the continuous-time state space representation being expressed as

$$\begin{aligned} \dot{x}(t) &= Ax(t) + Bu(t) \\ y(t) &= Cx(t) + Du(t), \end{aligned} \tag{5.9}$$

where u is the actuation (current); y is the observable (voltage); $\dot{x}$ is an internal state vector; and matrices A, B, C, and D are parameters that determine the system response. The zero-order hold equivalent discrete-time state-space representation is

$$\begin{aligned} x_{k+1} &= \Phi x_k + \Gamma u_k \\ y_k &= Cx_k + Du_k, \end{aligned} \tag{5.10}$$

where

$$\begin{bmatrix} \Phi & \Gamma \\ 0 & I \end{bmatrix} = \exp\left(\begin{bmatrix} A & B \\ 0 & 0 \end{bmatrix} T_s \right), \tag{5.11}$$

and exp is the matrix exponential.

When Φ has no eigenvalues at zero, the inverse is calculated as

$$\begin{bmatrix} A & B \\ 0 & 0 \end{bmatrix} = \frac{\log\begin{pmatrix} \Phi & \Gamma \\ 0 & I \end{pmatrix}}{T_s}, \tag{5.12}$$

where log is the matrix natural logarithm.

The frequency response of the continuous equivalent is then the desired estimate of the battery EIS, $\hat{Z}(\omega)$, given by

$$\hat{Z}(\omega) = C(j\omega I - A)^{-1}B + D. \tag{5.13}$$

The process of obtaining the continuous time frequency response is summarized as follows. From the system-identification procedure, parameters θ are identified as the estimated model. The identified model

$$y_k = \frac{B(q)}{F(q)}u_k = \frac{b_0 + b_1q^{-1} + \cdots + b_nq^{-n}}{1 + a_1q^{-1} + \cdots + a_nq^{-n}}u_k, \tag{5.14}$$

is converted to a state-space representation using control canonical form

$$\begin{aligned} \Phi &= \begin{bmatrix} 0 & 1 & 0 & \cdots & 0 \\ 0 & 0 & 1 & \cdots & 0 \\ \vdots & \vdots & & \ddots & \vdots \\ 0 & 0 & 0 & \cdots & 1 \\ -a_n & -a_{n-1} & \cdots & -a_2 & -a_1 \end{bmatrix}, \quad \Gamma = \begin{bmatrix} 0 \\ 0 \\ \vdots \\ 0 \\ 1 \end{bmatrix}, \\ C &= \begin{bmatrix} b_n - b_0a_n & b_{n-1} - b_0a_{n-1} & \cdots & b_1 - b_0a_1 \end{bmatrix}, \\ D &= b_0. \end{aligned} \tag{5.15}$$

Via (5.12), this is converted to a continuous-time state-space representation, and by using (5.13), the frequency response at frequency ω is evaluated. The composition of these steps is represented as the function f, i.e.

$$\hat{Z}(\omega) = f(\theta), \tag{5.16}$$

where the domain of f is the θ such that mapping (5.12) is valid.

We now turn to assessing the uncertainty of this estimate. Under the assumption that the true system lies within the model structure and that the unmeasured signal e_k is a random sequence that is zero mean, independent and identically distributed, with variance λ_0 and bounded fourth moments, the scaled error between the true and estimated parameters approaches a Gaussian distribution with covariance [24]

$$P = \frac{\lambda_o}{N}E\left[\frac{d\hat{y}_k}{d\theta}^T\frac{d\hat{y}_k}{d\theta}\right]^{-1}, \tag{5.17}$$

where $d\hat{y}_k/d\theta$ is a row vector, the length of θ containing the derivative of $\hat{y}$ with respect to each element of θ. This suggests an estimate of the parameter covariance given by

$$\hat{P} = \frac{\hat{\lambda}}{N}\left[\frac{1}{N}\sum_{t=1}^{N}\frac{d\hat{y}_k}{d\theta}^T\frac{d\hat{y}_k}{d\theta}\right]^{-1}, \tag{5.18}$$

$$\hat{\lambda} = \frac{1}{N}\sum_{k=1}^{N}(y_k - \hat{y}_k)^2. \tag{5.19}$$

Using a Taylor series expansion of f, the estimated variance in the frequency estimate in the real and imaginary directions is

$$\hat{\sigma}^2_{\mathrm{Re}}(\omega) = \frac{d\Re f}{d\theta}\hat{P}\frac{d\Re f}{d\theta}^T, \tag{5.20}$$

$$\hat{\sigma}^2_{\mathrm{Im}}(\omega) = \frac{d\Im f}{d\theta}\hat{P}\frac{d\Im f}{d\theta}^T, \tag{5.21}$$

where $\Re f$ is the real part of f and $\Im f$ is the magnitude of the imaginary part.

Figure 5.7(a) illustrates the computed variance of the frequency response for a given test. In this case, the frequency response and uncertainty calculated for the system identified in Section 5.2.2, which has the fit as shown in Figure 5.5. For this system, the parameter vector θ contains 15 elements, 8 for the numerator polynomial, and 7 for the denominator. The sensitivities $(d\Re f/d\theta)$ and $(d\Im f/d\theta)$ are calculated numerically by perturbing each parameter in turn by 0.001%, and using the resulting frequency response estimate, and nominal estimate, to calculate the sensitivity. To compare, the true frequency response for the simulated battery is plotted. In this plot, the error bars represent $3\hat{\sigma}_{\mathrm{Re}}(\omega)$ and $3\hat{\sigma}_{\mathrm{Im}}(\omega)$ in the real and imaginary

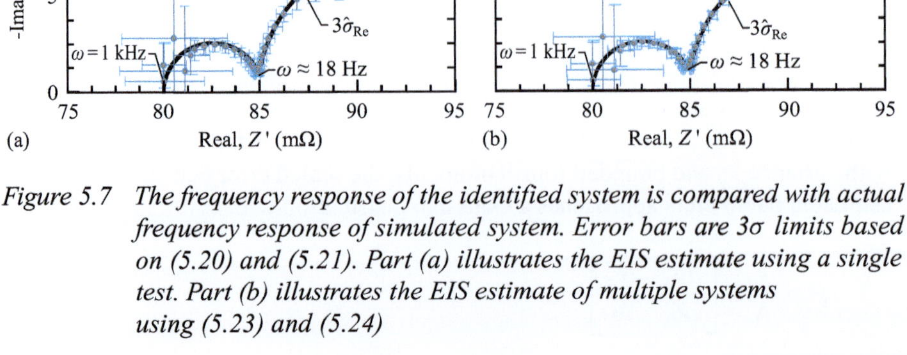

Figure 5.7 The frequency response of the identified system is compared with actual frequency response of simulated system. Error bars are 3σ limits based on (5.20) and (5.21). Part (a) illustrates the EIS estimate using a single test. Part (b) illustrates the EIS estimate of multiple systems using (5.23) and (5.24)

directions, respectively. Note that the estimate matches the true frequency response within the estimated uncertainty bounds. The uncertainty grows at both the higher and lower frequency ends of the estimate. Frequency estimates are also available at lower frequencies but were removed for clarity as the error bars become quite large.

5.2.4 Combined frequency response estimate

The estimates of the frequency response for each identified model need to be combined into a single estimate. Because the models are identified over different frequency ranges, it is difficult to do this combination using the parameters themselves, so the process outlined here combines the estimates frequency by frequency.

Let the frequency of interest be ω. Each model provides an estimate of the system frequency response, but given the sampling rate and length of experiment, some models can be immediately discounted because they inaccurately represent the system dynamics at particular frequencies. For example, if the data used to identify the model was obtained at sampling rate t_s, and the length of the experiment was N samples, then (with proper anti-aliasing filtering) the model is not informative if ω is greater than the Nyquist sampling frequency, i.e., $\omega > \pi/t_s$, or if ω is so small that the experiment length is on the order of or less than one period, i.e. $\omega < \pi/(Nt_s)$. Often, anti-aliasing filters may have nonzero phase response near the Nyquist frequency, so the high-frequency cutoff may be chosen even lower.

The first step in combining estimates is to find the systems for which the sampling rate t_s satisfies the upper and lower frequency bounds given above. Without loss of generality, let those systems be indexed $i = 1, \ldots, p$ and let the estimate of the impedance for each system be $\hat{Z}_i(\omega)$, with

$$\hat{Z}_i(\omega) = a_i(\omega) + jb_i(\omega) \tag{5.22}$$

and denote the estimate of the variance for a_i and b_i as $\hat{\sigma}^2_{\mathrm{Re},i}$ and $\hat{\sigma}^2_{\mathrm{Im},i}$, respectively.

We make the following assumptions:

- The error in the frequency response estimate is dominated by the noise (rather than bias due to an insufficiently flexible model structure).
- The error in the frequency response between models is independent.
- The estimate of the parameter error covariance (cf. (5.20) and (5.21)) is accurate.

While each of these assumptions is debatable in practice, the resulting process still makes engineering sense when the assumptions are not fully met.

Under these assumptions, the real and imaginary frequency response estimates are observations of a random variable with mean equal to the true frequency response and variance given by $\hat{\sigma}^2_{\mathrm{Re},i}$ and $\hat{\sigma}^2_{\mathrm{Im},i}$. The minimum variance estimate of the real and imaginary parts of the frequency response is then $\hat{Z}(\omega) = a_{\mathrm{tot}}(\omega) + jb_{\mathrm{tot}}(\omega)$, where

$$a_{\mathrm{tot}}(\omega) = \frac{\sum_{i=1}^{p} a_i(\omega)/\hat{\sigma}^2_{\mathrm{Re},i}}{\sum_{i=1}^{p} 1/\hat{\sigma}^2_{\mathrm{Re},i}}, \tag{5.23}$$

$$b_{\mathrm{tot}}(\omega) = \frac{\sum_{i=1}^{p} b_i(\omega)/\hat{\sigma}^2_{\mathrm{Im},i}}{\sum_{i=1}^{p} 1/\hat{\sigma}^2_{\mathrm{Im},i}}, \tag{5.24}$$

and the variance of these estimates are

$$\sigma_a^2 = \frac{1}{\sum_{i=1}^{p} 1/\hat{\sigma}_{\mathrm{Re},i}^2}, \quad \sigma_b^2 = \frac{1}{\sum_{i=1}^{p} 1/\hat{\sigma}_{\mathrm{Im},i}^2}. \tag{5.25}$$

This method was applied to combine the results of six different experiments on the simulated battery. Each experiment consisted of a PRBS perturbation sequence identical to that shown in Figure 5.4, but with a different sampling times. The selected sampling times t_s were 2.5 ms, 7.9 ms, 25 ms, 79 ms, 0.25 s, and 0.79 s, with hold time $h = 4$. The length of each experiment was 1,845 samples, except for the last two sampling rates, which were capped at a length of 300 s. The results of combining the individual frequency response estimates using (5.2) and (5.24) is shown in Figure 5.7(b). The bars represent 3σ calculated using (5.25). Note that the estimate at higher frequencies (lower left hand side of the figure) is similar to Figure 5.7(a), but the estimates and uncertainties are improved as the frequency decreases (moving toward the upper right hand side of the figure).

5.2.5 Review of frequency identification method

The procedure to accurately identify the frequency response of a battery using a passive cell balancing system discussed in Sections 5.2.1–5.2.4 is recapitulated:

1. Generate a PRBS signal with minimum switching time of t_s.
2. Identify the number of samples taken in each minimum switching time h.
3. Implement the PRBS on/off signal into the solid-state switch.
4. Measure the discrete voltage response $y_k = v(kt_s)$.
5. Determine the discrete current response u_k (cf. (5.2)).
6. Find a best fit ARX model with $n = 6$ (cf. (5.4)–(5.6)).
7. Obtain fitted, smoothed voltage estimate $\overline{v}_k$ (cf. (5.5)).
8. Obtain smoothed current estimate $\overline{u}_k$ (cf. (5.2)).
9. Determine OE models with $2 \leq n \leq 10$ using v_k and $\overline{u}_k$ for the first two-thirds of the data collected (cf. (5.5)).
10. Use the OE model with the lowest mean-squared one-step ahead prediction error in predicting the last one-third of the data.
11. Convert the chosen OE model to a discrete-time state-space model (cf. (5.10), (5.14), and (5.15)).
12. Convert the discrete-time state-space model to a continuous-time state-space model (cf. (5.12)).
13. Identify the frequency response of the continuous-time state-space model (cf. (5.13)).
14. Determine covariance of the identified frequency response (cf. (5.20) and (5.21)).
15. Repeat the above procedure for multiple sampling times t_s.
16. Combine the frequency responses of multiple identified systems (cf. (5.22)–(5.24)).
17. Determine the covariance of the combined system (cf. (5.25)).

5.3 Example experimental results

As an example experimental result, this method was applied to the EIS estimation of a lithium polymer battery, model GNB15004S120A, manufactured by GaoNeng Battery Limited. This is a power battery back of four cells in series, with a nominal capacity of 1,500 mA h and pack voltage of 14.8 V (cell voltage of 3.7 V). The package allows access to the voltage across each cell, which will be measured using an Analog Devices LTC6811 stack monitor. This battery monitor can measure up to 12 battery cells in series with a 1.2 mV accuracy, although the results reported here are for a single cell. The stack monitor includes passive cell balancing with a 33 Ω balancing resistor. The fastest sample rate for a single cell achievable by the monitor is 2.5 ms. For comparison, the EIS was also estimated using sinusoidal perturbations. This data was collected using a National Instruments NI-USB 6211 board. The battery was fully charged before the experiments began. The sinusoidal perturbation experiments were applied first, followed by the PRBS perturbation experiments.

5.3.1 Experimental conditions for PRBS perturbation

Seven PRBS perturbation experiments are implemented, with sampling times t_s of 2.5 ms, 7.9 ms, 25 ms, 79 ms, 250 ms, and 1 s, with the PRBS signal determining if the balancing resistor is active or inactive. The PRBS signal is held for $h = 4$ samples. The experiment length is 1,845 samples.

5.3.2 Experimental conditions for sinusoidal perturbation

A voltage controlled current source is used to apply sinusoidal perturbations applied at 35 frequencies, with the resulting voltage perturbation recorded at a rate of at least 1,000 samples per cycle. At least five cycles are recorded at each frequency, with the relative magnitude and phase determined using cross correlation.

5.3.3 Results

The EIS estimates are shown in Figures 5.8 and 5.9. Figure 5.8 illustrates the magnitude and phase estimates vs. frequency. Note that the magnitude scale is a linear scale with units of mΩ, rather than dB. Figure 5.9 illustrates the estimate plotted in the complex plane with the imaginary axis reversed. The 3σ uncertainty for the EIS estimate using PRBS input is also included as error bars. Because two different measurement systems were used, with different internal resistances, there is an unavoidable uncertain constant contact resistance added to the real part of the EIS for both estimates. To compare the results, 0.11 Ω was subtracted from the EIS estimate obtained from PRBS perturbations. It is apparent from the plots that the two methods for obtaining the EIS are equivalent. It is worth noting that the battery EIS contains the expected two-semicircles at frequencies greater than 1 rad/s. There is a larger discrepancy between the two methods at lower frequencies. This could mean the uncertainty estimates are too conservative in this region, perhaps due to a bias error. Additional

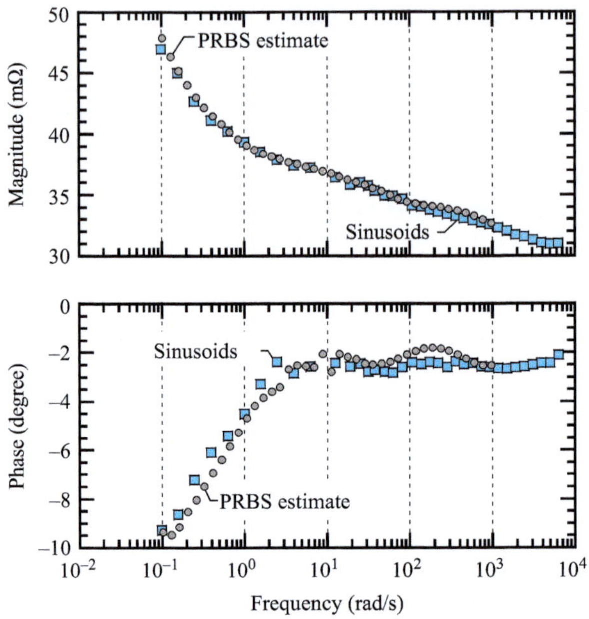

Figure 5.8 *Comparison of EIS estimates from two different types of experimental data. Circles indicate estimate using system identification and PRBS perturbations. Squares indicate estimates from sinusoidal perturbations. Data is plotted vs. frequency*

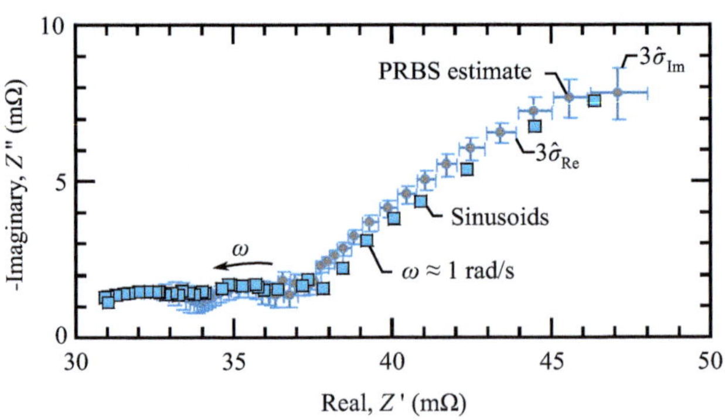

Figure 5.9 *Comparison of EIS estimates from two different types of experimental data. Circles indicate estimate using system identification and PRBS perturbations. Squares indicate estimates from sinusoidal perturbations. Data is plotted as real impedances vs. negative imaginary impedance*

PRBS experiments at lower sampling rates would also perhaps improve the estimate in this region.

Acknowledgments

This work was supported in part by the Office of Naval Research via grant N00014-14-1-0059 and grant N00014-17-1-2697. The authors also thank Analog Devices for supplying the stack monitor hardware used to obtain experimental data.

References

[1] Macdonald DD. Reflections on the history of electrochemical impedance spectroscopy. Electrochim Acta. 2006;51(8):1376–1388.
[2] Waag W, Käbitz S, Sauer DW. Experimental investigation of the lithium-ion battery impedance characteristic at various conditions and aging states and its influence on the application. Appl Energy. 2013;102:885–897.
[3] Pattipati B, Sankavaram C, Pattipati KR. System identification and estimation framework for pivotal automotive battery management system characteristics. IEEE Trans Syst Man Cybern C Appl Rev. 2011;41(6):869–884.
[4] Piao T, Park SM, Doh CH, *et al.* Intercalation of lithium ions into graphite electrodes studied by AC impedance measurements. J Electrochem Soc. 1999;146(8):2794–2798.
[5] Tippmann S, Walper D, Balboa L, *et al.* Low-temperature charging of lithium-ion cells Part I: Electrochemical modeling and experimental investigation of degradation behavior. J Power Sources. 2014;252:305–316.
[6] Ferguson TR, Bazant MZ. Nonequilibrium thermodynamics of porous electrodes. J Electrochem Soc. 2012;159:A1967–A1985.
[7] Smith RB, Bazant MZ. Multiphase porous electrode theory. J Electrochem Soc. 2017;11:E3291–E3310.
[8] Fuller TF, Doyle M, Newman J. Simulation and optimization of the dual lithium insertion cell. J Electrochem Soc. 1994;141:1–10.
[9] Kim GH, Smith K, Lee K, *et al.* Multi-domain modeling of lithium-ion batteries encompassing multi-physics in varied length scales. J Electrochem Soc. 2011;158:A955–A969.
[10] Newman J, Thomas-Alyea KE. Electrochemical Systems. Hoboken, NJ: Wiley; 2004.
[11] Colclasure AM, Kee RJ. Thermodynamically consistent modeling of elementary electrochemistry in lithium-ion batteries. Electrochim Acta. 2010;55: 8960–8973.
[12] Orazen ME, Tribollet B. Electrochemical Impedance Spectroscopy. Hoboken, NJ: Wiley; 2008.
[13] Bard AJ, Faulkner LR. Electrochemical Methods Fundamentals and Applications. Hoboken, NJ: Wiley; 2001.

[14] Reid MA. Impedance studies of nickel/cadmium and nickel/hydrogen cells using the cell case as a reference electrode. J Power Sources. 1990;29:467–476.
[15] Huet F. A review of impedance measurements for determination of the state-of-charge or state-of-health of secondary batteries. J Power Sources. 1998;70: 59–69.
[16] Hammouche A, Karden E, Doncker RWD. Monitoring state-of-charge of Ni–MH and Ni–Cd batteries using impedance spectroscopy. J Power Sources. 2004;127:105–111.
[17] Eddahech A, Briat O, Bertrand N, *et al.* Behavior and state-of-health monitoring of Li-ion batteries using impedance spectroscopy and recurrent neural networks. Int J Electr Power Energy Syst. 2012;42:487–494.
[18] Love CT, Swider-Lyons K. Impedance diagnostic for overcharged lithium-ion batteries. J Electrochem Soc. 2012;15:A53–A56.
[19] Love CT, Virji MVB, Rocheleau RE, *et al.* State-of-health monitoring of 18650 4S packs with a single-point impedance diagnostic. J Power Sources. 2014;266:512–519.
[20] Zhang QC, Qui XY, Xu SD, *et al.* Diagnosis of electrochemical impedance spectroscopy in lithium ion batteries. Prog Chem. 2010;22:1044–1057.
[21] Chang BY, Park SM. Electrochemical impedance spectroscopy. Annu Rev Anal Chem. 2010;3:207–229.
[22] Lee YD, Park SY, Han SB. Online embedded impedance measurement using high-power battery charger. IEEE Trans Ind Appl. 2015;51(1):498–508.
[23] Chang BY, Mark SM. Electrochemical impedance spectroscopy. Annu Rev Anal Chem. 2010;3:207–229.
[24] Ljung L. System Identification: Theory for the User. 2nd ed. Upper Saddle River, NJ: Prentice Hall PTR; 1999.
[25] Hu Y, Yurkovich S, Guezennec Y, *et al.* Electro-thermal battery model identification for automotive applications. J Power Sources. 2011;196(1):449–457.
[26] Zhao X, de Callafon RA. Data-based modeling of a lithium iron phosphate battery as an energy storage and delivery system. In: American Control Conference. IEEE; 2013. p. 1908–1913.
[27] Lin X, Perez HE, Mohan S, *et al.* A lumped-parameter electro-thermal model for cylindrical batteries. J Power Sources. 2014;257:1–11.
[28] Rahimi-Eichi H, Baronti F, Chow MY. Online adaptive parameter identification and state-of-charge coestimation for lithium-polymer battery cells. IEEE Trans Ind Electron. 2014;61(4):2053–2061.
[29] Thunot A, Robertson P. Battery diagnostic system and complex impedance measurement algorithm. In: International Electric Vehicle Symposium and Exhibition; 2015.
[30] Zou Y, Li SE, Shao B, *et al.* State-space model with non-integer order derivatives for lithium-ion battery. Appl Energy. 2016;161:330–336.
[31] Hu Y, Wang YY. Two time-scaled battery model identification with application to battery state estimation. IEEE Trans Control Syst Technol. 2015; 23(3):1180–1188.

[32] Dai H, Xu T, Zhu L, *et al.* Adaptive model parameter identification for large capacity Li-ion batteries on separated time scales. Appl Energy. 2016;184: 119–131.

[33] Weddle PJ, Kee RJ, Vincent TL. A stitching algorithm to identify wide-bandwidth electrochemical impedance spectra for Li-ion batteries using binary perturbations. J Electrochem Soc. 2018;9:A1679–A1684.

[34] Lu L, Han X, Li J, *et al.* A review on the key issues for lithium-ion battery management in electric vehicles. J Power Sources. 2013;226:272–288.

[35] Moore SW, Schneider PJ. A review of cell equalization methods for lithium ion and lithium polymer battery systems. In: SAE Technical Paper; 2001.

[36] Söderström T, Stoica P. System Identification. Upper Saddle River, NJ: Prentice-Hall, Inc.; 1988.

[37] Ljung L. System identification. In: Signal Analysis and Prediction. Boston, MA: Birkhäuser; 1998. p. 163–173.

[38] Franklin GF, Powell JD, Workman MK. Digital Control of Dynamic Systems. 3rd ed. Menlo Park, CA: Addison-Wesley; 1998.

Part II

Data-driven filtering and control

Chapter 6
Dynamic measurement

Ivan Markovsky[1]

In metrology, a given measurement technique has fundamental speed and accuracy limitations imposed by physical laws. Data processing allows us to overcome these limitations by using prior knowledge about the sensor dynamics. The prior knowledge considered in this paper is a model class to which the sensor dynamics belongs. We present methods that are applicable to linear time-invariant processes and are suitable for real-time implementation on a digital signal processor.

6.1 Introduction

The accuracy of measurement devices is reduced by measurement noise and transient response. The transient response of the measurement device decays exponentially, so that measuring longer has the effect of (eventually) reducing this source of error. Measuring longer also yields more data, which when used effectively reduces the error due to the measurement noise. The objective, however, is to achieve faster as well as more accurate measurement: there is a trade-off between speed and accuracy determined by the physical properties of the measurement device.

Our goal is to design methods that bring the performance of the measurement device to the theoretically optimal one. To do this, we take into account the dynamical properties of the sensor and reduce the measurement time needed for a specified measurement accuracy. Instead of waiting for the sensor's natural transient response to die out, the methods proposed *predict* the steady state value. The prediction is possible thanks to the assumption that the sensor is a low-order linear time-invariant system. At the same time as reducing the effect of the transient, the methods take into account the measurement noise, reducing also its effect.

How realistic is the low-order linear time-invariance assumption? The answer depends on the application. For example, in temperature and mass measurement, the assumptions are satisfied. In more complicated measurement processes, such as the ones in biochemical reactors, they are not.

[1]Department ELEC, Vrije Universiteit Brussel (VUB), Brussels, Belgium

The subfield of metrology that takes into account the dynamical properties of the measurement devices is called *dynamic measurement*. A system theoretic way of viewing the dynamic measurements problem is the problem of designing a *compensator* that reduces the transient process of the sensor. Then, the procedure is (1) design off-line a dynamical system—the compensator—so that the series connection of the measurement process with the compensator behaves as close as possible to a static gain and (2) process online the measurements of the sensor by the compensator.

Note 6.1 (Link to deadbeat control) *The question of making the transient response as short as possible occurs also in control. It turns out that a reachable discrete-time system can be steered to a steady-state in a finite number of samples. Moreover, the number of samples is less than or equal to the order of the system. The corresponding control is called* deadbeat. *The equivalent of the reference signal in control is the measured value in metrology and the equivalent of the controller is a compensator that implements the prediction method.*

Despite the apparent similarity between deadbeat control and dynamic measurement, however, there are important differences: (1) the reference signal is known, while the measured value is not, (2) the controller is typically implemented in feedback, while the compensator can be applied only in open loop, (3) in control, the plant's dynamics is typically assumed known, while in metrology, it is often unknown. These differences suggest that if deadbeat control methods are used in metrology, they have to be open loop and adaptive.

6.1.1 Literature review

Most authors assume that a linear time-invariant model of the measurement process is a priori given. In this case, the problem of designing a compensator reduces to the classical problem of designing an inverse system [1]. In the presence of noise, however, compensators that take into account the noise are needed. A theoretically optimal solution in the case of Gaussian noise is given in [2]. It is shown that in this case, the problem reduces to a state-estimation problem for a suitably defined autonomous linear time-invariant system. As a consequence the maximum likelihood solution is given by the Kalman filter.

The design problem is more complicated in the case of unknown measurement process dynamics. In this case, adaptive filtering methods are used [3,4], i.e., the compensator is tuned on-line by a parameter estimation algorithm. The solutions proposed in [3,4], however, are tailored to the mass-measurement problem, which is a second-order process.

In this paper, we present two methods that bypass the real-time model-identification step done in adaptive filtering methods. The first method is a maximum-likelihood method that requires nonconvex optimization. The second one belongs to the class of the subspace methods [5] and is computationally simple to implement; however, it is not statistically optimal in the maximum-likelihood sense.

6.2 Problem setup

The sensor is a dynamical system with input the unknown to-be-measured value and output the sensor's reading, corrupted by additive noise. We assume that

1. the sensor dynamics is linear time-invariant;
2. the order $\mathtt{n}$ is known;
3. the input u is a step function $u = \bar{u}s$, where $\bar{u}$ is the to-be-measured value;
4. the measurement noise is zero mean, white, Gaussian, with variance σ^2; and
5. in steady-state, $\bar{u}$ can be determined from y.

The linear time-invariance assumption is essential for the methods considered. The known order assumption can be relaxed. Order selection can be done, for example, by the methods presented in [6]. The assumption that the input is a step function implies that after an initial moment of time (start of the measurement process), the to-be-measured value is constant. This is a common assumption in metrology that is satisfied in many practical situations due to a "short" measurement time and the fact that the measurement process does not affect "significantly" the environment. The assumption that the output is corrupted by additive noise that is zero mean, white, Gaussian (output error setup) is appropriate for applications in metrology, where the mismatch between the observed data and the model is due to measurement noise rather than process noise and model uncertainty. Finally, assumption 5 implies that the DC-gain G of the sensor is known. The known DC-gain assumption means in practice that the sensor is calibrated.

Under the assumption that the input is a step function, the corresponding output is of the form

$$y_{\mathrm{d}} = y + e = G\bar{u} + y_0 + e. \tag{6.1}$$

where y_{d} is the observed noisy output, y is the noise-free output signal, e is the measurement noise, $G\bar{u}$ is the steady-state value of y, and y_0 is the transient response of the sensor that we aim to eliminate by the compensator. The assumption that the process is linear time-invariant implies that there are matrices $A \in \mathbb{R}^{\mathtt{n}\times\mathtt{n}}$, $c \in \mathbb{R}^{1\times\mathtt{n}}$, and a vector $x_0 \in \mathbb{R}^{\mathtt{n}}$, such that

$$\begin{aligned} x(t+1) &= Ax(t), \qquad x(0) = x_0 \\ y_0(t) &= cx(t). \end{aligned} \tag{6.2}$$

Therefore, the sensor's output is given by

$$\underbrace{\begin{bmatrix} y_{\mathrm{d}}(1) \\ y_{\mathrm{d}}(2) \\ \vdots \\ y_{\mathrm{d}}(T) \end{bmatrix}}_{y_{\mathrm{d}}} = \underbrace{\begin{bmatrix} G \\ G \\ \vdots \\ G \end{bmatrix}}_{\mathscr{G}} \bar{u} + \underbrace{\begin{bmatrix} c \\ cA \\ \vdots \\ cA^{T-1} \end{bmatrix}}_{\mathscr{O}} x_0 + \underbrace{\begin{bmatrix} e(1) \\ e(2) \\ \vdots \\ e(T) \end{bmatrix}}_{e}. \tag{6.3}$$

6.3 Model-based vs data-driven approaches

The model-based estimation procedure requires solving approximately the overdetermined system of equations

$$[\mathscr{G} \quad \mathscr{O}]\begin{bmatrix}\widehat{u}\\ \widehat{x}_0\end{bmatrix} \approx y_\mathrm{d}.$$

Under assumption 4, the maximum-likelihood estimator for $\bar{u}$ in (6.3) is defined by the standard least squares problem

$$\begin{aligned} &\text{minimize} \quad \text{over } \widehat{y}, \widehat{u}, \widehat{x}_0 \quad \|y_\mathrm{d} - \widehat{y}\| \\ &\text{subject to} \quad [\mathscr{G} \quad \mathscr{O}]\begin{bmatrix}\widehat{u}\\ \widehat{x}_0\end{bmatrix} = \widehat{y}, \end{aligned} \tag{6.4}$$

where $\|\cdot\|$ is the two norms. Correspondingly, $\operatorname{cov}(\widehat{u}) = \sigma^2 V_{11}$, where

$$V = \left([\mathscr{G} \quad \mathscr{O}]^\top [\mathscr{G} \quad \mathscr{O}]\right)^{-1}.$$

The covariance matrix of $\widehat{u}$ does not depend on the data y_d. However, it depends on the noise variance σ^2, which affects the data. If σ^2 is unknown, it can be estimated as

$$\widehat{\sigma}^2 = \frac{1}{T}\left\|y_\mathrm{d} - [\mathscr{G} \quad \mathscr{O}]^\top \begin{bmatrix}\widehat{u}\\ \widehat{x}_0\end{bmatrix}\right\|^2.$$

The matrix $\mathscr{O}$ depends on the model parameters A and c, so that the dynamic measurement method based on (6.4) requires a priori given model. The data-driven model-free approach considered next is based on the observation that, in the noise-free case, under an additional assumption (persistency of excitation), the matrix $\mathscr{O}$ can be replaced by a matrix that depends on the observed data only.

The difference signal

$$\Delta y(t) := y(t) - y(t-1) = y_0(t) - y_0(t-1)$$

satisfies the same dynamics as y_0, i.e., there is an initial condition $\Delta x \in \mathbb{R}^n$, such that

$$\begin{aligned} x(t+1) &= Ax(t), \qquad x(0) = \Delta x \\ \Delta y(t) &= cx(t). \end{aligned}$$

Let

$$\mathscr{H}(\Delta y) := \begin{bmatrix} \Delta y(1) & \Delta y(2) & \cdots & \Delta y(\mathrm{n}) \\ \Delta y(2) & \Delta y(3) & \cdots & \Delta y(\mathrm{n}+1) \\ \Delta y(3) & \Delta y(4) & \cdots & \Delta y(\mathrm{n}+2) \\ \vdots & \vdots & & \vdots \\ \Delta y(T-\mathrm{n}) & \Delta y(T-\mathrm{n}) & \cdots & \Delta y(T-1) \end{bmatrix}, \tag{6.5}$$

be the Hankel matrix with n columns, constructed from Δy and let

$$y|_t := \begin{bmatrix} y(1) & \cdots & y(t) \end{bmatrix}^\top.$$

Under the assumption that Δy is persistently exciting of order n [7], we have that

$$\text{image}\left(\begin{bmatrix} c \\ cA \\ \vdots \\ cA^{T-n-1} \end{bmatrix}\right) = \text{image}\,(\mathscr{H}(\Delta y)).$$

Therefore, for noise free data, there is a vector $\ell \in \mathbb{R}^{\mathrm{n}}$, such that

$$\begin{bmatrix} \mathscr{G} & \mathscr{H}(\Delta y) \end{bmatrix} \begin{bmatrix} \bar{u} \\ \ell \end{bmatrix} = y|_{T-\mathrm{n}}. \tag{6.6}$$

As long as $(1, \ldots, 1) \notin \text{image}\,(\mathscr{H}(\Delta y))$, the solution of (6.6) is unique. Therefore, solving (6.3) allows us to estimate $\bar{u}$ directly from the observed exact data y (without knowing the model).

With noisy data y_{d}, the data-driven estimation procedure aims at an approximate solution of the system of equations

$$\begin{bmatrix} \mathscr{G} & \mathscr{H}(\Delta y_{\mathrm{d}}) \end{bmatrix} \begin{bmatrix} \widehat{u} \\ \ell \end{bmatrix} \approx y_{\mathrm{d}}|_{T-\mathrm{n}}. \tag{6.7}$$

The noise e affects both the left-hand side of (6.7) (through Δy_{d}) as well as the right-hand side of (6.7) (through y_{d}), i.e., solving (6.7) is a total least squares problem [8]. Moreover, Δy_{d} enters in a structured way in the left-hand side of the equation, so that the maximum-likelihood estimator for the parameter $\bar{u}$ is the structured total least squares problem

$$\begin{aligned} &\text{minimize} \quad \text{over } \widehat{y}, \widehat{u}, \widehat{\ell} \quad \|y_{\mathrm{d}} - \widehat{y}\| \\ &\text{subject to} \quad \begin{bmatrix} \mathscr{G} & \mathscr{H}(\Delta \widehat{y}) \end{bmatrix} \begin{bmatrix} \widehat{u} \\ \widehat{\ell} \end{bmatrix} = \widehat{y}|_{T-\mathrm{n}}. \end{aligned} \tag{6.8}$$

The uncertainty $\text{cov}\,(\widehat{u})$ of the estimate $\widehat{u}$ is given by the Jacobian matrix and the optimal value of the cost function [9]. The Jacobian matrix and the cost-function value are computed as by-products of the local optimization solution method [10].

6.4 Maximum-likelihood data-driven estimation method

As shown in Section 6.3, the maximum-likelihood model-based estimation problem (6.4) is an ordinary least squares problem. A special feature of the problem however is the "shift structure" of the matrix $\begin{bmatrix} \mathscr{G} & \mathscr{O} \end{bmatrix}$. Taking into account this structure in an efficient real-time algorithm leads to the Kalman filter [11].

The maximum-likelihood model-based estimation problem (6.8) is a nonconvex optimization problem. Existing solution methods [8,12] use local optimization algorithms. Such algorithms require an initial approximation and are susceptible to convergence to a local minimum.

The initial approximation for the local optimization method can be computed by a convex relaxation of (6.8). This yields a method that is of independent interest.

One approach of "convexifying" the problem is to ignore the structure of the matrix $[\mathscr{G} \quad \mathscr{H}(\Delta\widehat{y})]$, which turns the structured total least squares problem into an ordinary total linear squares problem. The latter admits an analytical solution in terms of the singular value decomposition. Moreover, there are algorithms that recursively update the singular value decomposition [13,14]. These algorithms allow real-time implementation of the data-driven input estimation method.

Another approach of relaxing the structured total least squares problem (6.8) is to substitute the optimization variable $\widehat{y}$ in $[\mathscr{G} \quad \mathscr{H}(\Delta\widehat{y})]$ with the observed data y_d, which results in an ordinary least squares problem (similar to the maximum-likelihood model-based estimation problem (6.4)). This solution approach leads to the subspace method of [2,15].

Independent of the approach being used to compute the initial approximation, the initial estimate can be improved by local optimization. The methods developed in [16,17] can be used for this purpose; however, they operate off-line. Development of recursive methods, suitable for online solution of the maximum-likelihood model-based estimation problem (6.8) is a topic of current research.

6.5 Examples

As an illustration of the methods presented in the paper, we consider temperature and mass measurement, where the process dynamics is linear time-invariant first and second order, respectively. The data is simulated and the methods are implemented in MATLAB®. We show the actual code used to generate the numerical results in a literate programming style [18], using the no-web syntax [19] and the org-babel system [20]. This makes the results reproducible in the sense of [21]. The code is available from http://homepages.vub.ac.be/ imarkovs/software/sensor-uncer.tar.

6.5.1 Methods and evaluation criterion

The methods compared are

- y_d/G—the estimate obtained directly from the scaled raw measurements,
- `stepid_kf`—model-based maximum-likelihood method, solving (6.4) using the Kalman filter [2],
- `stepid_ml`—data-driven maximum-likelihood method, solving (6.8) using the SLRA package [17],
- `stepid_dd`—data-driven subspace method, using recursive least squares [15].

The estimates are computed for `ne` repetitions of the experiment with independent noise realizations (Monte Carlo simulation):

```
% mc simulation
ne = 1000;
G = dcgain(sys);
[a, b, c, d] = ssdata(sys);
n = size(sys, 'order');
```

```
for i = 1:ne
    <<noisy-output>>
    uh_nv(:, i)          = yd / G;
    [uh_dd(:, i), V_dd] = stepid_dd(yd, G, n);
    [uh_ml(i),    V_ml] = stepid_ml(yd, G, n);
    [uh_kf(:, i), V_kf] = stepid_kf(yd, a, b, c, d, sigma^2);
end
```

Let $\widehat{u}^k$ be the estimate computed in the kth repetitions of the experiment. The results reported are the empirical bias $(1/\mathrm{ne})\sum_{k=1}^{\mathrm{ne}} \bar{u} - \widehat{u}^k$, the empirical variance $(1/\mathrm{ne})\sum_{k=1}^{\mathrm{ne}} (\bar{u} - \widehat{u}^k)^2$, and the theoretically computed variance by the methods.

```
% show results
[mean(ub - uh_nv(T, :)) var(uh_nv(T, :)) sigma^2
 mean(ub - uh_dd(T, :)) var(uh_dd(T, :)) sigma^2 * V_dd
 mean(ub - uh_ml(:))    var(uh_ml(:))    sigma^2 * V_ml
 mean(ub - uh_kf(T, :)) var(uh_kf(T, :)) V_kf           ]
```

6.5.2 Example of temperature measurement

The setup of the temperature measurement problem is as follows. A thermometer is placed in an environment with temperature $\bar{u}$. The measured temperature y satisfies the Newton's law of cooling

$$\frac{\mathrm{d}}{\mathrm{d}t} y = a(y - \bar{u}). \tag{6.9}$$

where a is a negative constant that depends on the thermometer and the environment. The differential equation (6.9) defines a first-order linear time-invariant system with input, the environmental temperature $\bar{u}$, and output, the thermometer's reading y. The system is described by an input/state/output model

$$\frac{\mathrm{d}}{\mathrm{d}t} x = Ax + Bu, \qquad y = Cx + Dx,$$

where

$$A = -a, \quad B = a, \quad C = 1, \quad \text{and} \quad D = 0.$$

```
% temperature measurement process parameters
a = 0.4; A = -a; B = a; C = 1; D = 0;
```

The input is a step function $u = \bar{u}s$ and the initial condition is $x(0) = x_{\mathrm{ini}}$.

```
ub = 1; xini = 0.1;
```

The continuous-time model is discretized using the input zero-order hold method with a sampling period =ts=. The data

$$y_{\mathrm{d}} = (y_{\mathrm{d}}(1), \ldots, y_{\mathrm{d}}(T))$$

is generated in the output-error setup, using the discrete-time model and zero-mean, white Gaussian noise with noise-to-signal ratio `s`.

```
ts = 0.4; T = 10; s = 0.02;

% true system and true output
sys = c2d(ss(A, B, C, D), ts);
u = ub * ones(T, 1);
y = lsim(sys, u, [], xini);

% noisy output
yt = randn(T, 1);
sigma = s * norm(y) / norm(yt);
yd = y + sigma * yt;
```

6.5.3 Example of mass measurement

The setup of the mass-measurement problem is as follows: an object with mass M is placed on a scale with mass m. At the time of placing the object, the scale is in a specified (in general nonzero) initial condition. The object placement has the effect of a step input as well as a step change of the total mass of the system—scale and object. The sensor is the scale. It is modeled as a mass, spring, damper system

$$(M+m)\frac{\mathrm{d}^2}{\mathrm{d}t^2}y = -ky - d\frac{\mathrm{d}}{\mathrm{d}t}y - Mg, \tag{6.10}$$

where y is the scale's reading, g is the gravitational constant, k is the elasticity constant, and d is the damping constant. In this case, the sensor is a second-order linear time-invariant dynamical system with input the measured mass M.

```
% mass measurement process parameters
m = 1; M = 5; k = 1; d = 1; g = 9.81;
A = [0, 1; -k / (m + M), -d / (m + M)];
B = [0; -g / (m + M)]; C = [1, 0]; D = 0;
T = 50; ts = 1; s = 0.05; ub = M; xini = 0.1 * [1; 1];
```

6.5.4 Results

From theoretical considerations—use of prior knowledge about the true data generating system and the optimal vs suboptimal nature of the methods—we expect that the model-based maximum likelihood estimation method `stepid_kf` achieves the best performance (smallest bias and variance); the data-driven maximum likelihood estimation method `stepid_ml`, the second best; and the data-driven subspace method `stepid_dd` has the worst performance of the three methods, which, however is still better than the one of the naive estimator `yd/G`. This conjecture is confirmed by the simulation results shown in Table 6.1 for the temperature-measurement setup and Table 6.2 for the mass-measurement setup.

The theoretical variance of the estimators is based on asymptotic results that are not satisfied in the present measurement setups (transient response with $T = 10$ and $T = 50$ samples, respectively). Moreover, the theoretical variance of the `stepid_dd` estimator is based on results for the ordinary least square method, i.e., classical regression, while the actual setup in the application of dynamic measurements leads a

Table 6.1 *The results for the temperature measurement Monte Carlo simulation empirically confirm that the model-based maximum likelihood estimation method (*`stepid_kf`*) gives the best results in the sense of smallest bias and variance. The data-driven maximum likelihood estimation method (*`stepid_ml`*) leads to some loss of performance due to the weaker prior knowledge used (model class instead of a model). The data-driven subspace method (*`stepid_dd`*) incurs further loss of performance in comparison with* `stepid_ml` *due to its heuristic nature (solving the structured total least squares problem as an ordinary least squares problem). The theoretical variances predicted by the* `stepid_kf` *and* `stepid_ml` *estimators matches well the empirical variances.*

	Bias	**Empirical variance**	**Theoretical variance**
Naive	0.2129	0.0001	0.0001
`stepid_dd`	0.0920	0.0016	0.0001
`stepid_ml`	−0.0055	0.0017	0.0012
`stepid_kf`	0.0001	0.0001	0.0001

Table 6.2 *The same observations as the ones for temperature measurement (Table 6.1) hold true for the mass-measurement example. In this case, the difference in performance between model-based maximum-likelihood estimation method (*`stepid_kf`*) and the data-driven maximum likelihood estimation method (*`stepid_ml`*) is less significant. Also, the prediction of the empirical variance of the data-driven subspace estimator (*`stepid_dd`*) by the theoretical variance is good.*

	Bias	**Empirical variance**	**Theoretical variance**
Naive	0.0790	0.0636	0.0636
`stepid_dd`	−0.0265	0.0014	0.0016
`stepid_ml`	0.0006	0.0013	0.0016
`stepid_kf`	0.0005	0.0013	0.0015

structured total least squares problem, i.e., errors-in-variables regression. These considerations put doubt on how useful the theoretical variance formulas will be in practice. The simulation results, however, show that for the maximum-likelihood methods `stepid_kf` and `stepid_ml`, the theoretical variance give reasonably good prediction of the empirical variances. For the subspace method `stepid_dd`, the match between the theoretical and empirical variance is good in the case of mass measurement but wrong in the case of temperature measurement. A more rigorous approach to the uncertainty analysis of the subspace method is done in [22].

6.6 Conclusions and discussion

Metrology aims at producing faster and more accurate measurement devices. This can be done by the development of new measurement principles and corresponding sensors or by improvement of existing ones with signal processing. In this paper, we considered the latter approach: improvement of existing sensors by building a computational layer, typically implemented in a digital signal processor, which processes the raw-signal obtained from the sensor.

The signal-processing goal is to improve the given sensor's speed and accuracy characteristics. We presented both model-based and data-driven suboptimal and optimal in the maximum-likelihood sense methods for dynamic measurements. The model based-method is a classical least squares problem, which efficient implementation is the Kalman filter. In the data-driven case, the problem is a structured total least squares, which is a nonconvex optimization problem. Off-line solution methods performing nonlinear optimization exist; however, their recursive real-time implementation is a topic of current research. A computationally cheap alternative to the data-driven maximum likelihood method is a subspace method that can also be viewed as a convex relaxation of the structured total least squares problem.

The full parameter vectors $(\bar{u}, x_0)$ or $(\bar{u}, \ell)$ of the model-based and data-driven estimation problems, respectively, cannot be estimated consistently. This is due to the transient nature of the process (asymptotically decaying true signal). We conjecture that the parameter of interest $\widehat{u}$ can nevertheless be estimated consistently, due to the special structure of the regressors matrix: constant block $\mathscr{G}$ corresponding to $\widehat{u}$.

The bias of the maximum likelihood estimators, viewed as a function of the number of samples T, is the transient response of the compensated system, i.e., the bias is the remaining transient response. The subspace method introduces an additional bias, which is quantified in [22]. The variance of the maximum likelihood estimators is obtained as a by-product of the estimation algorithms: Kalman filter in the model-based case and the structured total least squares algorithm in the data-driven case. The variance of the subspace method is quantified in [22].

Acknowledgments

The research leading to these results has received funding from the European Research Council (ERC) under the European Union's Seventh Framework Programme (FP7/2007–2013) / ERC Grant agreement number 258581 "Structured low-rank approximation: Theory, algorithms, and applications" and Fund for Scientific Research Vlaanderen (FWO) projects G028015N "Decoupling multivariate polynomials in nonlinear system identification" and G090117N "Block-oriented nonlinear identification using Volterra series"; and Fonds de la Recherche Scientifique

(FNRS)—FWO Vlaanderen under Excellence of Science (EOS) Project no 30468160 "Structured low-rank matrix/tensor approximation: numerical optimization-based algorithms and applications."

References

[1] Sain M, Massey J. Invertibility of linear time-invariant dynamical systems. IEEE Trans Autom Control. 1969;14:141–149.
[2] Markovsky I. An application of system identification in metrology. Control Eng Pract. 2015;43:85–93.
[3] Shu W. Dynamic weighing under nonzero initial conditions. IEEE Trans Instrum Meas. 1993;42(4):806–811.
[4] Jafaripanah M, Al-Hashimi B, White N. Application of analog adaptive filters for dynamic sensor compensation. IEEE Trans Instrum Meas. 2005;54: 245–251.
[5] Van Overschee P, De Moor B. Subspace identification for linear systems: Theory, implementation, applications. Boston: Kluwer; 1996.
[6] Stoica P, Selén Y. Model-order selection: A review of information criterion rules. IEEE Signal Process Mag. 2004;21:36–47.
[7] Willems JC, Rapisarda P, Markovsky I, *et al.* A note on persistency of excitation. Control Lett. 2005;54(4):325–329.
[8] Markovsky I. Recent progress on variable projection methods for structured low-rank approximation. Signal Process. 2014;96(PB):406–419.
[9] Guillaume P, Pintelon R. A Gauss–Newton-like optimization algorithm for "weighted" nonlinear least-squares problems. IEEE Trans Signal Process. 1996;44(9):2222–2228.
[10] Markovsky I, Van Huffel S, Pintelon R. Block-Toeplitz/Hankel structured total least squares. SIAM J Matrix Anal Appl. 2005;26(4):1083–1099.
[11] Kailath T, Sayed AH, Hassibi B. Linear estimation. Upper Saddle River, NJ: Prentice Hall; 2000.
[12] Markovsky I. Structured low-rank approximation and its applications. Automatica. 2008;44(4):891–909.
[13] Rhode S, Usevich K, Markovsky I, *et al.* A recursive restricted total least-squares algorithm. IEEE Trans Signal Process. 2014;62(21):5652–5662.
[14] Yang B. Projection approximation subspace tracking. IEEE Trans Signal Process. 1995;43:95–107.
[15] Markovsky I. Comparison of adaptive and model-free methods for dynamic measurement. IEEE Signal Process Lett. 2015;22(8):1094–1097.
[16] Usevich K, Markovsky I. Variable projection for affinely structured low-rank approximation in weighted 2-norms. J Comput Appl Math. 2014;272:430–448.
[17] Markovsky I, Usevich K. Software for weighted structured low-rank approximation. J Comput Appl Math. 2014;256:278–292.
[18] Knuth D. Literate programming. Oxford: Cambridge University Press; 1992.
[19] Ramsey N. Literate programming simplified. IEEE Softw. 1994;11:97–105.

[20] Dominik C. The org mode 7 reference manual. Network Theory Ltd; 2010. Available from: http://orgmode.org/.
[21] Buckheit J, Donoho D. WaveLab and reproducible research. In: Wavelets and statistics. New York: Springer-Verlag; 1995.
[22] Carapia GQ, Markovsky I, Pintelon R, *et al.* Uncertainty analysis of the subspace step input estimation. Comput Stat Data An. 2019.

Chapter 7

Multivariable iterative learning control: analysis and designs for engineering applications

Lennart Blanken[1], Jurgen van Zundert[1], Robin de Rozario[1], Nard Strijbosch[1], and Tom Oomen[1]

Iterative learning control (ILC) enables high control performance through learning from measured data using limited model knowledge, typically in the form of a nominal parametric model. Robust stability requires robustness to modeling errors, often due to deliberate undermodeling. The aim of this chapter is to outline a range of design approaches for multivariable ILC that is suited for engineering applications, with specific attention to addressing interaction using limited model knowledge. The proposed methods either address the interaction in the nominal model or as uncertainty, i.e., through robust stability. The result is a range of techniques, including the use of the structured singular value (SSV) and Gershgorin bounds, that provide a different trade-off between modeling requirements, i.e., modeling effort and cost, and achievable performance. This allows control engineers to select the approach that best fits the modeling budget and control requirements. This trade-off is demonstrated in case studies on industrial printers. Additionally, two learning approaches are presented that are compatible with, and provide extensions to, the developed multivariable design framework: model-free iterative learning and ILC for varying tasks.

7.1 Introduction

7.1.1 ILC for complex engineering applications

Iterative learning control (ILC) [1–5] can significantly improve the control performance of systems that perform tasks which are mostly repeating in nature. After each iteration, or trial, the control action is improved by learning from past trials using an approximate model of the system. Many successful applications are reported, including additive manufacturing [6], microscopic imaging [7,8], printing systems [9], semiconductor bonding equipment [10], and wafer stages [11–13].

[1]Control Systems Technology Group, Department of Mechanical Engineering, Eindhoven University of Technology, Eindhoven, The Netherlands

Increasing requirements on product quality and throughput in high-precision engineering applications necessitate faster operation with increased accuracy. These developments impose strict specifications regarding the control performance, which lead to an ongoing development of increasingly complex systems in terms of flexible dynamics, and required actuators and sensors to actively control these dynamics, see, e.g., [14]. This has significant implications for learning control design, especially in view of the multivariable aspect, which is explored in this chapter.

7.1.2 Design requirements for high-precision applications

Successful implementation of ILC algorithms on complex engineering applications imposes several modeling and design requirements. First, many multivariable high-precision applications inherently exhibit strong interaction between inputs and outputs, since the flexible dynamical behavior is generally not aligned with the motion degrees of freedom [14]. Second, low-complexity design procedures are typically favored in industrial practice, particularly in view of algorithmic complexity and required user effort. Third, many industrial systems perform slightly varying tasks, including atomic force microscopes [8], bonding equipment [15] and lithographic equipment [16], in which case the application of traditional ILC may yield disastrous performance, see e.g., [10,12,13]. These observations lead to the following requirements for ILC design:

R1. robust stability of the multivariable system w.r.t. unmodeled dynamics, e.g., ignored interaction;
R2. limited user effort, i.e., intuitive design tools and low modeling requirements;
R3. high performance, possibly for varying tasks.

7.1.3 Robust multivariable ILC design: the importance of (under) modeling (R1–R2)

Related to requirement R1, the observation that many ILC applications are inherently multivariable, i.e., are multi-input multi-output (MIMO), has led to the development of ILC theory for multivariable systems. Most design algorithms are developed in the so-called lifted or supervector framework [2], where the ILC controller follows from a norm-based optimization problem over a finite-time horizon, see, e.g., [17].

Robust convergence w.r.t. modeling errors, i.e., robust stability in the iteration domain, is a key property of ILC algorithms to enable achieving extremely high performance using only approximate models. In particular, since the ILC control action often is effective up to the Nyquist frequency, model accuracy requirements are imposed over the entire frequency range [18], which is in sharp contrast to classical feedback control [14]. Norm-based optimization approaches have been further extended for robust monotonic convergence in, e.g., [11,19–22]. These approaches rely on a detailed specification of a nominal model and its uncertainty in some prescribed form. Despite being very systematic, this imposes a large burden on the model requirements, since the modeling of uncertainty requires substantial effort of the user [23], in addition to non-convex optimization algorithms, e.g., *D–K* iterations.

Related to requirement R2, although robust MIMO ILC has been significantly developed, especially from a theoretical perspective, these approaches are often not employed due to stringent requirements on uncertainty modeling. In many applications, ILC algorithms are designed in the frequency domain [1,2,10,24–26]. Compared to the norm-optimal framework, frequency-domain ILC enables a systematic and inexpensive robust design in the sense of modeling requirements, especially regarding the model uncertainty. Indeed, confidence intervals around frequency response function (FRF) estimates [27] can be utilized to model the uncertainty, see, e.g., [11,26,28], which are typically readily obtained for FRF estimates that result from experimental data. In addition, frequency-domain design allows for manual loop shaping, which is often preferred by control engineers. However, since such manual design approaches are mainly focused on single-input–single-output (SISO) systems, design for MIMO systems typically involves their application to multiple SISO loops, see, e.g., [2,29]. As such, interaction is typically ignored, which can potentially lead to robust stability issues in the sense of non-convergent algorithms [30].

Crucially, the seemingly drastic increase in modeling effort required to enforce robust stability of multivariable ILC algorithms must be justified by the performance requirements imposed on the considered system. Interestingly, interaction is typically accounted for through centralized, or full MIMO, ILC design. Successful MIMO design approaches include standard $\mathscr{H}_\infty$ synthesis without preview [11,31,32], and P-type ILC [33–35]. Centralized techniques enable robust convergence and superior performance, yet require a full MIMO parametric model of the system, including the interaction terms. These models can be difficult and expensive to obtain due to complex dynamics [36,37] and numerical issues [38]. In this chapter, it is shown that a MIMO parametric model is not necessarily justified by the performance requirements: through systematic decentralized design using only SISO design tools and SISO parametric models, robust convergence and high performance can be achieved.

7.1.4 Model-free iterative learning (R2)

Related to requirement R2, and as an alternative to the approaches outlined in Section 7.1.3, model-free learning approaches have been developed driven by the seemingly required user effort to obtain parametric models for ILC. In iterative inversion-based control (IIC) [39], the updating of the input signal is explicitly performed in the frequency domain by using the discrete Fourier transform (DFT). The key advantage of this approach is threefold. First, IIC merely requires an estimate of the FRF of the system, which are typically accurate, inexpensive and fast to obtain [27]. Second, under the assumption that the system is approximately in steady state, convergence can be assessed for each individual frequency of the DFT grid. This leads to significantly increased design freedom since the robustness and learning filters can be tuned as a set of frequency-dependent coefficients instead of dynamic filters [40]. Third, it can be shown that under mild conditions, this approach automatically results in bounded signals that may be noncausal in case the system possesses zeros outside the unit disc [41].

7.1.5 ILC for varying tasks (R3)

Related to requirement R3, independent of the pursued ILC or IIC approach, a key underlying assumption is that the considered systems perform exactly repeating tasks. This assumption is often violated in industrial systems [8,10,15,16], which can result in severe performance degradations. In fact, a small deviation from exactly repeating tasks can lead to worse performance compared to traditional control without ILC, see, e.g., [12,13,42].

To enhance the robustness to varying tasks, ILC approaches have been extended with basis functions in, e.g., [42–49]. However, since these approaches are developed in the norm-based optimization framework, the large burden imposed on the modeling requirements may prevent industrial implementation (R2).

Recent developments for ILC with basis functions [10] enable a manual loop-shaping design approach in the frequency-domain, facilitating industrial adoption where loop-shaping is commonly employed. However, the developed approach is yet restricted to SISO systems. In this chapter, the frequency-domain design approach for ILC with basis functions is generalized toward MIMO systems, and it is unified with the MIMO frequency-domain design framework introduced above, enabling decentralized designs to cope with trade-offs between modeling requirements (R2) and performance (R1).

7.1.6 Contributions

The main contribution of this chapter is an overview of a systematic and unifying framework for analysis and design of MIMO ILC in the frequency domain, both for exactly repeating and varying tasks, with or without (MIMO) parametric models. A step-by-step procedure is presented that gradually extends SISO manual ILC design to the MIMO case. This procedure consists of (1) interaction analysis, (2) decoupling, (3) independent SISO design for MIMO systems, (4) robust decentralized MIMO design and (5) full MIMO design. Optionally, if desired by the user, the designed ILC can be combined and implemented with basis functions for flexibility to task variations. The framework enables control engineers to explicitly address the design trade-offs between modeling requirements and performance. The developed solutions are hence all suitable alternatives, and depending on the specific application at hand, an appropriate choice can be made. To illustrate and substantiate the outlined trade-offs in the design framework, throughout the chapter the presented design solutions are applied to the high-precision printing systems shown in Figure 7.1.

The outline of the chapter is as follows. The general ILC design problem is formalized in Section 7.2. Frequency-domain design of SISO ILC is reviewed in Section 7.3, which is subsequently extended and generalized to a MIMO ILC design framework in Section 7.4. Extensions of this frequency-domain framework are presented in Section 7.5 toward model-free iterative learning, and in Section 7.6 toward the use of basis functions for task variations. Finally, conclusions and an outlook on ongoing developments are provided in Section 7.7.

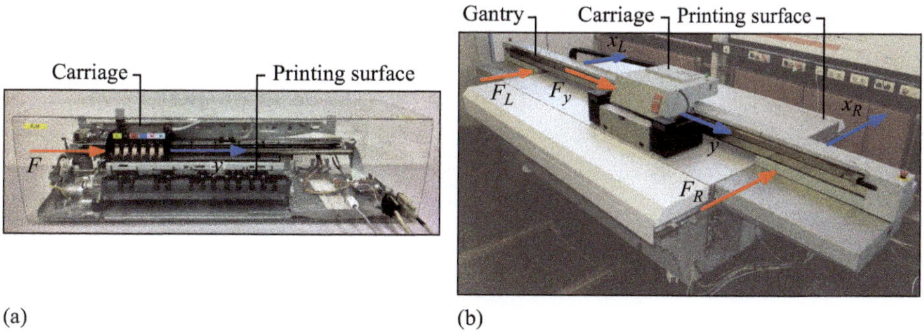

Figure 7.1 *SISO (left) and MIMO (right) printing systems, used throughout the chapter to illustrate design implications of the outlined framework for ILC design. (a) Photograph of single-input single-output A3 desktop printer, with input force F and output position y. (b) Photograph of multivariable Océ Arizona 550 GT flatbed printer. The considered variables for control are the translation $x = (x_L + x_R)/2$, and the rotation φ of the carriage in the horizontal plane, while the position of the carriage along the gantry is fixed*

7.1.7 Notation

The following notation is used throughout. All systems are assumed to be discrete-time, MIMO and linear time-invariant. The continuous-time case follows analogously. $\mathscr{RL}_\infty$ denotes the set of real rational functions bounded on the unit circle, $\mathscr{RH}_\infty$ is the subset of $\mathscr{RL}_\infty$ analytic in $|z| > 1$. The space ℓ_2^n is the set of n-dimensional sequences on the time interval $\mathbb{Z}$ with finite inner product $\langle x(t), y(t)\rangle_{\ell_2} = \sum_{t\in\mathbb{Z}} x(t)^* y(t)$, where $x(t), y(t) \in \ell_2^n$, and associated induced norm $\|x\|_{\ell_2} = \sqrt{\langle x(t), x(t)\rangle_{\ell_2}}$. The dimension n is omitted when it is clear from the context, e.g., ℓ_2^n is written ℓ_2.

Consider the response $y(t) = H(q)u(t)$, where $y(t) = \sum_{k=-\infty}^{\infty} H_k u(t-k)$ and $H(q) = \sum_{k=-\infty}^{\infty} H_k q^{-k}$, with H_k the infinite-time Markov parameters, and q denotes the forward time shift operator, i.e., $qu(t) = u(t+1)$. Using frequency domain representations, this response is described by $Y(\omega)e^{\iota\omega t} = H(e^{\iota\omega})U(\omega)e^{\iota\omega t}$, or $Y(\omega) = H(e^{\iota\omega})U(\omega)$, where $Y(\omega) = \sum_{k=-\infty}^{\infty} y(k)e^{-\iota\omega k}$, and $H(e^{\iota\omega}) = \sum_{k=-\infty}^{\infty} H_k e^{-\iota\omega k}$, with ι denoting the imaginary unit, i.e., $\iota^2 = -1$. That is, the notation $Y(\omega)$ refers to the sinusoidal vector-valued signal $Y(\omega)e^{\iota\omega t}$; the frequency response of system H is denoted as complex matrix $H(e^{\iota\omega})$. In the remainder, the arguments t, q, z, ω and $e^{\iota\omega}$ are often omitted when they are clear from the context.

7.2 System description and problem formulation

7.2.1 ILC framework

Consider the control configuration in Figure 7.2, consisting of the plant $G \in \mathscr{R}^{n_y \times n_u}(z)$ and internally stabilizing feedback controller $C \in \mathscr{R}^{n_u \times n_y}(z)$. The exogenous

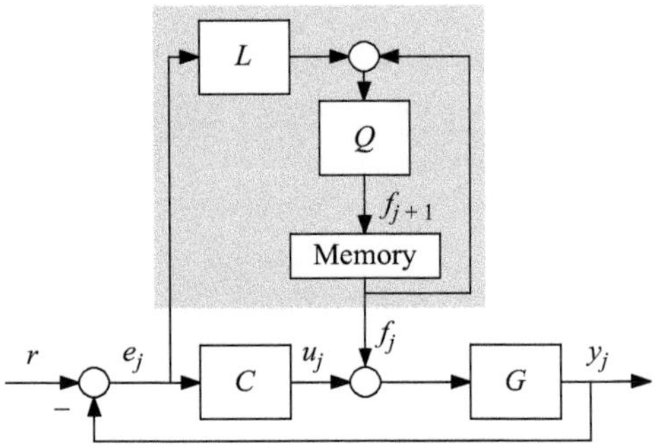

Figure 7.2 ILC control configuration

disturbance $r \in \ell_2$ is trial-invariant unless stated otherwise. The aim is to minimize the tracking error e in the presence of r using feedforward f. The error is given by

$$e_j = Sr - Jf_j, \tag{7.1}$$

with stable and causal sensitivity function $S = (I + GC)^{-1} \in \mathscr{RH}_\infty^{n_y \times n_y}$, and process sensitivity function $J = SG \in \mathscr{RH}_\infty^{n_y \times n_u}$, i.e., the transfer function matrix mapping f_j to e_j, where the index $j \in \mathbb{N}_{\geq 0}$ denotes the trial number. Zero initial conditions are assumed without loss of generality [2]. Note that if G is stable, then $C = 0$ is admissible in which case $S = I$ and $J = G$ in (7.1).

The objective of ILC is to iteratively improve control performance over trials j by selecting the command input f_{j+1} for the next trial $j + 1$. Typically an update law of the following form is invoked:

$$f_{j+1} = Q(f_j + Le_j), \tag{7.2}$$

where the learning filter $L \in \mathscr{RL}_\infty^{n_u \times n_y}$ and robustness filter $Q \in \mathscr{RL}_\infty^{n_u \times n_u}$ are to be designed, which can be noncausal, since the signals f_j and e_j in (7.2) are available off-line. In Figure 7.2, the ILC update is placed in parallel to C; alternatively and nonrestrictively it can be placed in a serial configuration, see, e.g., [1]. In the design, it is often assumed that the trial length is infinite, such that the ILC can essentially be implemented for any trial length, see [50] for a detailed analysis.

Clearly, assuming perfect system knowledge and J invertible, selecting the update law $f_{j+1} = f_j + J^{-1}e_j$, i.e., $Q = I$ and $L = J^{-1}$, gives $e_{j+1} = Sr - Jf_j - JJ^{-1}e_j = Sr - Jf_j - (Sr - Jf_j) = 0$; hence, perfect tracking is achieved in a single iteration. However, in practice modeling errors are inevitable, i.e., $LJ \neq I$, and robustness through Q may be required to enforce convergence of the algorithm (7.2) subject to dynamics (7.1). These are the key problems addressed in this chapter.

7.2.2 Convergence and performance

In this subsection, convergence and performance properties are investigated, both transient and asymptotic, of the closed-loop ILC system.

Since the ILC algorithm (7.2) operates from trial to trial, ILC can be interpreted as feedback control in the trial domain. A closed-loop system description can be obtained by combining (7.1) and (7.2), yielding the linear iterative systems:

$$f_{j+1} = Q(I - LJ)f_j + QLSr, \tag{7.3}$$

$$e_{j+1} = JQ(I - LJ)J^{-1}e_j + (I - JQJ^{-1})Sr, \tag{7.4}$$

where (7.4) requires J invertible. Convergence of the ILC can be interpreted as asymptotic stability of (7.3) and (7.4), and performance is typically expressed based on their transient and asymptotic behavior, i.e., in the limit $j \to \infty$. Considering (7.3) for instance, f_j converges in a given norm $\| \cdot \|$ if and only if for all r, f_0, there exists an asymptotic signal $f_\infty = \lim_{j\to\infty} f_j$ such that

$$\lim_{j\to\infty} \|f_\infty - f_j\| = 0.$$

The asymptotic signals f_∞ and e_∞ are obtained through $f_\infty = \lim_{j\to\infty} f_j = \lim_{j\to\infty} f_{j+1}$ and substitution into (7.1), i.e.,

$$f_\infty = (I - Q(I - LJ))^{-1}\, QLSr,$$

$$e_\infty = \left(I - J\,(I - Q(I - LJ))^{-1}\, QL\right) Sr. \tag{7.5}$$

Two observations are made. First, convergence does not imply that e_∞ is smaller than e_0, i.e., it is possible that the addition of ILC to the control system degrades tracking performance. Second, if and only if $Q = I$ and (7.3) converges, then $e_\infty = 0$ for all r and general $L, J \neq 0, \forall\omega$.

Transient performance can be expressed in terms of monotonic convergence of the ILC algorithm. Typically in practice, monotonic convergence of f_j under a given norm $\| \cdot \|$ is desired to avoid actuator limitations, in the sense that for all r, f_0, there exists an f_∞ for which

$$\|f_\infty - f_{j+1}\| \leq \gamma \|f_\infty - f_j\| \quad \forall j,$$

where $0 \leq \gamma < 1$ is the convergence rate.

7.2.3 Design conditions for convergence and performance

Design procedures of L and Q for (possibly monotonic) convergence typically aim to satisfy the following well-known necessary and sufficient conditions. The iterations (7.3) and (7.4) are convergent if and only if [30, Theorem 1]

$$\rho\left(Q(e^{\iota\omega})(I - L(e^{\iota\omega})J(e^{\iota\omega})\right) < 1 \quad \forall\omega \in [0, \pi], \tag{7.6}$$

and (7.3) is monotonically convergent in the ℓ_2 norm of f_j with $0 \leq \gamma_1 < 1$ if and only if [30, Theorem 2]

$$\|Q(I - LJ)\|_\infty \leq \gamma_1, \tag{7.7}$$

where $\|H(z)\|_\infty = \sup_{\omega\in[0,\pi]} \bar{\sigma}(H(e^{\iota\omega}))$ is the $\mathscr{L}_\infty$-norm, and $\bar{\sigma}$ denotes the maximum singular value. Hence, condition (7.7) is equivalent to requiring

$$\bar{\sigma}\left(Q(e^{\iota\omega})(I - L(e^{\iota\omega})J(e^{\iota\omega})\right) \leq \gamma_1 \quad \forall\omega \in [0, \pi]. \tag{7.8}$$

Three observations are made. First, (7.7) explicitly allows for noncausal filters L and Q since an $\mathscr{L}_\infty$-norm is used. Second, monotonic convergence of (7.3) and (7.4) is not equivalent in the MIMO case: a condition for monotonic convergence in the ℓ_2 norm of e_j is $\left\|JQ(I-LJ)J^{-1}\right\|_\infty \leq \gamma_1$. Third, monotonic convergence is a stronger property than convergence, i.e., (7.8) implies (7.6).

7.2.4 Modeling considerations

The theoretical results on convergence and performance have direct consequences for the design of $L(z)$ and $Q(z)$. For optimal asymptotic tracking performance, i.e., $e_\infty = 0$, Q must equal I, and to achieve optimal transient performance, i.e., monotonic convergence rate $\gamma_1 = 0$, L should equal the inverse of $J = (I + GC)^{-1}G$. However, this optimal design may not be achievable in practice due to the following reasons:

1. A parametric model $\hat{J}(z)$ of $J(z)$ is required for inversion-based construction of $L = \hat{J}^{-1}(z)$, which inevitably contains modeling errors, i.e., $\hat{J}(z) \neq J(z)$.
2. The (approximate) inversion of parametric models $\hat{J}(z)$ may introduce additional errors, e.g., if $\hat{J}(z)$ is non-invertible or has zeros outside the usual stability region, i.e., $\hat{J}^{-1}(z)$ has poles outside the unit disk, see [51].
3. For MIMO systems, systematic procedures to deal with structure in parametric models $\hat{J}(z)$, e.g., resulting from deliberately ignoring the off-diagonal interaction terms during the parametric modeling step, are largely undeveloped.

Regardless of the underlying reason, if $LJ \neq I$ for any frequency, robustness may be required to enforce (monotonic) convergence of the algorithm. For the SISO case, practical procedures for robust ILC design are well developed, where inexpensive FRF models of $J(e^{\iota\omega})$ can be used to evaluate robust convergence in the presence of undermodeling. This is reviewed in Section 7.3.

For the MIMO case, however, such practical design procedures are mostly undeveloped. Crucial yet largely unexplored aspects herein are (i) modeling requirements, especially regarding the seemingly required MIMO parametric model of J for inversion-based design of L, see, e.g., [11,19–22] and (ii) dealing with robustness to (often deliberate) modeling errors.

Summarizing, at present, a gap is present between SISO and MIMO ILC designs in view of modeling requirements (R2) and resulting performance (R1). Thus, the aim of the present manuscript is to extend and generalize procedures for SISO manual ILC design to the MIMO case, with particular attention to modeling requirements, e.g., alleviating or completely circumventing the requirement for MIMO parametric models through the use of FRFs. In view of this, frequency-domain design procedures for SISO ILC are reviewed in the next section.

7.3 ILC design—the SISO case

In the SISO case, frequency-domain design procedures are well developed. Typically, the two-step approach in Procedure 7.1 is followed, see, e.g., [24–26,28]. The steps in Procedure 7.1, i.e., design of L and Q, are investigated in forthcoming subsections, including modeling requirements and noncausal implementations.

Procedure 7.1. Design procedure for SISO ILC

---

1. Design of L:
   i. Obtain SISO parametric model $\hat{J}(z)$ of $J(z)$.
   ii. Inversion-based design of L based on $\hat{J}(z)$, such that $L\hat{J} \approx 1$.
2. Design of Q:
   i. Obtain a SISO nonparametric FRF model $\hat{J}_{\mathrm{FRF}}(e^{\iota\omega})$ of J.
   ii. For robust convergence, design Q such that $|Q(e^{\iota\omega})(1 - L(e^{\iota\omega})\hat{J}_{\mathrm{FRF}}(e^{\iota\omega}))| < 1, \forall\omega$.

Remark 7.1. *In view of robust convergence w.r.t. plant uncertainty, the evaluation of* (7.6) *and* (7.8) *based on FRF models $\hat{J}_{FRF}(e^{\iota\omega})$ straightforwardly allows to include confidence intervals, which are typically readily determined for FRF estimates that result from experimental data, see, e.g., [27]. This is a distinctive advantage compared to alternative robust norm-based ILC frameworks, especially for systems where FRF measurements are accurate and inexpensive.*

7.3.1 Manual design in the frequency domain

In the SISO case $n_u = n_y = 1$ and infinite time setting, convergence and monotonic convergence are equivalent, i.e., conditions (7.6) and (7.7) are equivalent. That is, monotonic convergence in the ℓ_2 norms of f_j and e_j is achieved with convergence rate $0 \leq \gamma_1 < 1$ if and only if

$$|Q(e^{\iota\omega})(1 - L(e^{\iota\omega})J(e^{\iota\omega}))| \leq \gamma_1, \quad \forall\omega \in [0, \pi]. \tag{7.9}$$

In the next subsections, the design of L and Q is investigated based on models $\hat{J}$ and $\hat{J}_{FRF}$, as outlined in Procedure 7.1. The proposed manual design procedure is illustrated on the desktop printer, see Figure 7.1(a), with particular attention to the required models, i.e., parametric and nonparametric, and design, i.e., model inversion and loop-shaping in the frequency domain.

7.3.1.1 Design of L by inverting $\hat{J}$

Typically, the SISO filter $L(z)$ is designed by inverting an approximate low-order parametric model $\hat{J}(z)$ of J, such that $L\hat{J} \approx 1$, see, e.g., [1,2,26,29]. For the considered printing example, $\hat{J}(z)$ is obtained as $\hat{J} = \hat{G}/(1 + \hat{G}C)$, where $\hat{G}(z)$ is a low-order parametric approximation of a measured FRF model $\hat{G}_{\mathrm{FRF}}$, see Figure 7.3(a).

Regarding the inversion step, in case $\hat{J}$ is minimum phase, L can be constructed using straightforward inversion, i.e., $L(z) = 1/\hat{J}(z)$. In case J is non-minimum phase, i.e., has zeros outside the unit disk, directly taking the inverse $1/\hat{J}(z)$ leads to an unbounded solution and hence cannot be used. In Section 7.3.2, a range of methods is

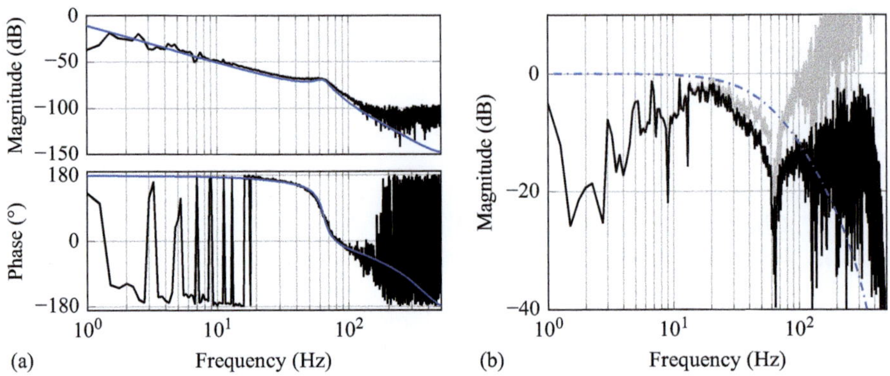

Figure 7.3 Manual SISO ILC design in the frequency domain: using models $\hat{G}(z)$ and $\hat{G}_{FRF}(e^{\iota\omega})$ (left), L is designed by inverting $\hat{J} = \hat{G}/(1+\hat{G}C)$, and Q is designed such that $|Q(1 - L\hat{J}_{FRF})| < 1, \forall\omega$ (right). (a) Bode diagram of frequency response function model $\hat{G}_{FRF}$ and low-order parametric model $\hat{G}$ of the A3 desktop printer, see Figure 7.1(a). (b) SISO design: Q (-.-), $(1-L\hat{J}_{FRF})$ (—), $Q(1-L\hat{J}_{FRF})$ (—)

outlined to compute a stable (approximate) inverse of a non-minimum phase (NMP) model. Each technique requires a parametric model $\hat{J}(z)$, yet differs in terms of computational complexity and approximation accuracy.

7.3.1.2 Design of Q based on FRF measurements

In step 2 of Procedure 7.1, robust convergence of the ILC algorithm is enforced by designing a filter Q. The key point is that (7.9) can be verified using a nonparametric FRF model $\hat{J}_{\mathrm{FRF}}$ of J, which are for mechatronic systems often accurate, inexpensive and fast to obtain [52,53], and enables an evaluation of robust convergence in the presence of undermodeling. For the considered printing example, an identified FRF model is shown in Figure 7.3(a).

In particular, given L from step 1, robustness is required in case $|1 - LJ| \geq 1$, see Figure 7.3(b), for instance due to modeling errors, i.e., $\hat{J} \neq J$, and inversion errors, i.e., $L \neq 1/\hat{J}$. The robustness filter Q is typically designed as a zero-phase filter, since its sole aim is to guarantee $|Q(1 - LJ)| < 1$. The zero phase behavior can be obtained through noncausal filtering, i.e., an operation with $Q(z)$ and its adjoint $Q^*(z) = Q(1/z)$, where the adjoint can be implemented by filtering backwards in time, see, e.g., [54].

For the considered printing example, Q is designed as a low-pass filter, since the robustness is mainly required at high frequencies. The convergence condition (7.9) is verified on the basis of $\hat{J}_{\mathrm{FRF}} = 1/(1 + G_{\mathrm{FRF}}C)$, as is shown in Figure 7.3(b). Using this manual ILC design, the experimental results depicted in Figure 7.4(a) show that tracking performance can significantly be improved.

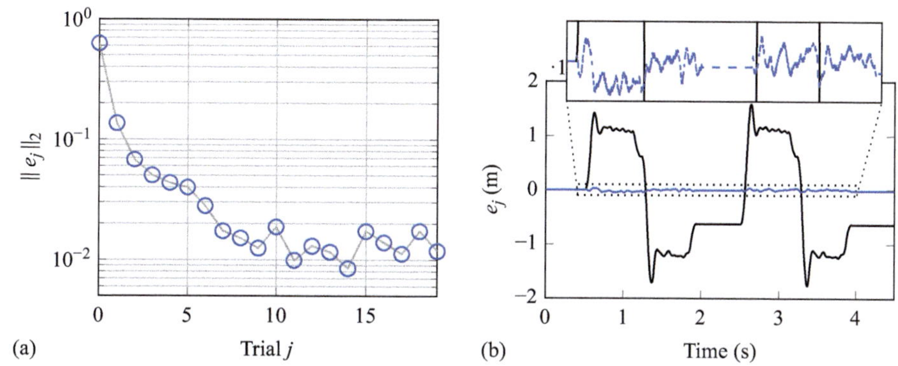

Figure 7.4 Experimental results using SISO ILC: the resulting performance $\|e_j\|_2$ is improved by a factor 30: (a) $\|e_j\|_2$ as function of trials j and (b) initial error e_0 and final error e_{19} as function of time

7.3.2 Design of learning filter: SISO inversion techniques

Next, several techniques for inversion of a NMP parametric model $\hat{J}$ are investigated. The aim is to construct a filter L such that the transfer function $I - L\hat{J}$ is small, where the ILC command input f_{j+1} must remain bounded.

Let $\hat{J}(z)$ have relative degree $d \in \mathbb{N}$ and $p \in \mathbb{N}$ NMP zeros. In view of inverting $\hat{J}$, there are two main issues:

1. Delay: for $d > 0$, $\hat{J}^{-1}$ is improper.
2. NMP zeros: for $p > 0$, $\hat{J}^{-1}$ has poles outside the usual stability region which, when solved forward in time, yields unbounded outputs.

The first aspect can be overcome by inverting the bi-proper system $\bar{J} = z^d\hat{J}$, where the anticausal z^d is implemented as a time shift on the input signal, see, e.g., [55]. Note that this approach is exact for an infinite-time horizon, whereas this may lead to boundary errors on a finite-time horizon.

The second issue can be overcome by the use of inversion techniques for NMP systems. In the forthcoming subsections, several inversion techniques are presented that are suitable for ILC, see also [51] for an overview. An overview of these techniques is presented in Table 7.1.

7.3.2.1 Approximate inversion

Let $\hat{J}$ be decomposed as

$$\hat{J}(z) = \frac{B_s(z)B_u(z)}{A(z)}, \tag{7.10}$$

with $B_s(z)$ containing all minimum-phase zeros and $B_u(z)$ the p NMP zeros. A key issue is that $B_u^{-1}(z)$ is unstable. Several techniques have been proposed to approximate the inverse of $B_u(z)$, including non-minimum phase zeros ignore (NPZ-Ignore) [56],

Table 7.1 Overview of the inversion techniques and their characteristics

Technique	Preview	I/O dimensions	Aim
NPZ-Ignore	Finite, $p+d$	SISO	$\hat{J}^{-1}$ approximation
ZPETC	Finite, $p+d$	SISO	$\hat{J}^{-1}$ approximation
ZMETC	Finite, d	SISO	$\hat{J}^{-1}$ approximation
$\mathscr{H}_\infty$-preview control	Finite, user-defined	Non-square	$\min_L \|I - L\hat{J}\|_\infty$
Stable inversion	Infinite	Square	$\hat{J}^{-1}$ exact

Table 7.2 Overview of NPZ-Ignore, ZPETC, and ZMETC for decomposition (7.10). *The DC gain is compensated by* $\beta = B_u(1)$

Technique	*L(z)*	Preview	*L(z)J(z)*
NPZ-Ignore	$A(z)/\beta B_s(z)$	$p{+}d$	$B_u(z)/\beta$
ZPETC	$(z^{-p}A(z)B_u^*(z))/\beta^2 B_s(z)$	$p{+}d$	$(z^{-p}B_u(z)B_u^*(z))/\beta^2$
ZMETC	$A(z)/(B_s(z)B_u^*(z))$	d	$B_u(z)/B_u^*(z)$

zero-phase-error tracking control (ZPETC) [57], and zero-magnitude-error tracking control (ZMETC). The results for these approaches are summarized in Table 7.2. Importantly, all approaches compute an approximate inverse in a heuristic manner that is stable and utilize a fixed and finite amount of preview.

7.3.2.2 $\mathscr{H}_\infty$-optimal synthesis with preview

Alternatively to the heuristic approaches listed in Table 7.2, where the preview is directly determined by the model $\hat{J}$, one may attempt to compute an approximate inverse with predefined and user-chosen preview that is optimal with respect to the $\mathscr{L}_\infty$-condition for monotonic convergence (7.7). The key idea is to optimize the causal part of L in $\mathscr{H}_\infty$ for a specified amount of preview.

Consider the system shown in Figure 7.5(a), where L is decomposed as $L = z^{p_l}L_c$ with user-defined preview $p_l \in \mathbb{N}$ and causal $L_c \in \mathscr{RH}_\infty$, and Q is decomposed as $Q = z^{p_q}Q_c$ with causal $Q_c \in \mathscr{RH}_\infty$ and user-defined preview $p_q \in \mathbb{N}$. Considering Figure 7.5(b), note that the multiplications with anticausal parts z^{p_l} and z^{p_q} are not part of generalized plant P to ensure $P \in \mathscr{RH}_\infty$. Then, the noncausal filter L follows from the following optimization problem over $\mathscr{H}_\infty$:

$$L = z^{p_l}L_c = z^{p_l} \arg \min_{L_c \in \mathscr{RH}_\infty} \|Q_c \left(z^{-p_l}I - L_c\hat{J}\right)\|_{\mathscr{H}_\infty},$$

where it is emphasized that an $\mathscr{H}_\infty$-norm is considered, which is in sharp contrast with, e.g., (7.7). Note also that the a priori chosen $Q(z)$ effectively acts as an output

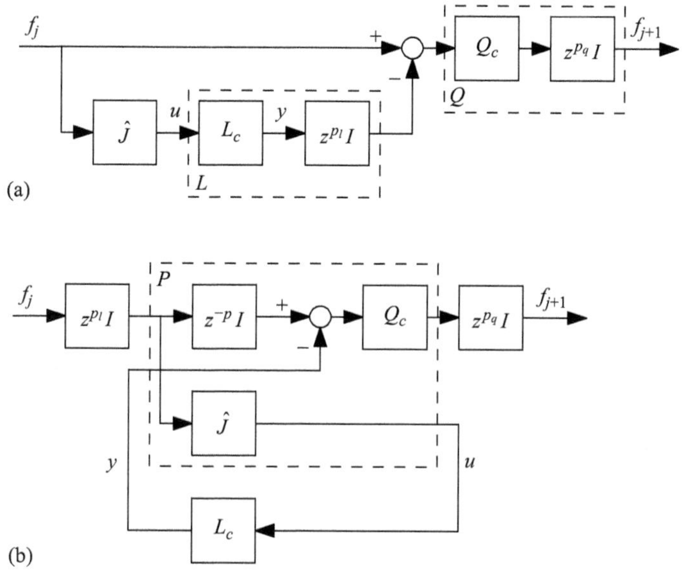

Figure 7.5 *$\mathcal{H}_\infty$-optimal synthesis with preview: the objective is minimization of $\|Q(I - LJ)\|_{\mathcal{L}_\infty}$: (a) block diagram corresponding to transfer $f_j \mapsto f_{j+1}$, which is in $\mathcal{RL}_\infty$ and (b) standard plant formulation with $P \in \mathcal{RH}_\infty$*

weighting and is restricted to have finite preview, e.g., $Q(z)$ can be chosen as a noncausal zero-phase finite impulse response (FIR) filter. For the special case without preview, i.e., $p_l, p_q = 0$, the standard $\mathcal{H}_\infty$ approach is recovered, see, e.g., [11,31,32]. In addition, the preview terms can be optimized, e.g., through bisections over p_l and p_q.

7.3.2.3 Stable inversion

The techniques presented in the previous subsections are all based on finite-preview approximations of the inverse system, which have all poles strictly inside the unit disk. In contrast, stable inversion aims to compute an exact solution by interpreting the inverse as a noncausal operator in $\mathcal{L}_2$, i.e., with poles both strictly inside and outside of the unit disk, rather than a causal operator in $\mathcal{H}_2$, see, e.g., [58, Section 1.5]. This interpretation directly fits into the considered frequency-domain ILC framework, since indeed an $\mathcal{L}_\infty$ norm is used in (7.7), rather than an $\mathcal{H}_\infty$ norm.

Next, the required steps for computing $f_{e,j} = Le_j$ are presented, see (7.2) with $f_{j+1} = Q(f_j + f_{e,j})$. Let the bi-proper hyperbolic system $\hat{J}$ be described by a state-space realization (A, B, C, D), such that $\hat{J}^{-1}$ is described by

$$\hat{J}^{-1} \stackrel{s}{=} \left[\begin{array}{c|c} A - BD^{-1}C & BD^{-1} \\ \hline -D^{-1}C & D^{-1} \end{array}\right].$$

The state of $\hat{J}^{-1}$ is divided into a strictly stable and unstable part by applying the state transformation $x(t) = T\begin{bmatrix} x_s(t) \\ x_u(t) \end{bmatrix}$, where T contains the eigenvectors of $A - BD^{-1}C$ such that

$$\begin{bmatrix} x_s(t+1) \\ x_u(t+1) \end{bmatrix} = \begin{bmatrix} A_s & 0 \\ 0 & A_u \end{bmatrix} \begin{bmatrix} x_s(t) \\ x_u(t) \end{bmatrix} + \begin{bmatrix} B_s \\ B_u \end{bmatrix} e_j(t),$$

$$f_{e,j}(t) = \begin{bmatrix} C_s & C_u \end{bmatrix} \begin{bmatrix} x_s(t) \\ x_u(t) \end{bmatrix} + De_j(t),$$

with $|\lambda(A_s)| < 1$ and $|\lambda(A_u)| > 1$, i.e., all stable poles are contained in A_s and all unstable poles in A_u. Then, the bounded states are obtained by solving

$$x_s(t+1) = A_s x_s(t) + B_s e_j(t),$$

forward in time with suitable initial conditions, e.g., $x_s(-\infty) = 0$, and

$$x_u(t+1) = A_u x_u(t) + B_u e_j(t),$$

backward in time with suitable final conditions, e.g., $x_u(\infty) = 0$. The bounded and noncausal output $f_{e,j}$ follows from

$$f_{e,j}(t) = C_s x_s(t) + C_u x_u(t) + De_j(t).$$

In practice, the boundary conditions are imposed on finite time instants, which introduces boundary errors. For motion systems, if the system is at rest before and after the motion task, setting $x_s(0) = 0$, $x_u(N) = 0$ typically leads to sufficiently small errors, see, e.g., [51].

7.3.3 Toward MIMO ILC design: naive SISO design for MIMO systems

Since the tuning and design algorithms for SISO ILC are well developed and relatively intuitive to use, see the preceding subsections, it is tempting to apply multiple SISO controllers to MIMO systems. However, when considerable interaction is present in the system, naive application of multiple SISO ILC controllers may lead to robust stability issues in the sense of non-convergent systems. Considering (7.6) and (7.9), it is clear that

$$|q_i(e^{\iota\omega})(1 - l_i(e^{\iota\omega}) J_{ii}(e^{\iota\omega}))| < 1, \quad \forall i \tag{7.11}$$

$$\not\Rightarrow$$

$$\rho\left(Q(e^{\iota\omega})(I - L(e^{\iota\omega}) J(e^{\iota\omega})\right) < 1, \tag{7.12}$$

hence, convergence is not guaranteed. This can be observed by applying the SISO design techniques, developed in this section, to the multivariable Arizona printer, see Figure 7.1(b). Although the SISO designs for $i = 1, 2$ satisfy (7.11), see Figure 7.6, the multivariable condition for convergence (7.12) is not satisfied, see Figure 7.7. This motivates to develop systematic design procedures for MIMO ILC.

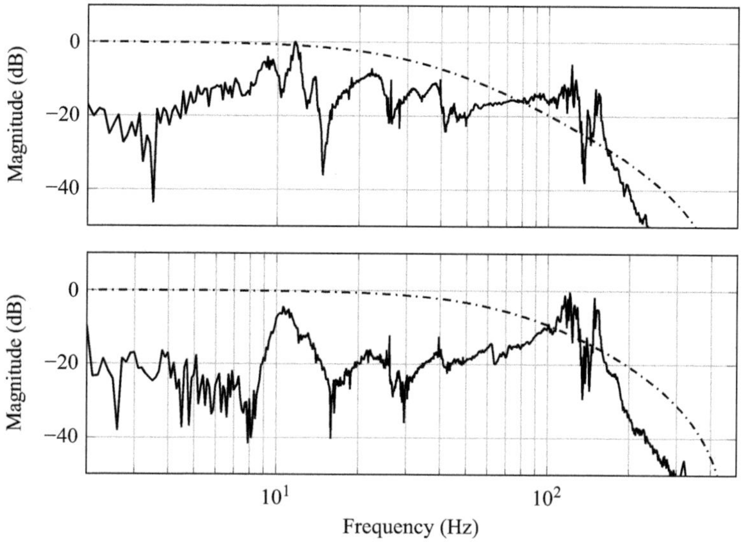

Figure 7.6 *Interaction-ignoring SISO design using* (7.11)*: filters q_i (–·–) (top: $i = 1$; bottom: $i = 2$) are designed such that $|q_i(1 - l_i J_{ii})| < 1, \forall\omega$ (—) for both loops $i = 1,2$*

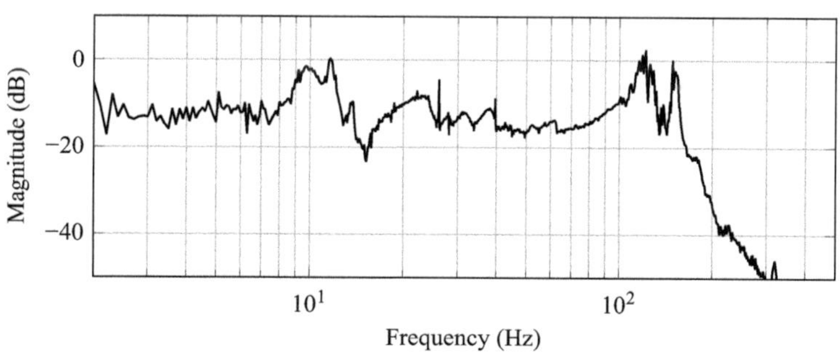

Figure 7.7 *Naive SISO designs for a MIMO system: although the interaction-ignoring SISO designs satisfy* (7.11)*, see Figure 7.6, the condition for convergence $\rho(Q(I - LJ)) < 1, \forall\omega$ (—) is not satisfied*

7.4 ILC Design—the MIMO case

To ensure robust convergence for MIMO systems, in this section, a systematic framework is developed for MIMO ILC design in the frequency domain. Manual frequency-domain design for MIMO systems is much less trivial than for SISO systems, see, e.g., Figure 7.4, which is often a key limiting factor for industrial application of MIMO ILC algorithms. For that reason, the proposed design framework comprises a step-by-step procedure, which gradually extends well-known design techniques from the SISO case to the MIMO case. Crucially, the design complexity in terms of modeling requirements and design tools is only increased if justified by the problem at hand, which enables control engineers to explicitly balance user effort with performance. Note that this is a standard approach in feedback control design for motion systems, see, e.g., [59,60]. Further details on the presented framework for MIMO ILC design can be found in [30].

Procedure 7.2. Design procedure for multivariable ILC

---

1. Interaction analysis (Section 7.4.1). Decoupled?
   - i. Yes: independent SISO design (Section 7.3).
   - ii. No: proceed to next step.
2. Decoupling transformations (Section 7.4.2). Decoupled?
   - i. Yes: independent SISO design (Section 7.3).
   - ii. No: proceed to the next step.
3. Robust multi-loop SISO design.
   - i. Obtain SISO parametric models $\widehat{J_{ii}}$ of the diagonal elements J_{ii}, $i = 1, \ldots, n$;
   - ii. Multi-loop SISO design of L using $\widehat{J_{ii}}$ (Section 7.3.2);
   - iii. Robust SISO design of Q using a MIMO nonparametric model of J (Section 7.3.3).

   Performance not satisfactory? Proceed to next step.
4. Robust decentralized MIMO design.
   - i. Given L, e.g., from step 3, robust decentralized design of Q w.r.t. ignored interaction and modeling errors, using a MIMO nonparametric model of J (Section 7.4.4).

   Performance not satisfactory? Proceed to next step.
5. Centralized MIMO design.
   - i. Obtain MIMO parametric model $\hat{J}$ of J, including interaction;
   - ii. MIMO design of L using $\hat{J}$ such that $I - L\hat{J}$ is small (Section 7.4.5.1);
   - iii. Robust design of Q w.r.t. modeling errors, e.g., through decentralized design using a MIMO nonparametric model of J (Section 7.4.4).

The design techniques constituting steps 4 and 5 are applicable to non-square systems unless stated otherwise, while the techniques in steps 1–3 assume that the considered system is square, i.e., $n \equiv n_u = n_y$, possibly after a squaring-down process, see, e.g., [51,61].

In view of modeling requirements, notice that all except for the last step requires only SISO parametric models, while step 5 requires a MIMO parametric model, which can be expensive to obtain in practice.

Throughout this section, the required models and design approaches are investigated step-by-step, and exemplified on a case study of an industrial flatbed printer. The considered Océ Arizona 550 GT is depicted in Figure 7.1(b). In contrast to standard consumer printers, the printer can handle a large range of media, including flexible and rigid media such as paper, plastics, wood and metals. As a result, the medium is fixed on the printing surface, while the carriage, which contains the printheads, moves in the horizontal plane. This makes the system inherently multivariable. In combination with significant interaction and complex dynamics, see Figure 7.8, this demands the development of multivariable ILC techniques to achieve requirements R1, R2, and R3. An overview of the approaches and resulting designs is presented in Table 7.3.

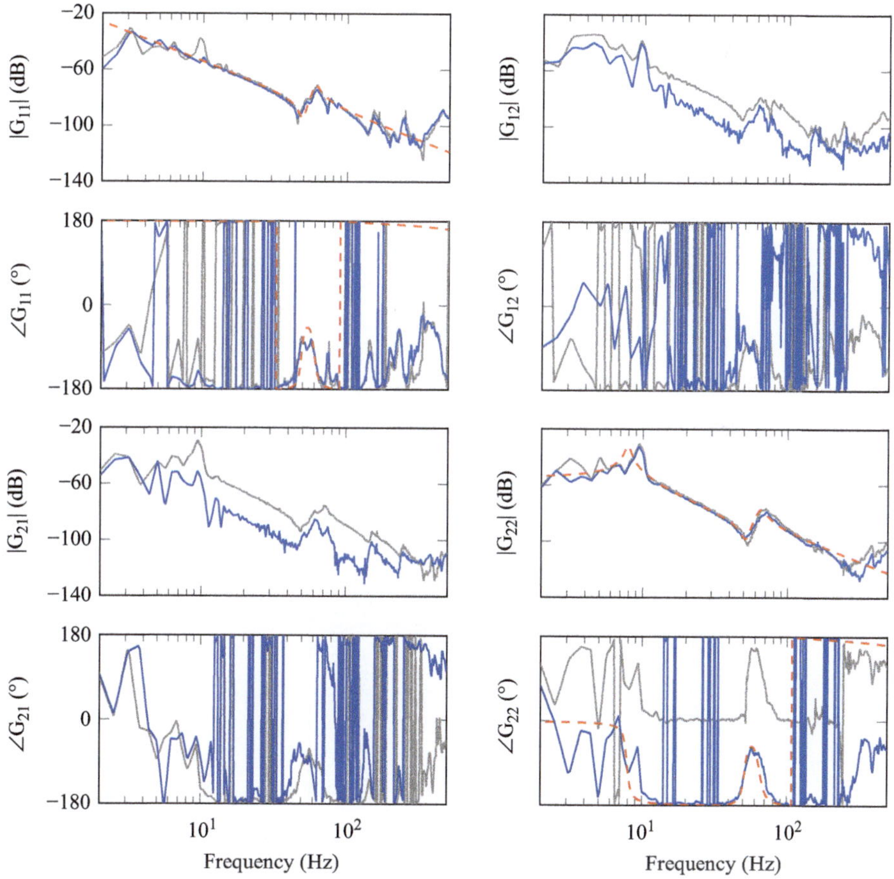

Figure 7.8 *Bode diagram of non-decoupled FRF model $G_o(e^{\iota\omega})$ (—), FRF model $G(e^{\iota\omega})$ (—) after decoupling transformations in step 2 of Procedure 7.2, and SISO parametric models of decoupled plant $\widehat{G}_{ii}(z)$ (- -) used for ILC design. The closed-loop systems of interest for ILC design are $J = (I + GC)^{-1}G$ and $\hat{J}_{ii}(z) = (I + \hat{G}_{ii}C_{ii})^{-1}\hat{G}_{ii}$*

Table 7.3 Overview of approaches in design framework: modeling requirements and designs

Design step	***L*-filter**	**Required parametric models**	***Q*-filter**	**Cutoff f_c [Hz]**		**Robust stability (R1)**	**User effort (R2)**	**Performance (R3)**
				q_1	q_2			
2	*n* Times SISO	*n* Times SISO	SISO, §7.4.3	35	73	−	+	−
3	*n* Times SISO	*n* Times SISO	Robust SISO, §7.4.3	27	27	+	+	−
4	*n* Times SISO	*n* Times SISO	Decentralized, §7.4.4	21	60	+	+/−	+/−
5	MIMO, §7.4.5	Full MIMO	Robust SISO, §7.4.3	30	30	+	−	+

7.4.1 Interaction analysis

The goal of interaction analysis is to identify two-sided interactions in J. If there is no two-sided interaction, then ILC design becomes a standard multi-loop SISO problem. If there is substantial interaction, this should be accounted for in subsequent design steps.

The frequency-dependent relative gain array (RGA) [62,63] is a useful interaction measure, defined as

$$\Lambda(J(e^{\iota\omega})) = J(e^{\iota\omega}) \circ J^{-\top}(e^{\iota\omega}),$$

where $\circ$ denotes element-wise multiplication. Two important observations are made: (i) if $\Lambda(J(e^{\iota\omega})) = I$, then no two-sided interaction is present at frequency ω, whereas (ii) significant non-zero off-diagonal elements indicate interaction.

For the considered case study, the RGA is depicted in Figure 7.9. It is observed that the system is approximately decoupled for frequencies below 30 Hz, while considerable interaction is present around the first resonance frequency at 60 Hz and beyond. Hence, to enable successful decentralized design, decoupling transformations are needed.

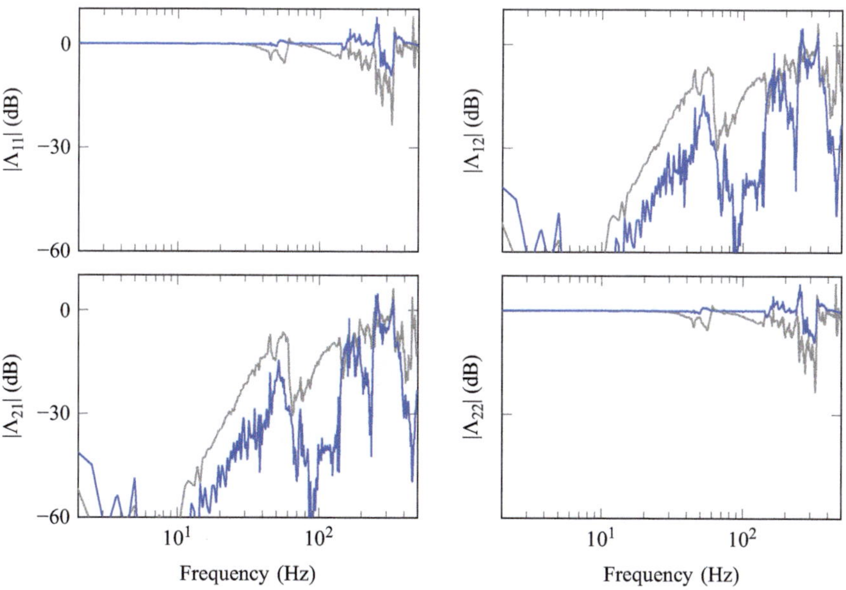

Figure 7.9 Interaction analysis: the frequency-dependent relative gain array (RGA) of $J_o = (I + G_oC)^{-1}G_o$ (— (gray thick line)) and $J = (I + GC)^{-1}G$ (— (blue thick line)) shows that (i) significant interaction is present before decoupling, which is (ii) considerably reduced using a static input transformation, enabling a successful decentralized ILC design

7.4.2 Decoupling transformations

To reduce interaction and facilitate decentralized design, the input and output of the system may be redefined. For motion systems, often linear combinations of actuator variables and sensor variables are defined, such that the transformed inputs and outputs act in orthogonal directions at the center of gravity. Typically these static transformations are based on an inversion of a kinematic model of the system, resulting in an approximately decoupled system in frequency ranges where it behaves as a rigid body, see, e.g., [59,60,64].

The considered system is decoupled using a static input transformation matrix T_u, see Figures 7.8 and 7.9. It can be observed that the effective plant $G = G_o T_u$ is diagonally dominant in the frequency ranges 10–40 Hz and 70–130 Hz, i.e., where the system behaves as a rigid body.

7.4.3 Robust multi-loop SISO design

For systems where interaction is sufficiently low, a SISO design procedure may be pursued. In the first step, multiple independent SISO ILC algorithms may be designed as if no interaction is present, see Section 7.3.3 and its application to the Arizona printer in Figure 7.6. In the second step, to ensure convergence of the multi-loop SISO design, robustness can be added a posteriori by designing a SISO filter $q_d \in \mathcal{RL}_\infty$, where $Q(z) = q_d(z)I$, according to

$$|q_d(e^{\iota\omega})|\rho\,(I - L(e^{\iota\omega})J(e^{\iota\omega})) < 1 \quad \forall\omega \in [0, \pi], \tag{7.13}$$

for convergence, see Figure 7.10, and for monotonic convergence according to

$$|q_d(e^{\iota\omega})|\bar{\sigma}\,(I - L(e^{\iota\omega})J(e^{\iota\omega})) < 1 \quad \forall\omega \in [0, \pi].$$

This approach is attractive due to its simplicity and may yield satisfactory results for the particular problem at hand, yet it is restrictive in the sense that the a posteriori added robustness limits performance. Indeed, as can be observed in Figure 7.10 for the case study, q_d accounts for the worst case modeling errors and interaction over all loops combined: the low-pass filter q_d is mainly required in view of the modeling error in loop 1 at 11 Hz, while this is seemingly over-restrictive for loop 2. This motivates to develop systematic design procedures for each SISO loop independently.

7.4.4 Robust decentralized MIMO design

The objective is to design a robust decentralized MIMO controller by solving an independent set of SISO design problems. By addressing the required robustness for each SISO loop separately, performance of the MIMO system can potentially be improved. The developed design techniques for diagonal $Q = \text{diag}\{q_1, q_2, \ldots, q_{n_u}\} \in \mathcal{RL}_\infty^{n_u \times n_u}$ are applicable to the general non-square case $J \in \mathcal{RH}_\infty^{n_y \times n_u}$. Note however that the results are most often applied to square systems, since this enables decentralized design also of diagonal learning filter $L = \text{diag}\{l_1, l_2, \ldots, l_n\}$ using the techniques presented in Section 7.3.2.

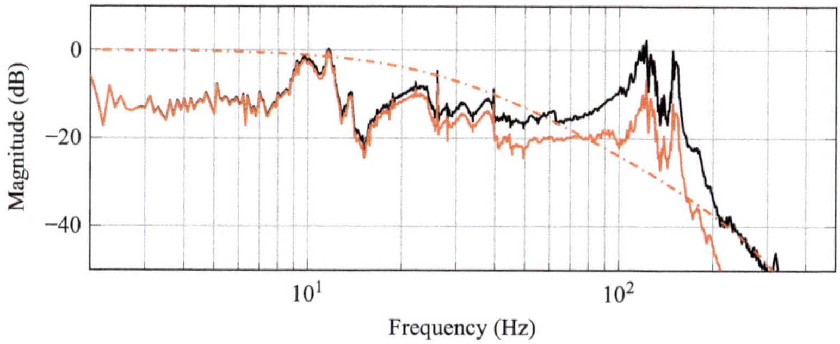

Figure 7.10 *Robust SISO design: although the interaction-ignoring SISO designs satisfy* (7.9)*, see Figure 7.6, the condition for convergence* $\rho(Q(I-LJ)) < 1, \forall\omega$ *is not satisfied (—). In contrast, through additional robustness* q_d *(- -) according to* (7.13)*, convergence can be guaranteed (—)*

To enable decentralized Q-filter design, the interaction in (7.3) is separated using the following factorization

$$f_{j+1} = QMf_j + \tilde{w} = QM_{\mathrm{d}}(I+E)f_j + \tilde{w},$$

where $M = I - LJ$, $\tilde{w} = QLSr$, and $E = M_{\mathrm{d}}^{-1}(M - M_{\mathrm{d}})$ contains the normalized interaction in M, where M_{d} is the diagonal part of the transfer function M. Then, the diagonal term QM_{d} and interaction term $I + E$ can be used to analyze and design for robust stability. Particularly, with introduction of conservatism, the following sufficient conditions guarantee convergence:

$$\rho\left(QM_{\mathrm{d}}(I+E)\right) \leq \|QM_{\mathrm{d}}(I+E)\|_{ip} < 1 \quad \forall\omega \in [0, \pi],$$

and

$$\rho\left(QM_{\mathrm{d}}(I+E)\right) \leq \bar{\sigma}(QM_{\mathrm{d}})\mu_d(I+E) < 1 \quad \forall\omega \in [0, \pi],$$

where μ_d is the SSV [65] with respect to the diagonal structure of QM_{d}. As $\|A\|_{i1} = \max_i\left(\sum_j |a_{ji}|\right)$, $\|A\|_{i\infty} = \max_i\left(\sum_j |a_{ij}|\right)$ and $\bar{\sigma}(QM_{\mathrm{d}}) = \max_i |q_i M_{ii}|$, the following sufficient conditions for convergence follow, see [30, Theorems 4, 5]:

$$|q_i(e^{\iota\omega})M_{ii}(e^{\iota\omega})| < \frac{1}{\sum_j |I + E(e^{\iota\omega})|_{ij}} \quad \forall i, \omega \in [0, \pi], \tag{7.14}$$

$$|q_i(e^{\iota\omega})M_{ii}(e^{\iota\omega})| < \frac{1}{\sum_j |I + E(e^{\iota\omega})|_{ji}} \quad \forall i, \omega \in [0, \pi], \tag{7.15}$$

$$|q_i(e^{\iota\omega})M_{ii}(e^{\iota\omega})| < \frac{1}{\mu_{\mathrm{d}}(I + E(e^{\iota\omega}))} \quad \forall i, \omega \in [0, \pi], \tag{7.16}$$

where, in this case, $|\cdot|$ denotes the element-wise absolute value. Similarly, design conditions for monotonic convergence can be derived. Using that

$$\begin{aligned}\bar{\sigma}(QM_{\mathrm{d}}) = \sqrt{\rho(QM(QM)^H)} &= \sqrt{\rho(QM_{\mathrm{d}}(I+E)(I+E)^H M_{\mathrm{d}}^H Q^H)} \\ &= \sqrt{\rho(M_{\mathrm{d}}^H Q^H Q M_{\mathrm{d}}(I+E)(I+E)^H)},\end{aligned}$$

since the products of two square matrices A, B have identical spectra $\{\lambda_i(AB)\} = \{\lambda_i(BA)\}$, the following sufficient conditions for monotonic convergence in the ℓ_2 norm of f_j are obtained:

$$|q_i(e^{\iota\omega})M_{ii}(e^{\iota\omega})| < \frac{1}{\sqrt{\sum_j |(I+E(e^{\iota\omega}))(I+E(e^{\iota\omega})^H|_{ij}}} \quad \forall i, \omega \in [0, \pi], \tag{7.17}$$

$$|q_i(e^{\iota\omega})M_{ii}(e^{\iota\omega})| < \frac{1}{\sqrt{\mu_{\mathrm{d}}((I+E(e^{\iota\omega}))(I+E(e^{\iota\omega}))^H)}} \quad \forall i, \omega \in [0, \pi], \tag{7.18}$$

Importantly, the developed sufficient bounds (7.14)–(7.16), respectively (7.17) and (7.18), complement each other in the sense that the ordering of their tightness may vary over different frequency ranges. Hence, they should be considered jointly during design. Notice however that they cannot be mixed over the different SISO loops i per evaluated frequency: at least one condition must be satisfied for all loops i simultaneously.

The main advantage of a decentralized design of Q is that it allows for SISO loop-shaping, while explicitly addressing interaction, using only SISO parametric models for the design of L. This is illustrated by Figure 7.11, where the SISO filters q_i are designed according to joint evaluation of design conditions (7.14)–(7.16) based on a nonparametric FRF model $\hat{J}_{\mathrm{FRF}}(e^{\iota\omega})$. As can be observed, in loop 1, robustness is required to a modeling error at 11 Hz, hence limiting the cutoff frequency of low-pass filter q_1. Yet, q_2 can be designed independent of the required robustness in loop 1 and is in fact limited by interaction at significantly higher frequencies, see also Table 7.3. This potentially leads to increased performance compared to robust SISO design.

If the decentralized approach does not lead to satisfactory performance, and in case increased modeling requirements are justified, then a full multivariable design approach may be pursued.

7.4.5 Centralized MIMO design

A full multivariable design approach is suitable in case the decentralized approaches in the previous subsections do not lead to satisfactory performance, and the increased modeling effort is justified by the performance requirements. In this subsection, a range of techniques is investigated for multivariable learning filter design, i.e., step 5(ii) of Procedure 7.2, using a MIMO parametric model of the system.

The main idea is that the performance is potentially increased compared to robust decentralized approaches, since the ILC is explicitly designed for interaction rather

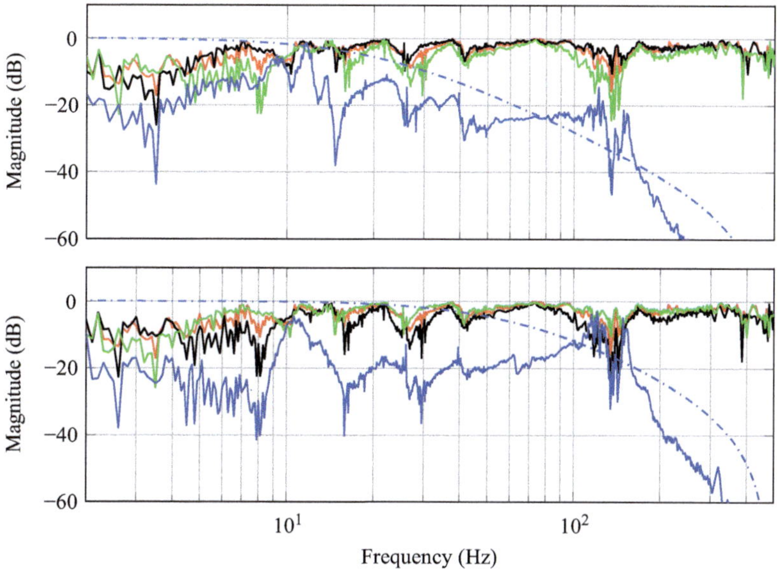

Figure 7.11 *Decentralized design for robust convergence: the filters q_i (-.-) (top: $i = 1$; bottom: $i = 2$) are chosen such that, for each frequency, $|q_i(1 - l_iJ_{ii})| = |q_iM_{ii}|$ (—) are simultaneously upper bounded by at least one of the right-hand sides of* (7.14), (7.15) *and* (7.16) *(—,—,—, respectively), which can be verified using inexpensive nonparametric FRF models of J*

than dealing with interaction through robustness. That is, by explicitly designing $L \in \mathscr{RL}_\infty^{n_u \times n_y}$ such that $I - LJ$ is small, the need for robustness through Q is alleviated.

7.4.5.1 Design of learning filter: MIMO inversion techniques

A range of algorithms is available for (approximate) inversion of MIMO parametric models $\hat{J}(z)$. The applicability of each inversion technique presented in Section 7.3.2 to MIMO systems is summarized in Table 7.1.

For NPZ-Ignore, ZPETC, and ZMETC, the extension to multivariable systems is nontrivial. In [30,66], an approach for multivariable ZPETC is proposed which applies multiple times SISO ZPETC to the Smith form of the system. Similarly, the Smith form can be used to construct MIMO inverses using NPZ-Ignore and ZMETC. It is noted however that numerical reliability for computing Smith forms of transfer matrices is challenging [67], and constructing them manually is a cumbersome task. Hence, the practical application of these techniques currently seems to be limited to SISO systems.

In view of the inversion algorithms that are applicable to multivariable systems, a third requirement is identified in addition to the two design requirements posed in Section 7.3.2:

3. non-square systems: for $n_y \neq n_u$, direct inversion of $J(z) \in \mathcal{RH}_\infty^{n_y \times n_u}$ is not possible.

In view of exact inversion, depending on the excess of inputs or outputs, infinitely many or no solutions may exist to the problem $LJ = I$. If J is underactuated, i.e., $n_y \geq n_u$, but left invertible, then stable inversion can directly be applied and achieve $I - L\hat{J} = 0$, given a state-space model of the left inverse $\hat{J}_l^{-1}(z)$ with $\hat{J}_l^{-1}\hat{J} = I$. For overactuated systems J, i.e., with $n_u \geq n_y$, a squaring-down process may be pursued such that a square system is obtained. Moreover, the additional degrees of freedom can be exploited to satisfy additional requirements, such as obtaining minimum-phase square systems, see, for example, the approach in [51, Section 6]. $\mathcal{H}_\infty$-preview control can directly be used for general non-square systems.

Regardless of the pursued approach, if L is designed such that $\rho(I - LJ) < 1$ for all frequencies, then perfect asymptotic tracking $e_\infty = 0$ can be achieved, see (7.5). Finally, note that non-square systems in general do not have NMP zeros [51], hence the developed techniques are foreseen to be most often applied to square systems.

7.5 Iterative inversion-based control: avoiding the need for parametric models

In this section, a model-free learning approach is presented called IIC [39,68–71], in the sense that no parametric models are required. IIC is closely related to the inverse model-based ILC approaches as presented in the previous sections, with the key difference that the input updating is explicitly performed by means of an inverse FRF model in the frequency domain. The advantage of this approach is that merely a nonparametric model of the system is required, hence significantly alleviating the required user effort (R2) in terms of modeling requirements.

7.5.1 System description and procedure

The frequency-domain approach to learning control is facilitated by the DFT and its inverse DFT (IDFT), which are defined as follows:

$$X(\omega_k) = \mathscr{F}_N\{x(n)\} \triangleq \frac{1}{\sqrt{N}}\sum_{n=0}^{N-1} x(n)e^{-\iota\omega_k n}, \; x(k) = \mathscr{F}_N^{-1}\{X(\omega_n)\} \triangleq \frac{1}{\sqrt{N}}\sum_{n=0}^{N-1} X(\omega_n)e^{\iota\omega_k n},$$

$$\omega_k \in \Omega, \quad \Omega \triangleq \left\{\omega_i \in \mathbb{R} \mid \omega_i = \tfrac{2\pi i}{N}, \quad i = 0, \quad 1, \ldots, N-1\right\},$$

where Ω is referred to as the DFT grid. After transformation, the output of $J(z)$ is described by

$$Y_j(\omega_k) = J(e^{\iota\omega_k})F_j(\omega_k) + D_j(\omega_k),$$

where the trial-varying disturbance $D_j(\omega_k)$ contains the lumped contributions of exogenous disturbances, nonlinear perturbations and leakage. Under the assumptions that the system is in steady-state, i.e., its states are periodic with N samples, and the remaining disturbances are trial-invariant, i.e., $D_j(\omega_k) = D(\omega_k)$, the error is described by

$$E_{j+1}(\omega_k) = E_j(\omega_k) - J(e^{\iota\omega_k})(F_{j+1}(\omega_k) - F_j(\omega_k)).$$

Similar to ILC, the aim of IIC is to minimize $E_{j+1}(\omega_k)$ in an appropriate sense, by iteratively updating $F_j(\omega_k)$ in the frequency-domain as follows, see [71]:

$$F_{j+1}(\omega_k) = Q(\omega_k)\left(F_j(\omega_k) + L(\omega_k)E_j(\omega_k)\right), \quad \forall k. \tag{7.19}$$

where $Q(\omega_k), L(\omega_k) \in \mathbb{C}^{n\times n}$ are here referred to as the robustness- and learning coefficient matrices, respectively. Note that although this approach is similar to ILC, an essential difference is that $Q(\omega_k)$ and $L(\omega_k)$ are not restricted to represent the FRF of a low order filter but can instead be tuned independently per frequency ω_k to achieve convergence [71].

The following procedure is proposed for design and implementation of IIC:

Procedure 7.3. Design and implementation of iterative inversion-based control

---

1. Frequency-domain IIC design:
   i. Design matrices $L(\omega_k) \in \mathbb{C}^{n_u\times n_y}$, $Q(\omega_k) \in \mathbb{C}^{n_u\times n_u}$ for each $\omega_k \in \Omega$ according to (7.20), (7.21) using a nonparametric FRF model $\hat{J}_{\mathrm{FRF}}(e^{\iota\omega_k})$.
   ii. Select an initial input f_0.
2. Set the task index $j = 0$, and perform the following steps:
   i. Implement f_j on the system and measure e_j.
   ii. Input updating in the frequency domain:
      a. Compute $F_j(\omega_k)$ and $E_j(\omega_k)$ by applying the DFT to $f_j(k)$ and $e_j(k)$.
      b. Construct frequency-domain input $F_{j+1}(\omega_k) = Q(\omega_k)(F_j(\omega_k) + L(\omega_k)E_j(\omega_k))$, see (7.19).
      c. Compute the time domain input $f_{j+1}(k)$ by applying the IDFT to $F_{j+1}(\omega_k)$.
   iii. Set $j \to j+1$ and return to step 2(i).

7.5.2 *Convergence analysis, modeling requirements and design*

Equivalently to (7.6) and (7.8), convergence, respectively monotonic convergence w.r.t. the ℓ_2 norm of f_j, is achieved if and only if

$$\rho(Q(\omega_k)\,(I - L(\omega_k)J(e^{\iota\omega_k}))\,) < 1, \quad \forall k, \tag{7.20}$$

$$\bar{\sigma}(Q(\omega_k)\,(I - L(\omega_k)J(e^{\iota\omega_k}))\,) < 1, \quad \forall k. \tag{7.21}$$

Design approaches for $Q(\omega_k)$ and $L(\omega_k)$ are facilitated by an FRF model of $J(z)$. Hence crucially, no parametric models are required. Clearly, if a model $\hat{J}_{\mathrm{FRF}}(\omega_k)$ is available that is regular on the DFT grid, a suitable choice is $Q(\omega_k) = I$ and $L(\omega_k) = \hat{J}_{\mathrm{FRF}}^{-1}(\omega_k)$, since this renders $\lim_{j\to\infty} E_j = 0$. Note that in contrast to norm-optimal ILC, regularity of $J(\omega_k)$ is implied if $J(z)$ is nonzero and has no zeros on the unit disc for all frequencies ω_k, hence delay is allowed. This does not require that the system is minimum phase and it can in fact be shown that under these conditions, see [71],

$$f(k) = \mathscr{F}_N^{-1}\{\hat{J}_{\mathrm{FRF}}^{-1}(\omega_k)\mathscr{F}_N\{r(k)\}\},$$

is identical to stable inversion as discussed in Section 7.3.2.3 and results in bounded but possibly noncausal inputs $f(k)$.

7.6 ILC with basis functions: enhancing flexibility to varying tasks

In this section, a frequency-domain design approach is presented for MIMO ILC with basis functions, which enables high-tracking performance also for varying tasks. Hence, the pursued approach fits into the developed frequency-domain design framework as described in Section 7.4. As a special case, the results recover the work of [10] for SISO systems.

7.6.1 Flexibility in ILC—case study on a flatbed printer

A key assumption underlying conventional ILC is that the system performs exactly repeating tasks. Violating this assumption can significantly affect performance, as is demonstrated using the experimental results in Figure 7.12. In the next subsection, basis functions are incorporated in ILC for high performance for varying tasks.

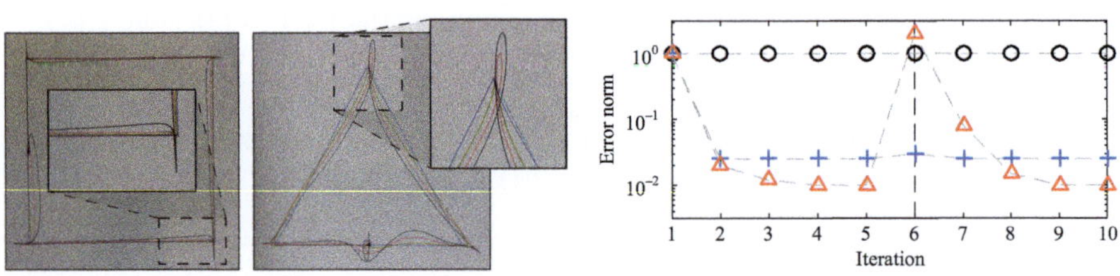

(a) (b)

Figure 7.12 Flexibility to varying tasks: by equipping the Arizona flatbed printer, see Figure 7.1(b), with a whiteboard marker, control performance can be visualized. After learning to accurately draw a square (iterations 1–5), the performance of standard ILC severely deteriorates when the motion task is changed to a triangle (iterations 6–10). By incorporating basis functions in ILC, performance can be maintained upon task variations. (a) Drawn output of standard ILC during iterations 1–5 (left) and iterations 6–10 (right). Each iteration is drawn using a different color. (b) Tracking performance without ILC (○), standard ILC (□) and ILC with basis functions (+) as function of iterations

7.6.2 Basis functions in ILC

To enhance the flexibility of ILC to varying tasks, ILC approaches have been extended with basis functions in, e.g., [42,44,48]. Typically, the input signal is parametrized on a finite time interval as an explicit function of r_{j+1}, i.e.,

$$f_{j+1} = F(\theta_{j+1})r_{j+1}, \tag{7.22}$$

where the transfer function $F(\theta_{j+1})$ is a linear combination of weighted basis functions $\psi_i(z)$, i.e., $F(\theta) = \sum_{i=1}^{n_\theta} \theta_i \psi_i(z)$. In contrast to standard ILC, see (7.2), the aim of ILC with basis functions is now to iteratively determine the parameters $\theta_j \in \mathbb{R}^{n_\theta}$ such that the tracking error e_j is small. Substituting (7.22) into (7.1) and rearranging terms yields

$$e_{j+1} = S(I - PF(\theta_{j+1}))r_{j+1},$$

which shows that optimal tracking performance, i.e., $e_{j+1} = 0$, is achieved independent of r_{j+1} if $PF(\theta) = I$. This leads to the following two-step design procedure for ILC with basis functions:

1. To enable optimal tracking performance, the basis functions should be chosen such that $P^{-1}(z)$ is included in the linear span $\{\psi_i(z)\}$.
2. The iterative updating algorithm for corresponding parameters θ should be designed, ensuring robust convergence and performance of the iterative system.

In view of step 1, recent developments on the selection of basis functions can be found in, e.g., [44,72] for FIR filters and [49] for rational functions with fixed poles.

Step 2, i.e., the design of the parameter update algorithm, is typically performed using norm-based ILC approaches, see, e.g., [48,73,74]. This imposes a large burden on the model requirement, since these approaches rely on detailed specifications of a nominal model and its uncertainty to guarantee robust stability. Recent developments in [10] aim to enable frequency-domain design of parameter update algorithms, hence enabling manual loop-shaping as is often preferred by control engineers. However, currently the approach is restricted to SISO systems. In the next subsection, the frequency-domain design approach of [10] is generalized toward MIMO systems and unified with the MIMO frequency-domain design framework as developed in Section 7.4.

Remark 7.2. *Alternatively to the linear combination of basis functions utilized here, rational (nonlinear) combinations of basis functions can be employed, see, e.g., [13,42,47,75]. Although these parametrizations potentially enable performance improvements, the associated optimization problem becomes non-convex, and often nonlinear algorithms are required for which, in general, convergence to global minima cannot be guaranteed.*

7.6.3 Projection-based MIMO ILC with basis functions: frequency-domain design

Next, an ILC approach is developed that enables high performance for varying tasks. Crucially, the design can be done in the frequency-domain, using for instance the techniques developed in Section 7.4. The flexibility to varying tasks is achieved by projecting the ILC update onto basis functions in a suitable manner. In prospect of forthcoming analyses, the following procedure is proposed for design and implementation of projection-based ILC with basis functions.

Procedure 7.4. Design and implementation of projection-based MIMO ILC with basis functions

---

1. Frequency-domain ILC design and selection of basis functions:
   i. Design MIMO filters $L(z)$, $Q(z)$ in the frequency-domain, e.g., using Procedure 7.2.
   ii. Select a set of basis functions $\{\psi_i(z)\}$, initialize θ_0, and set $f_0^{\text{proj}} = \Psi_{r0}\theta_0$.
2. Set the task index $j = 0$, and perform the following steps
   i. Implement f_j^{proj} and measure e_j.
   ii. Construct $f_{j+1} = Q(q)(f_j^{\text{proj}} + L(q)e_j$.
   iii. Projection step:
      a. Compute $\hat{\theta}_{j+1} = (\Psi_{Jr_{j+1}}^\top \Psi_{Jr_{j+1}})^{-1}\Psi_{Jr_{j+1}}^\top Jf_{j+1}$.
      b. Construct $f_{j+1}^{\text{proj}} = \Psi_{r_{j+1}}\hat{\theta}_{j+1}$.
   iv. Set $j \to j+1$ and return to step 2(i).

Step 2 of the procedure is developed in the remainder of this section. The required notation is introduced first; second the projection step is investigated, and finally convergence of the projection-based ILC algorithm is analyzed.

7.6.3.1 Basis functions for MIMO ILC

The feedforward (7.22) is linearly parametrized as $F(z,\theta) = \sum_{i=1}^{n_\theta} \theta_i^\top \psi_i(z)$, where in the MIMO case either $\psi_i(z) \in \mathcal{RL}_\infty^{n_u \times n_y}$ and $\theta_i \in \mathbb{R}^{n_u \times n_u}$, i.e., MIMO basis functions, or $\psi_i(z) \in \mathcal{RL}_\infty$ and $\theta_i \in \mathbb{R}^{n_y \times n_u}$, i.e., SISO basis functions. The feedforward signal at time k is constructed as

$$f(k) = F(q,\theta)r(k) = \underbrace{\left[\theta_1^\top,\, \theta_2^\top,\, \ldots, \theta_{n_\theta}^\top\right]}_{\theta^\top} \underbrace{\begin{bmatrix} \psi_1(q) \\ \psi_2(q) \\ \vdots \\ \psi_{n_\theta}(q) \end{bmatrix} r(k)}_{\Psi_r(k)} = \theta^\top \Psi_r(k), \tag{7.23}$$

where if $\{\psi_i(q)\}$ are matrices, i.e., MIMO basis functions, then $\theta_i \in \mathbb{R}^{n_u \times n_u}$, $\theta \in \mathbb{R}^{n_u n_\theta \times n_u}$, $\Psi_r(k) \in \mathbb{R}^{n_u n_\theta \times 1}$, and if $\{\psi_i(q)\}$ are scalars, i.e., SISO basis functions, then

$\theta_i \in \mathbb{R}^{n_y \times n_u}$, $\theta \in \mathbb{R}^{n_y n_\theta \times n_u}$, $\Psi_r(k) \in \mathbb{R}^{n_y n_\theta \times 1}$, and in all cases $r(k) \in \mathbb{R}^{n_y \times 1}$ and $f(k) \in \mathbb{R}^{n_u \times 1}$. The finite-time response is conveniently described by adopting the vectorized notation

$$f^\top = [f(0), f(1), \ldots, f(N-1)], \quad \Psi_r^\top = [\Psi_r(0), \Psi_r(1), \ldots, \Psi_r(N-1)],$$

such that (7.23) can be expressed as $f = \mathscr{F}(\theta)r = \Psi_r \theta$.

7.6.3.2 Projection step

Given a set of basis functions $\{\psi_i(q)\}$, a weighted projection step is performed to approximate f_{j+1} in (7.2) with $f_{j+1}^{\text{proj}} = \mathscr{F}(\theta_{j+1})r_{j+1}$. From (7.1), observe that $e_j = Sr_j - Jf_j$ and $e_{j+1} = Sr_{j+1} - Jf_{j+1}$. Using these expressions, a prediction of the error in task $j+1$, given input f_{j+1}, is described by

$$\hat{e}_{j+1} = e_j - J(f_{j+1} - f_j) + S(r_{j+1} - r_j),$$

and similarly define $\hat{e}_{j+1}^{\text{proj}}(\theta_{j+1}) = e_j - J(f_{j+1}^{\text{proj}}(\theta_{j+1}) - f_j) + S(r_{j+1} - r_j)$. The parameters θ_{j+1} are determined by projecting $\hat{e}_{j+1}(k)$ onto $\hat{e}_{j+1}^{\text{proj}}(k)$, as

$$\begin{aligned} \hat{\theta}_{j+1} &= \arg\min_{\theta_{j+1}} \frac{1}{N} \left\| \hat{e}_{j+1} - \hat{e}_{j+1}^{\text{proj}}(\theta_{j+1}) \right\|_F^2 \\ &= \arg\min_{\theta_{j+1}} \frac{1}{N} \left\| Jf_{j+1} - J\Psi_{r_{j+1}}\theta_{j+1} \right\|_F^2, \end{aligned} \tag{7.24}$$

where $\|\cdot\|_F$ denotes the Frobenius norm. Here, it is crucial to note that the weighting J in (7.24) implies that the projection is performed such that $\hat{e}_{j+1}^{\text{proj}}$ approximates $\hat{e}_{j+1}$. Assuming that $\Psi_{Jr_{j+1}}^\top \Psi_{Jr_{j+1}}$ is nonsingular with $\Psi_{Jr_{j+1}} = J\Psi_{r_{j+1}}$, which imposes a mild persistence of excitation condition on r_{j+1}, see [10], the unique solution to (7.24) is

$$\hat{\theta}_{j+1} = (\Psi_{Jr_{j+1}}^\top \Psi_{Jr_{j+1}})^{-1} \Psi_{Jr_{j+1}}^\top Jf_{j+1}.$$

The projection of f_{j+1} onto the basis functions is hence given by $f_{j+1}^{\text{proj}} = \Psi_{r_{j+1}}\hat{\theta}_{j+1}$, which is the command input to be implemented in trial $j+1$.

Regarding convergence of the projected ILC algorithm, the following key result is proven [10, Theorem 1] for the SISO case. The MIMO case follows straightforwardly in a similar manner. If $L(z), Q(z), J(z), \psi_i(z) \in \mathscr{RH}_\infty$, then f_j^{proj} converges monotonically w.r.t. its ℓ_2 norm if

$$\bar{\sigma}\left(Q(e^{\iota\omega})(I - L(e^{\iota\omega})J(e^{\iota\omega})\right) < 1 \quad \forall \omega \in [0, \pi].$$

Here, it is emphasized that this is the same frequency-domain condition that guarantees monotonic convergence for standard ILC without basis functions. This is a crucial result, since it enables the use of frequency-domain design techniques as developed in Section 7.4 for ILC with basis functions.

7.7 Conclusion and ongoing work

In this chapter, a systematic and unifying framework is developed for analysis and design of multivariable ILC in the frequency-domain, which is often preferred by control engineers for complex engineering applications. A step-by-step design procedure is presented that gradually extends manual SISO ILC design toward the MIMO case, with the underlying rationale that design complexity and associated modeling requirements should only be increased if necessitated by performance requirements. The procedure consists of (1) interaction analysis, (2) decoupling, (3) robust SISO design for MIMO systems, (4) robust decentralized MIMO design and (5) full MIMO design. The framework enables control engineers to explicitly address the design trade-offs between performance and modeling requirements, i.e., parametric versus nonparametric models and SISO versus MIMO models. The developed solutions are hence all suitable alternatives, and depending on the specific application at hand, an appropriate choice can be made.

The frequency-domain design procedure is applicable to a range of learning control approaches, each addressing various control and design requirements in engineering applications. Relevant approaches include (i) traditional ILC for high performance to repeating tasks using approximate parametric models, (ii) FRF-based iterative learning, which solely requires nonparametric models and (iii) ILC with basis functions, which enables high performance also for varying tasks.

Ongoing research includes several extensions of the presented approaches. In [76], initial results are presented on the application of the developed design framework to multivariable repetitive control. In [68], the need for (nonparametric) models in IIC (Section 7.5) is completely removed by estimating the system FRF from trial data; in [71], it is shown that the algorithm is guaranteed for converge, albeit that additional measures are required to avoid excessive learning transients. The joint use and design of traditional feedback control with learning control is investigated in, e.g., [77] with particular attention to the inferential control setting [78], where an explicit distinction is made between measured variables that are available for real-time feedback control, and performance variables that are exploited for learning control. Finally, initial results on extending ILC with basis functions (Section 7.6) toward linear parameter-varying (LPV) systems are presented in [79]; basis function and sampling pattern selection are investigated in [80].

Acknowledgments

The authors would like to thank Abhishek Bareja, Frank Boeren, Joost Bolder, Sjirk Koekebakker, and Jeroen Willems for their contributions. This work is supported by Océ Technologies and is a part of the research programme VIDI with project number 15698, which is (partly) financed by the Netherlands Organisation for Scientific Research (NWO).

References

[1] Bristow D, Tharayil M, Alleyne A. A survey of iterative learning control. IEEE Control Systems Magazine. 2006;26(3):96–114.

[2] Moore K. Iterative Learning Control for Deterministic Dystems. Springer-Verlag, London, UK; 1993.

[3] Xu JX, Tan Y. Linear and Nonlinear Iterative Learning Control. vol. 291. Springer-Verlag, Berlin Heidelberg, Germany; 2003.

[4] Ahn HS, Moore K, Chen Y. Iterative learning control: Robustness and monotonic convergence for interval systems. Communications and Control Engineering Series. Springer-Verlag, London, UK; 2007.

[5] Owens DH. Iterative Learning Control: An Optimization Paradigm. Advances in Industrial Control. Springer-Verlag, London; 2016.

[6] Barton K, Hoelzle D, Alleyne A, Johnson A. Cross-coupled iterative learning control of systems with dissimilar dynamics: Design and implementation. International Journal of Control. 2011;84(7):1223–1233.

[7] Clayton GM, Tien S, Leang K, Zou Q, Devasia S. A review of feedforward control approaches in nanopositioning for high-speed SPM. Journal of Dynamic Systems, Measurements, and Control. 2009;131(6):061101.

[8] Wu Y, Zou Q. An iterative-based feedforward-feedback control approach to high-speed atomic force microscope imaging. Journal of Dynamic Systems, Measurements, and Control. 2009;131(6):061105.

[9] van Zundert J, Bolder J, Koekebakker S, Oomen T. Resource-efficient ILC for LTI/LTV systems through LQ tracking and stable inversion: Enabling large feedforward tasks on a position-dependent printer. Mechatronics. 2016;38: 76–90.

[10] Boeren F, Bareja A, Kok T, Oomen T. Frequency-domain ILC approach for repeating and varying tasks: With application to semiconductor bonding equipment. IEEE/ASME Transactions on Mechatronics. 2016;21(6):2716–2727.

[11] de Roover D, Bosgra OH. Synthesis of robust multivariable iterative learning controllers with application to a wafer stage motion system. International Journal of Control. 2000;73(10):968–979.

[12] Mishra S, Tomizuka M. Projection-based iterative learning control for wafer scanner systems. IEEE/ASME Transactions on Mechatronics. 2009;14(3): 388–393.

[13] Blanken L, Boeren F, Bruijnen D, Oomen T. Batch-to-batch rational feedforward control: From iterative learning to identification approaches, with application to a wafer stage. IEEE/ASME Transactions on Mechatronics. 2017; 22(2):826–837.

[14] Oomen T, van der Maas R, Rojas CR, Hjalmarsson H. Iterative data-driven $\mathscr{H}_\infty$ norm estimation of multivariable systems with application to robust active vibration isolation. IEEE Transactions on Control Systems Technology. 2014;22(6):2247–2260.

[15] Harman G. Wire Bonding in Microelectronics. McGraw-Hill, New York, NY, USA; 2010.

[16] Butler H. Position control in lithographic equipment. IEEE Control Systems Magazine. 2011;31(5):28–47.

[17] Owens DH, Freeman CT, Chu B. Multivariable norm optimal iterative learning control with auxiliary optimisation. International Journal of Control. 2013;86(6):1026–1045.

[18] Paszke W, Rogers E, Gałkowski K, Cai Z. Robust finite frequency range iterative learning control design and experimental verification. Control Engineering Practice. 2013;21(10):1310–1320.

[19] Li X, Huang D, Chu B, Xu JX. Robust iterative learning control for systems with norm-bounded uncertainties. International Journal of Robust and Nonlinear Control. 2016;26(4):697–718.

[20] Son T, Pipeleers G, Swevers J. Robust monotonic convergent iterative learning control. IEEE Transactions on Automatic Control. 2016;61(4):1063–1068.

[21] van de Wijdeven J, Donkers T, Bosgra O. Iterative learning control for uncertain systems: Robust monotonic convergence analysis. Automatica. 2009;45(10):2383–2391.

[22] Janssens P, Pipeleers G, Swevers J. A data-driven constrained norm-optimal iterative learning control framework for LTI systems. IEEE Transactions on Control Systems Technology. 2013;21(2):546–551.

[23] Smith R, Dahleh M. The Modeling of Uncertainty in Control Systems. vol. 192 of Lecture Notes in Control and Information Sciences. 1st ed. Springer-Verlag, Berlin, Heidelberg; 1994.

[24] De Luca A, Paesano G, Ulivi G. A frequency-domain approach to learning control: Implementation for a robot manipulator. IEEE Transactions on Industrial Electronics. 1992;39(1):1–10.

[25] Kavli T. Frequency domain synthesis of trajectory learning controllers for robot manipulators. Journal of Field Robotics. 1992;9(5):663–680.

[26] Steinbuch M, van de Molengraft M. Iterative learning control of industrial motion systems. In: 1st IFAC Conference on Mechatronic Systems. Darmstadt, Germany; 2000. p. 967–972.

[27] Pintelon R, Schoukens J. System Identification: A Frequency Domain Approach. John Wiley & Sons, Hoboken, NJ, USA; 2012.

[28] Strijbosch N, Blanken L, Oomen T. Frequency domain design of iterative learning control and repetitive control for complex motion systems. In: IEEJ International Workshop on Sensing, Actuation, Motion Control, and Optimization (SAMCON). Tokyo, Japan; 2018.

[29] Wallén J, Norrlof M, Gunnarsson S. Arm-side evaluation of ILC applied to a six-degrees-of-freedom industrial robot. In: Proceedings of the 17th IFAC World Congress. Seoul, Korea; 2008. p. 13450–13455.

[30] Blanken L, Oomen T. Multivariable iterative learning control design procedures: from decentralized to centralized, illustrated on an industrial printer. IEEE Transactions on Control Systems Technology. 2019.

[31] Amann N, Owens DH, Rogers E, Wahl A. An H_∞ approach to linear iterative learning control design. International Journal of Adaptive Control and Signal Processing. 1996;10(6):767–781.

[32] Wen JT, Potsaid B. An experimental study of a high performance motion control system. In: Proceedings of the 2004 American Control Conference. Boston, Massachusetts; 2004. p. 5158–5163.

[33] Fang Y, Chow TWS. Iterative learning control of linear discrete-time multivariable systems. Automatica. 1998;34(11):1459–1462.

[34] Olivares M, Albertos P, Sala A. Iterative learning controller design for multivariable systems. IFAC Proceedings Volumes. 2002;35(1):355–360.

[35] Tayebi A. Analysis of two particular iterative learning control schemes in frequency and time domains. Automatica. 2007;43(9):1565–1572.

[36] Gevers M. Identification for control: From the early achievements to the revival of experiment design. European Journal of Control. 2005;11(4):335–352.

[37] Hjalmarsson H. System identification of complex and structured systems. European Journal of Control. 2009;15(3):275–310.

[38] van Herpen R, Bosgra O, Oomen T. Bi-orthonormal polynomial basis function framework with applications in system identification. IEEE Transactions on Automatic Control. 2016;61(11):3285–3300.

[39] Tien S, Zou Q, Devasia S. Iterative control of dynamics-coupling-caused errors in piezoscanners during high-speed AFM operation. IEEE Transactions on Control Systems and Technology. 2005;13(6):921–931.

[40] de Rozario R, Fleming AJ, Oomen T. Iterative control for periodic tasks with robustness considerations, applied to a nanopositioning stage. IFAC-PapersOnLine. 2016;49(21):623–628.

[41] Bayo E. A finite-element approach to control the end-point motion of a single-link flexible robot. Journal of Field Robotics. 1987;4(1):63–75.

[42] Bolder J, Oomen T, Steinbuch M. Rational basis functions in Iterative learning control – with experimental verification on a motion system. IEEE Transactions on Control Systems Technology. 2015;23(2):722–729.

[43] Phan MQ, Frueh J. Learning control for trajectory tracking using basis functions. In: Proceedings of the 35th IEEE Conference on Decision and Control. Kobe, Japan; 1996. p. 2490–2492.

[44] van de Wijdeven J, Bosgra O. Using basis functions in iterative learning control: Analysis and design theory. International Journal of Control. 2010;83(4): 661–675.

[45] Hoelzle D, Alleyne A, Johnson A. Basis task approach to iterative learning control with applications to micro-robotic deposition. IEEE Transactions on Control Systems Technology. 2011;19(5):1138–1148.

[46] Hoelzle D, Johnson A, Alleyne A. Bumpless transfer filter for exogenous feedforward signals. IEEE Transactions on Control Systems Technology. 2014;22(4):1581–1588.

[47] van Zundert J, Bolder J, Oomen T. Optimality and flexibility in iterative learning control for varying tasks. Automatica. 2016;67:295–302.

[48] Blanken L, Isil G, Koekebakker S, Oomen T. Flexible ILC: Towards a convex approach for non-causal rational basis functions. IFAC-PapersOnLine. 2017;50(1):12107–12112. IFAC 2017 World Congress, Toulouse, France.

[49] Blanken L, Isil G, Koekebakker S, Oomen T. Data-driven feedforward learning using non-causal rational basis functions: Application to an industrial flatbed printer. In: 2018 IEEE American Control Conference. Milwaukee, Wisconsin; 2018.

[50] Ardakani MMG, Khong SZ, Bernhardsson B. On the convergence of iterative learning control. Automatica. 2017;78:266–273.

[51] van Zundert J, Oomen T. On inversion-based approaches for feedforward and ILC. IFAC Mechatronics. 2017;50:282–291.

[52] Voorhoeve R, van der Maas A, Oomen T. Non-parametric identification of multivariable systems: A local rational modeling approach with application to a vibration isolation benchmark. Mechanical Systems and Signal Processing. 2018;105:129–152.

[53] Geerardyn E, Oomen T. A local rational model approach for H_∞ norm estimation: With application to an active vibration isolation system. Control Engineering Practice. 2017;68:63–70.

[54] Bolder J, Kleinendorst S, Oomen T. Data-driven multivariable ILC: Enhanced performance by eliminating L and Q filters. International Journal of Robust and Nonlinear Control. 2018;28(12):3728–3751.

[55] Zou Q, Devasia S. Preview-based stable-inversion for output tracking of linear systems. Journal of Dynamic Systems, Measurements, and Control. 1999;121(4):625–630.

[56] Gross E MW Tomizuka M. Cancellation of discrete time unstable zeros by feedforward control. Journal of Dynamic Systems, Measurements, and Control. 1994;116(1):33–38.

[57] Tomizuka M. Zero phase error tracking algorithm for digital control. Journal of Dynamic Systems, Measurements, and Control. 1987;109(1):65–68.

[58] Vinnicombe G. Uncertainty and Feedback: $\mathscr{H}_\infty$ Loop-Shaping and the ν-Gap Metric. Imperial College Press, London, UK; 2001.

[59] Steinbuch M, Merry R, Boerlage M, Ronde M, van de Molengraft M. Advanced Motion Control Design. 2nd ed. Levine W, editor. The Control Handbook. CRC Press, Inc., Salem, MA; 2010.

[60] Oomen T. Advanced motion control for precision mechatronics: Control, identification, and learning of complex system. IEEJ Journal of Industry Applications. 2018;7(2):127–140.

[61] Sain MK, Schrader CB. The role of zeros in the performance of multiinput, multioutput feedback systems. IEEE Transactions on Education. 1990;33(3):244–257.

[62] Bristol E. On a new measure of interaction for multivariable process control. IEEE Transactions on Automatic Control. 1966;p. 133–134.

[63] Skogestad S, Postlethwaite I. Multivariable Feedback Control: Analysis and Design. 2nd ed. Wiley, New York; 2007.

[64] Stoev J, Ertveldt J, Oomen T, Schoukens J. Tensor methods for MIMO decoupling and control design using frequency response functions. Mechatronics. 2017;45:71–81.

[65] Zhou K, Doyle J, Glover K. Robust and Optimal Control. Prentice Hall, Upper Saddle River, NJ, USA; 1996.

[66] Blanken L, Willems J, Koekebakker S, Oomen T. Design techniques for multivariable ILC: Application to an industrial flatbed printer. IFAC-PapersOnLine. 2016;49(21):213–221. 7th IFAC Symposium on Mechatronic Systems, Loughborough, UK.

[67] Datta BN. Numerical Methods for Linear Control Systems: Design and Analysis. vol. 1. Academic Press, San Diego, CA, USA; 2004.

[68] Kim KS, Zou Q. A modeling-free inversion-based iterative feedforward control for precision output tracking of linear time-invariant systems. IEEE/ASME Transactions on Mechatronics. 2013;18(6):1767–1777.

[69] Yang L, Bechhoefer J. Feedforward control of a piezoelectric flexure stage for AFM. In: Proc. Amer. Control Conf. Seattle, WA. IEEE; 2008. p. 2703–2709.

[70] Wang Z, Zou Q. A modeling-free differential-inversion-based iterative control approach to simultaneous hysteresis-dynamics compensation: High-speed large-range motion tracking example. In: Proc. Amer. Control Conf. IEEE; 2015. p. 3558–3563.

[71] de Rozario R, Oomen T. Improving transient learning behavior in model-free inversion-based iterative control with application to a desktop printer. In: Proc. 15th Int. Workshop on Advanced Motion Control, Tokyo, Japan. IEEE; 2018.

[72] van der Meulen S, Tousain R, Bosgra O. Fixed structure feedforward controller design exploiting iterative trials: Application to a wafer stage and a desktop printer. Journal of Dynamic Systems, Measurements, and Control. 2008;130(5):0510061–05100616.

[73] Bolder J, Oomen T, Koekebakker S, Steinbuch M. Using iterative learning control with basis functions to compensate medium deformation in a wide-format inkjet printer. Mechatronics. 2014;24(8):944–953.

[74] Boeren F, Oomen T, Steinbuch M. Iterative motion feedforward tuning: A data-driven approach based on instrumental variable identification. Control Engineering Practice. 2015;37:11–19.

[75] Boeren F, Blanken L, Bruijnen D, Oomen T. Optimal estimation of rational feedforward control via instrumental variables: With application to a wafer stage. Asian Journal of Control. 2017;20(Special Issue: Theoretical and Practical Challenges in Learning Control):975–992.

[76] Blanken L, Hazelaar T, Koekebakker S, Oomen T. Multivariable repetitive control design framework applied to flatbed printing with continuous media flow. In: IEEE 56th Conference on Decision and Control. Melbourne, Australia; 2017. p. 4727–4732.

[77] van Zundert J, Oomen T. On optimal feedforward and ILC: The role of feedback for optimal performance and inferential control. IFAC-PapersOnLine. 2017;50(1):6093–6098. 20th IFAC World Congress.

[78] Bolder J, Oomen T. Inferential iterative learning control: A 2D-system approach. Automatica. 2016;71:247–253.

[79] de Rozario R, Pelzer R, Koekebakker S, Oomen T. Accommodating trial-varying tasks in iterative learning control for LPV systems, applied to printer sheet positioning. In: 2018 IEEE American Control Conference. Milwaukee, Wisconsin; 2018.

[80] Oomen T, Rojas CR. Sparse iterative learning control with application to a wafer stage: Achieving performance, resource efficiency, and task flexibility. Mechatronics. 2017;47:134–147.

Chapter 8
Algorithms for data-driven $\mathcal{H}_\infty$-norm estimation

Cristian R. Rojas[1] and Matias I. Müller[1]

In this chapter, the problem of estimating in a model-free manner the $\mathcal{H}_\infty$ norm of a linear dynamic system is discussed at a tutorial level. Two recently developed methods for addressing this problem are presented, namely the power iterations method and a class of multi-armed bandit (MAB) algorithms. Due to reasons of space, many details are omitted, but references are provided to complement this exposition.

8.1 Motivation and problem formulation

In automatic control, the design of industrial controllers is based on the available knowledge about the process to be controlled, usually condensed in the form of a mathematical model which approximates the dynamics of the system. Such a model is typically obtained using system identification techniques, based on empirical data collected from the process.

Since data is finite and may be contaminated with noise, models are at best the only approximations to the systems they aim to represent. Due to the approximate nature of a model, the task of designing a control law might be difficult due to different sources of uncertainty. Therefore, the mismatch between the model and the true process, or *modelling error*, must be taken into account when the closed-loop performance is aimed to be optimized. Previous works [1–3] have studied this topic under the name of robust control, and its solution involves spectral properties of the modelling error. These techniques require not only a nominal model of the system, but also a measure of its uncertainty, i.e. a bound on the modelling error. A standard bound for this purpose is the largest gain of the modelling error (or a weighted version of it), also known as its $\mathcal{H}_\infty$-norm.

There are several $\mathcal{H}_\infty$-norm estimation methods available in the literature (see e.g. [4,5]); however, most of these techniques require a parametric model. This requirement would involve determining an even more complex model of the difference between the true system and its nominal model (that is, of the modelling error). Since only one specific feature of such a model is sought, namely its largest gain, one

[1]Division of Decision and Learning, KTH Royal Institute of Technology, Stockholm, Sweden

could instead appeal to data-driven, or also called model-free, approaches to estimate the $\mathcal{H}_\infty$-norm of the modelling error. Under this setup, and provided with a nominal model, input–output data of the modelling error can be obtained by exciting the real system and its model with exactly the same input signals and then subtracting their outputs.

The idea behind data-driven $\mathcal{H}_\infty$-norm estimation methods is that this norm can be characterized as the maximum quotient between the Euclidean norms of the noiseless output and the input, and that, for linear and time-invariant (LTI) systems, the input that maximizes this ratio is a sinusoid whose frequency is equal to the one where the largest gain of the system is located (called the *peak frequency*) [6]. Therefore, an $\mathcal{H}_\infty$-norm estimation method should aim to produce an input that resembles a sinusoid at the peak frequency. To this end, the method would rely on a series of batch experiments, where the input sequence applied to the process during the ith experiment is designed based on the data collected from the previous $i-1$ experiments.

In this chapter, two classes of data-driven $\mathcal{H}_\infty$-norm estimation methods will be discussed. The first class of algorithms is based on the so-called power iterations method [7,8] from linear algebra, to estimate the largest eigenvalue of a matrix [9]. This method is easy to apply, and it delivers a good estimate of the $\mathcal{H}_\infty$-norm with just a few experiments. However, it can be shown to yield asymptotically biased estimates in the presence of noise [8].

To address the bias problem of the power iterations method, a second class of algorithms is then introduced. These methods are based on the *stochastic MAB* framework [10]. The MAB problem [11] is a sequential decision problem where, at each round, an agent selects an action that reveals a stochastic reward, whose distribution is unknown. Here, the aim is to design a policy that maximizes the expected cumulative reward, subject to a fundamental trade-off between exploration and exploitation. In some sense, the MAB framework is an adaptive control formulation where the process to be controlled is static (i.e. does not have memory). The MAB framework seems well suited for the data-driven $\mathcal{H}_\infty$-norm estimation problem, under the assumption that the plant is reset at the beginning of each experiment or, equivalently, that the time between consecutive experiments is sufficiently long so the process does not depend (directly) on its output from the previous batch experiment. In particular, we will introduce optimal MAB algorithms for the $\mathcal{H}_\infty$-norm estimation problem, in the sense of achieving the best rate of convergence for the estimation error of the $\mathcal{H}_\infty$-norm.

It is worth mentioning that some recent work [12] has focused on alternative identification schemes to estimate the $\mathcal{H}_\infty$-norm, based on fitting first a high-order finite impulse response model and then computing its gain. This line of research is, however, outside the scope of this chapter due to space limitations and the imposition of a different set of assumptions on the underlying true system.

The two classes of $\mathcal{H}_\infty$-norm estimation methods mentioned above have their own merits and weaknesses, so in the last part of this chapter, we will discuss their properties and future work needed to combine their strengths.

8.1.1 Problem formulation

Before presenting the two classes of estimation methods, we need to formalize our estimation problem. To this end, consider a linear discrete-time single-input–single-output stable causal and proper system of the form

$$y_k = G(q)u_k + e_k, \qquad k = 0, 1, \ldots,$$

where $\{u_k\}$ is an input signal, $\{y_k\}$ is the output of the system and $\{e_k\}$ is Gaussian, but not necessarily white, noise. q denotes the shift operator, $qx_k = x_{k+1}$, and G is a transfer function which can be described as

$$G(q) = g_0 + g_1 q^{-1} + g_2 q^{-2} + \cdots = \sum_{i=0}^{\infty} g_i q^{-i}.$$

The relation between the signals u, e and y is shown graphically in Figure 8.1.

If we let ℓ_2 denote the Hilbert space of real square-summable sequences (i.e. real sequences $\{x_k\}_{k=0}^{\infty}$ such that $\sum_{k=0}^{\infty} x_k^2 < \infty$), then the ℓ_2-gain of G is defined as

$$\|G\|_{\ell_2} := \sup_{u \in \ell_2 \setminus \{0\}} \frac{\|Gu\|_2}{\|u\|_2}. \tag{8.1}$$

Here, Gu denotes the sequence $\{G(q)u_k\}_{k=0}^{\infty}$. Because G is LTI, the ℓ_2-gain of G is equal to its $\mathscr{H}_\infty$-norm:

$$\|G\|_\infty := \sup_{\omega \in (-\pi, \pi]} |G(e^{j\omega})|,$$

where $G(e^{j\omega}) = g_0 + g_1 e^{-j\omega} + g_2 e^{-2j\omega} + \cdots$ is the frequency response of G. Therefore, in the sequel, the terms $\mathscr{H}_\infty$-norm and ℓ_2-norm will be used interchangeably when referring to linear systems.

The methods for $\mathscr{H}_\infty$-norm estimation to be presented next fall under the class of *iterative identification* procedures [13]. These procedures work with batches of input–output data: at round or iteration t, a finite input sequence $\{u_k^t\}_{k=1,\ldots,N}$ of length N is applied to the system, and its (noisy) output $\{y_k^t\}_{k=1,\ldots,N}$ is recorded. Based on the data collected from iterations 1 to t, the input for iteration $t+1$ is then designed. Furthermore, it is assumed that either the system is restarted with zero (or identically fixed) initial conditions at each round or that the batches are temporally separated widely enough so that the system has reached a stationary regime before the next round starts. Thanks to this assumption, the system G can be seen as a *static* (memoryless) but random mapping acting on $\{u_k^t\}_{k=1,\ldots,N}$ and returning $\{y_k^t\}_{k=1,\ldots,N}$. This way of

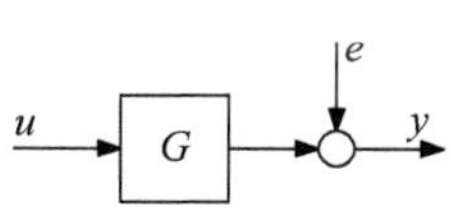

Figure 8.1 Relation between the signals u, e and y associated with the system G

interpreting the action of the system will be useful for the development of $\mathscr{H}_\infty$-norm estimation methods in the next two sections.

8.2 Power iterations

Assume that at iteration t, the system has been reset to zero initial condition. If there is no noise affecting the output of the system, the mapping from $\{u_k^t\}_{k=1,\dots,N}$ to $\{y_k^t\}_{k=1,\dots,N}$ can be written as

$$\mathbf{y}^t = \mathbf{G}\mathbf{u}^t,$$

with $\mathbf{u}^t = [u_1^t \ \cdots \ u_N^t]^T$, $\mathbf{y}^t = [y_1^t \ \cdots \ y_N^t]^T$, and where

$$\mathbf{G} = \begin{bmatrix} g_0 & 0 & 0 & \cdots & 0 \\ g_1 & g_0 & 0 & \cdots & 0 \\ g_2 & g_1 & g_0 & \cdots & 0 \\ \vdots & \vdots & \vdots & \ddots & \vdots \\ g_{N-1} & g_{N-2} & \cdots & g_1 & g_0 \end{bmatrix}$$

is a lower triangular Toeplitz matrix. In terms of this representation, the ℓ_2-gain can be approximated, for N sufficiently large, as follows [8, Lemma 1]:

$$\|G\|_{\ell_2}^2 \approx \sup_{\mathbf{u}^t \neq 0} \frac{(\mathbf{u}^t)^T \mathbf{G}^T \mathbf{G} \mathbf{u}^t}{(\mathbf{u}^t)^T \mathbf{u}^t}. \tag{8.2}$$

The right-hand side, known as the *Rayleigh quotient* of $\mathbf{G}^T\mathbf{G}$, corresponds to the largest eigenvalue of $\mathbf{G}^T\mathbf{G}$ (or, equivalently, the square of the largest singular value of $\mathbf{G}$) [14]. Thus, $\|G\|_{\ell_2}^2$ could be estimated using techniques from numerical linear algebra, such as the *power iterations method* [9].

8.2.1 Power iterations in linear algebra

The power iterations method for estimating the largest eigenvalue of a square matrix $\mathbf{A} \in \mathbb{R}^{n\times n}$ and its corresponding eigenvector[1] is presented in Algorithm 8.1.

Algorithm 8.1: Power iterations for matrices

1: Input: $\mathbf{A} \in \mathbb{R}^{n\times n}$
2: Choose $\mathbf{x}_0 \in \mathbb{R}^n$ randomly
3: **for** $i = 1, 2, \dots$ **do**
4:     Compute $\mathbf{x}_i \leftarrow \frac{1}{\|\mathbf{A}\mathbf{x}_{i-1}\|_2}\mathbf{A}\mathbf{x}_{i-1}$
5: **end for**

---

[1]For the power iterations method to work, the largest eigenvalue of $\mathbf{A}$ should be unique.

The largest eigenvalue of $\mathbf{A}$, at iteration i, can be estimated, e.g. as

$$\hat{\lambda}_i = \mathbf{x}_{i-1}\mathbf{A}\mathbf{x}_{i-1}.$$

(Notice that the product $\mathbf{A}\mathbf{x}_{i-1}$ is already computed at iteration i.)

The idea behind this method can be explained as follows. Let $\{\lambda_i\}$ denote the eigenvalues of $\mathbf{A}$ (where multiple eigenvalues are considered separately), sorted in descending order according to their magnitude, and denote by $\{\mathbf{v}_i\}$ the corresponding eigenvectors. Since the eigenvectors are linearly independent, one can represent $\mathbf{x}_{i-1}$ uniquely as

$$\mathbf{x}_{i-1} = \alpha_{i-1}^{(1)}\mathbf{v}_1 + \alpha_{i-1}^{(2)}\mathbf{v}_2 + \cdots + \alpha_{i-1}^{(n)}\mathbf{v}_n.$$

Therefore,[2]

$$\begin{aligned}\mathbf{x}_i &\propto \mathbf{A}\mathbf{x}_{i-1}\\ &\propto \alpha_{i-1}^{(1)}\mathbf{A}\mathbf{v}_1 + \alpha_{i-1}^{(2)}\mathbf{A}\mathbf{v}_2 + \cdots + \alpha_{i-1}^{(n)}\mathbf{A}\mathbf{v}_n\\ &\propto \alpha_{i-1}^{(1)}\lambda_1\mathbf{v}_1 + \alpha_{i-1}^{(2)}\lambda_2\mathbf{v}_2 + \cdots + \alpha_{i-1}^{(n)}\lambda_n\mathbf{v}_n,\end{aligned}$$

and by induction on i,

$$\begin{aligned}\mathbf{x}_i &\propto \alpha_0^{(1)}\lambda_1^i\mathbf{v}_1 + \alpha_0^{(2)}\lambda_2^i\mathbf{v}_2 + \cdots + \alpha_0^{(n)}\lambda_n^i\mathbf{v}_n\\ &\propto \lambda_1^i\left[\alpha_0^{(1)}\mathbf{v}_1 + \alpha_0^{(2)}\left(\frac{\lambda_2}{\lambda_1}\right)^i\mathbf{v}_2 + \cdots + \alpha_0^{(n)}\left(\frac{\lambda_n}{\lambda_1}\right)^i\mathbf{v}_n\right].\end{aligned}$$

Due to the random initialization of $\mathbf{x}_0$, we can assume that $\alpha_0^{(1)} \neq 0$ with probability 1. Therefore, since $|\lambda_1| > |\lambda_2| \geq \cdots \geq |\lambda_n|$, the normalization of $\mathbf{x}_i$ yields that $\mathbf{x}_i \approx \pm\mathbf{v}_1$ for i sufficiently large.

8.2.2 Power iterations for linear dynamical systems

In order to apply the power iterations method to estimate the gain of a linear dynamical system G based solely on experiments on the system, we need a way to compute $\mathbf{G}^T$ from experiments. This can be achieved [15,16] by noticing that

$$\mathbf{G}^T = \tau\,\mathbf{G}\,\tau,$$

where

$$\tau = \begin{bmatrix} 0 & \cdots & 0 & 1\\ \vdots & \iddots & 1 & 0\\ \vdots & \iddots & & \vdots\\ 1 & \cdots & 0 & 0 \end{bmatrix}$$

is a *time reversal* operator, such that $\tau\,[x_1 \;\cdots\; x_n]^T = [x_n \;\cdots\; x_1]^T$. Using this operator, we have that $\mathbf{G}^T\mathbf{G} = \tau\,\mathbf{G}\,\tau\,\mathbf{G}$, which leads to the *power iterations method for $\mathcal{H}_\infty$-norm estimation* [8], described in Algorithm 8.2.

[2]The notation $\mathbf{x} \propto \mathbf{y}$ means that there is a real scalar a such that $\mathbf{x} = a\mathbf{y}$.

Algorithm 8.2: Power iterations for $\mathscr{H}_\infty$-norm estimation

Choose $\mathbf{u}^0 \in \mathbb{R}^N$ at random, and normalize it to have norm equal to $\sqrt{n}$
for $i = 1, 2, \ldots$ **do**
  Apply $\mathbf{u}^{i-1}$ to the system, and record its output, $\mathbf{y}^i \in \mathbb{R}^N$
  Time reverse $\mathbf{y}^i$ and normalize it to have norm equal to $\sqrt{n}$: $\mathbf{u}_i \leftarrow \frac{\sqrt{n}}{\|\tau\,\mathbf{y}^i\|_2}\tau\,\mathbf{y}^i$
end for

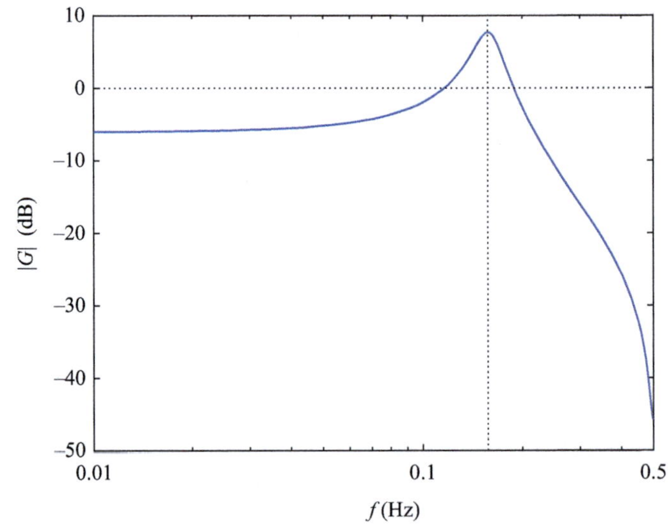

Figure 8.2 Bode magnitude diagram of G

The estimate of the $\mathscr{H}_\infty$-norm of G at iteration i is given by

$$\hat{\beta}^i = \frac{(\mathbf{u}^{i-1})^T\,\tau\,\mathbf{y}^i}{\sqrt{n}}.$$

8.2.3 An example

As an example of this procedure, consider a system with the frequency response depicted in Figure 8.2, with noise variance $\sigma_e = 9$. The $\mathscr{H}_\infty$-norm of this system is approximately 2.397, and its peak frequency is 0.159 (Hz). Each batch is an experiment of $N = 100$ samples. The evolution of $\hat{\beta}^{(i)}$ and its average over 1,000 realizations are shown in Figure 8.3. As this figure reveals, power iterations converges rather quickly, in about 30 iterations. However, the limit value of $\hat{\beta}^{(i)}$ has a negative bias, which is due to the presence of noise in the observations [8].

Figure 8.4 shows the input sequences generated after iterations 1 and 100. As expected, after a large number of iterations, the input sequence closely resembles a

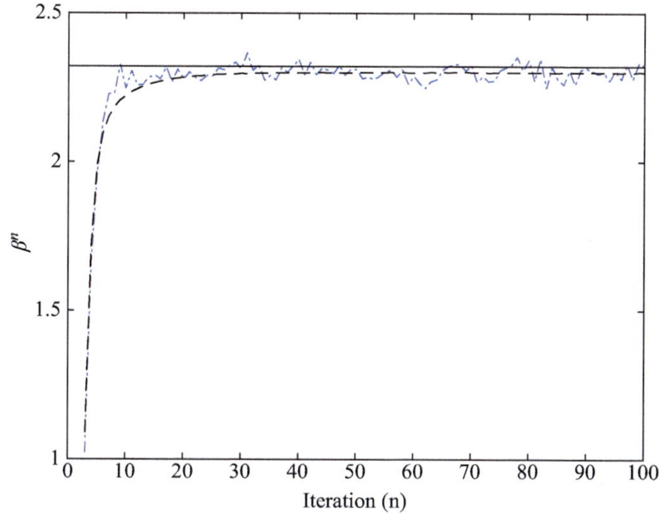

Figure 8.3 Estimate of the $\mathcal{H}_\infty$-norm of G (blue, dash-dotted), and average of 1,000 iterations (black, dashed). Also shown is the actual $\mathcal{H}_\infty$-norm of G (black, solid)

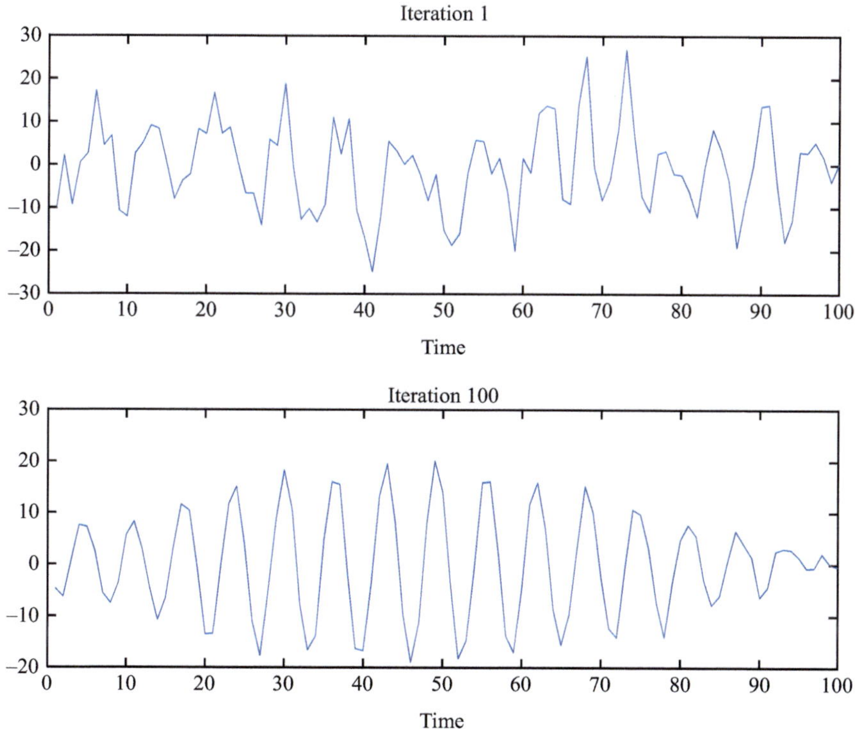

Figure 8.4 Input signal generated by the power iterations method after iterations 1 (top) and 100 (bottom)

sinusoid of period equal to $1/0.159 \approx 6.28$ samples (the period corresponding to the peak frequency of G).

This example illustrates that, while power iterations converges reasonably fast to a good estimate of the $\mathscr{H}_\infty$-norm, the estimate is in general biased due to the presence of noise, and in fact, this bias is roughly proportional to the noise variance σ_e (when σ_e is small). Furthermore, the spectrum of the limit input sequence generated by the method, when the data is noisy, is not that of a sinusoid, but it rather spreads power throughout the entire frequency range, which is the reason why the estimate of the $\mathscr{H}_\infty$-norm is negatively biased (i.e. the Rayleigh quotient (8.2) does not reach its maximum). These properties of the power iterations methods have been thoroughly analysed in [8].

The power iterations methods introduced here have been extended to multivariable systems in [17]. Also, in [16], other $\mathscr{H}_\infty$-norm estimation methods, inspired by standard techniques from numerical linear algebra such as Lanczos' method [9], are derived. Some of these algorithms may perform better than power iterations under some circumstances, but they also suffer from bias problems in the presence of noise.

In the next sections, an alternative class of $\mathscr{H}_\infty$-norm estimation methods will be covered, based on the stochastic MAB framework. These methods overcome the bias problem of power iterations and possess an optimal rate of convergence, even though their performance for small number of iterations can be slower than that of power iterations.

8.3 Multi-armed bandits

In this section, we discuss how the problem of $\mathscr{H}_\infty$ norm estimation can be posed within the MAB framework. As previously discussed, the problem of estimating efficiently the $\mathscr{H}_\infty$ norm of a system, in a model-free manner, involves finding the peak frequency as quick as possible, so the $\mathscr{H}_\infty$ norm can be then estimated from the data produced by the system when applying a sinusoid of such frequency as input signal. This argument serves as motivation to pose the $\mathscr{H}_\infty$-norm estimation as a two-step problem, one involving the exploration/exploitation dilemma and the other involving statistical inference. The solution of finding the optimal frequency is assessed by a regret function, while the performance of the solution to the norm estimation is the mean-squared error. The theory on MABs precisely provides a framework to solve the former problem efficiently.

8.3.1 Stochastic multi-arm bandits in a nutshell

MAB problems are sequential decision problems where, at each round/iteration/experiment, the decision maker (known as *agent*) selects an action (also known as *arm*) from a predefined set, observing a noisy measurement (known as *reward*) produced by the system/environment [10]. The agent does not know the reward distribution, and the final goal within this framework is to maximize the cumulative expected reward. At this point, we should remark that the final goal of this framework is not to identify how the rewards (from different arms) are drawn but only the average of each of them. This task requires the agent to balance between *exploitation* (the arms known to provide large rewards should be played more often) and *exploration* (all arms should be played enough so the agent actually learns each average reward).

Formally, the stochastic MAB setup is composed of a set of K different arms, from which at each round $t = 1, \ldots, T$ a stochastic reward X_{k_t} is revealed, where k_t denotes the arm chosen at round t. In a more general formulation, the goal is to find a policy $\boldsymbol{\pi} \in \boldsymbol{\Pi}$ minimizing the following notion of regret:

$$\mathscr{R}^{\pi}(T) := \mathbb{E}\left\{\sum_{t=1}^{T} f(X_{k^\star}) - f(X_{k_t^\pi})\right\}, \tag{8.3}$$

where f is a real-valued function and $k^\star := \max_{k=1,\ldots,K} \mathbb{E}\{f(X_k)\}$ denotes the optimal arm. In other words, we compare our performance to the one an oracle, knowing the best arm in hindsight, would achieve. Notice that this framework assumes that rewards are drawn independently over rounds. Other assumptions on how the rewards are generated derive on different bandit settings such as Markovian and Adversarial [10]. For an application of the Adversarial MAB framework to $\mathscr{H}_\infty$-norm estimation, the interested reader may check [18]. In what follows, we refer to MAB and stochastic MAB problems indistinctly.

8.3.2 $\mathscr{H}_\infty$-norm estimation as an MAB problem

The frequency axis $[0, \pi]$ is discretized into K frequencies, denoting all the possible arms. In this sense, the $\mathscr{H}_\infty$ norm of G is approximated as

$$\|G\|_\infty = \max_{\omega\in[0,\pi]} |G(e^{j\omega})| \approx \max_{k\in\{1,\ldots,K\}} |G(e^{j\omega_k})|, \tag{8.4}$$

where $\omega_k := 2\pi k/(2K+1)$.

Each round $t = 1, \ldots, T$ represents a different experiment, where input/output data in the frequency domain is obtained. More specifically, at each round, the agent receives the discrete Fourier transforms (DFT) $U^t(\omega_k), Y^t(\omega_k)$, $k = 1, \ldots, K$, of the input u and output y, respectively. The underlying number of data points in the time domain is $N = 2K + 1$. The power allocated at each frequency of the input signal is denoted as $p_k^t := |U^t(\omega_k)|^2$, and it will be assumed that at each round, without loss of generality, $\sum_{k=1}^{K} p_k^t = 1$. The (unknown) DFT of the noise, $E^t(\omega_k)$, $k = 1, \ldots, K$, is a circularly symmetric random variable whose real and imaginary components are statistically independent and of variance σ^2 [19]; this assumption will be denoted as $E^t(\omega_k) \sim \mathscr{N}_{\mathscr{C}}(0, \sigma^2 I)$. Moreover, the sequence $\{E^t(\omega_k)\}_{k=1}^{K}$ is white.

We allow the agent to choose policies within $\boldsymbol{\Pi}$ defined as the set of all adaptive policies specifying a power profile $p^t := \{p_k^t\}_{k=1}^{K}$ at each round based on previous input/output signals. That is, for $\pi \in \boldsymbol{\Pi}$, the power levels $p^{t,\pi}$ selected under $\boldsymbol{\Pi}$ at the tth experiment are $\mathscr{F}_t^\pi$-measurable, where $\mathscr{F}_t^\pi$ is the σ-algebra generated by $(p^{1,\pi}, X^1, \ldots, p^{t-1,\pi}, X^{t-1})$ [20].

After applying a power profile p^t, we observe a complex outcome sequence $X^t := \{X_k^t\}_{k=1}^K$ defined as

$$X_k^t = X^t(\omega_k) := \frac{Y^t(\omega_k)}{U^t(\omega_k)} = G(\mathrm{e}^{j\omega_k}) + \frac{E^t(\omega_k)}{U^t(\omega_k)}, \quad k = 1, \ldots, K.$$

Since $E^t(\omega_k) \sim \mathcal{N}_{\mathscr{C}}(0, \sigma^2 I)$, it follows that

$$\begin{aligned} \mathbb{E}\left\{X_k^t | U^t\right\} &= \mathbb{E}\left\{G(\mathrm{e}^{j\omega_k}) + \frac{E^t(\omega_k)}{U^t(\omega_k)}\right\} = G(\mathrm{e}^{j\omega_k}) =: \boldsymbol{\mu}_k, \\ \operatorname{cov}\left\{X_k^t | U^t\right\} &= \frac{\sigma^2}{p_k^t}, \end{aligned}$$

Additionally, the X_k^ts are independent for different ks, conditioned on U^t.

The problem of finding the optimal frequency can be reduced to an adaptive input design problem where the DFT of u^t is designed recursively at each experiment, aiming to maximize the ℓ_2 gain. In other words, the goal is to find a combination of *arms* p^t such that

$$p^t = \arg \max_{p^t \,:\, p^t \geq 0,\, \sum_{k=1}^K p_k^t = 1} \frac{\|y^t\|_2}{\|u^t\|_2}. \tag{8.5}$$

Since we assume that $|G(1)|, |G(-1)| \neq \|G\|_\infty$, and since p^t is restricted to lie in the simplex $\{p^t : p^t \geq 0,\, \sum_{k=1}^K p_k^t = 1\}$, the above problem can be casted as

$$\begin{aligned} p^t &= \arg \max_{p^t \,:\, p^t \geq 0,\, \sum_{k=1}^K p_k^t = 1} \|y\|_2^2 \\ &= \arg \max_{p^t \,:\, p^t \geq 0,\, \sum_{k=1}^K p_k^t = 1} \|Y^t\|_2^2 \\ &= \arg \max_{p^t \,:\, p^t \geq 0,\, \sum_{k=1}^K p_k^t = 1} \sum_{k=1}^K \left|U_k^t X_k^t\right|^2. \end{aligned}$$

This expression motivates our notion of reward at round t: $\sum_{k=1}^K \left|U_k^t X_k^t\right|^2$. It turns out that either the latter or $\sum_{k=1}^K p_k^t \left|X_k^t\right|^2$ achieve the same expected cumulative reward: $\sum_{k=1}^k \mathbb{E}\left\{p_k^{t,\pi}\right\} |G(\mathrm{e}^{j\omega_k})|^2 + K\sigma^2$, where the expectation is taken with respect to the noise and the randomness of $p_k^{t,\pi}$ in $\boldsymbol{\pi}$. Then, the regret minimization problem can be casted as

$$\mathscr{R}^\pi(\boldsymbol{\mu}, T) = T\,|\boldsymbol{\mu}_{k^\star}|^2 - \sum_{t=1}^T \sum_{k=1}^K \mathbb{E}\left\{p_k^{t,\pi}\right\} |\boldsymbol{\mu}_k|^2, \tag{8.6}$$

where $k^\star := \max_{k \in \{1,\ldots,K\}} |\mu_k|$, and the dependence of the regret on $\boldsymbol{\mu} := \{\boldsymbol{\mu}_k\}_{k=1}^K$ has been made explicit. The first term in the right-hand side of (8.6) corresponds to the expected cumulative reward collected by an oracle that assigns $p_{k^\star}^t = 1$ at each round, collecting $|\boldsymbol{\mu}_{k^\star}|^2 + K\sigma^2$ as expected reward.

The choice of $|\boldsymbol{\mu}_k|^2$, rather than $|\boldsymbol{\mu}_k|$, entails a proper and simpler notion of regret, namely $\sum_{k=1}^K p_k^{t,\pi} |X_k^t|^2$, corresponding to the squared norm of the output y^t, which we aim to maximize via the iterative-input-design algorithm defined by the MAB framework.

8.3.3 Regret lower bounds and optimal algorithms

Before stating the results of this approach, it is insightful to restrict our attention to the policies that concentrate all the available power at one single frequency, $\boldsymbol{\Pi}_{\text{SF}}$ (SF stands for 'single frequency'). That is, for $\boldsymbol{\pi} \in \boldsymbol{\Pi}_{\text{SF}}, p^{\pi,t} \in \{0,1\}^K, \sum_{k=1}^K p^{\pi,t} = 1$ for each $t = 1, \ldots, T$. Notice that $\boldsymbol{\Pi}_{\text{SF}} \subset \boldsymbol{\Pi}$.

8.3.3.1 Regret lower bound in $\boldsymbol{\Pi}_{\text{SF}}$

When restricting our attention to the single-frequency strategies in $\boldsymbol{\Pi}_{\text{SF}}$, we recover the classical stochastic MAB problem with a finite number of arms (here called frequencies), where $\min_{\pi \in \boldsymbol{\Pi}_{\text{SF}}} \mathscr{R}(\boldsymbol{\mu}, T)$ is the regret minimization introduced by Robbins [21], also described in the first part of this section. In their seminal paper, Lai and Robbins [11] derived a tight asymptotic (when T grows large) regret lower bound satisfied by any *uniformly efficient* algorithm in $\boldsymbol{\Pi}_{\text{SF}}$. An algorithm $\boldsymbol{\pi}$ is said to be uniformly efficient if, for any $\boldsymbol{\mu}$, the number of rounds $N_k^\pi(t)$ a suboptimal arm k is selected up to round t satisfies $\mathbb{E}\left\{N_k^\pi(t)\right\} = o(t^\alpha)$, for every $\alpha > 0$. Let ν_k denote the distribution $\mathscr{N}_{\mathscr{C}}(\boldsymbol{\mu}_k, \sigma^2 I_2)$. It is shown in [11] that under any uniformly efficient algorithm $\pi \in \boldsymbol{\Pi}_{\text{SF}}$:

$$\liminf_{T\to\infty} \frac{\mathbb{E}\left\{N_k^\pi(T)\right\}}{\log T} \geq \frac{1}{\mathbb{D}\left\{\nu_k || \nu_k^\star\right\}}, \tag{8.7}$$

where $\mathbb{D}\left\{\nu_k || \nu_k^\star\right\} := \int \ln(d\nu_k/d\nu_k^\star) d\nu_k$ denotes the Kullback–Leibler divergence between distributions ν_k and $\nu_k^\star$. In particular, the regret under algorithm $\boldsymbol{\pi}$ satisfies

$$\begin{aligned}\liminf_{T\to\infty} \frac{\mathscr{R}^\pi(\boldsymbol{\mu}, T)}{\log T} &\geq \sum_{k\neq k^\star} \frac{|\boldsymbol{\mu}_{k^\star}|^2 - |\boldsymbol{\mu}_k|^2}{\mathbb{D}\left\{\nu_k || \nu_k^\star\right\}} \\ &= \sum_{k\neq k^\star} \sigma \frac{|\boldsymbol{\mu}_{k^\star}|^2 - |\boldsymbol{\mu}_k|^2}{|\boldsymbol{\mu}_{k^\star} - \boldsymbol{\mu}_k|^2}. \end{aligned} \tag{8.8}$$

Algorithms achieving the regret lower bound in (8.8) are called *asymptotically optimal algorithms*.

8.3.3.2 Regret lower bound in $\boldsymbol{\Pi}$

The problem of finding a power-spreading policy can be interpreted as a linear stochastic MAB, where the observed reward corresponds to a noisy linear combination of the true means. In linear bandits, the regret lower bound usually scales as $\Theta(\sqrt{T})$, as T grows large, but as the following theorem shows, the minimal regret in our setup scales logarithmically. This is due to the fact that the agent receives much more detailed feedback, since it has access to noisy measurements of each mean, via the

DFT. To state the result, we need to redefine the notion of uniform efficiency: an algorithm $\pi \in \Pi$ is *uniformly efficient* if, for every μ, $\sum_{t=1}^{T} \mathbb{E}\{p_k^{\pi,t}\} = o(T^\alpha)$, for all $\alpha > 0$, where $p_k^{t,\pi}$ denotes the power allocated at frequency ω_k under π at round t.

Theorem 8.1 ([22]). *For every uniformly efficient algorithm $\pi \in \Pi$, the cumulative expected power allocated at some frequency ω_k satisfies*

$$\liminf_{T\to\infty} \frac{\sum_{t=1}^{T} \mathbb{E}\{p_k^{t,\pi}\}}{\log T} \geq \frac{\sigma}{|\mu_{k^\star} - \mu_k|^2}, \quad \text{for every } k \neq k. \tag{8.9}$$

This implies that the regret under π satisfies

$$\liminf_{T\to\infty} \frac{\mathscr{R}^\pi(\mu, T)}{\log T} \geq \sum_{k=1}^{K} \sigma \frac{|\mu_{k^\star}|^2 - |\mu_k|^2}{|\mu_{k^\star} - \mu_k|^2}. \tag{8.10}$$

By comparing the above theorem with (8.8), we conclude that to obtain an asymptotically optimal algorithm, one may restrict to strategies in Π_{SF}, i.e. selecting only one frequency at each experiment. It is important to note that the above regret lower bound is valid when T is very large only. When T is small, it might be much wiser to spread the power on different frequencies so as to quickly gather information on all frequencies and exploit this information to minimize the regret. This hand-waving argument motivates the development of a novel algorithm, introduced in [22], which is formally presented in the following subsection.

8.3.4 The weighted Thompson sampling (WTS) algorithm

Thompson sampling (TS) [23] is a well-known algorithm that is asymptotically optimal in the basic stochastic MAB setup, i.e. for single frequency strategies [24]. It belongs to the family of (randomized) Bayesian algorithms, and it starts with a prior distribution on the μ_ks, e.g. $\mu_k \sim \mathscr{N}_{\mathscr{C}}(0, \lambda^2 I)$, for some $\lambda > 0$. Based on this prior, the initial probability distribution ρ^1 of each arm being the optimal one is computed, i.e. ρ_k^1 :=Prob$\{k^\star = k\}$. At each round $t = 2, \ldots, T$, TS chooses an arm k randomly sampled from the posterior distribution ρ_k^t :=Prob$\{k^\star = k|\mathscr{F}_t\}$, where $\mathscr{F}_t$ represents the information collected in all previous experiments. The posterior distribution is then updated to ρ_k^{t+1}, accordingly. The procedure is summarized in Algorithm 8.3.

Algorithm 8.3: Thompson sampling

1: Input: $\rho^1 = (\rho_1^1, \ldots, \rho_K^1)$ (prior distribution for $k^\star$)
2: **for** $t = 1$ to T **do**
3: Sample $k^t \sim \rho^t$
4: Observe the outcome $X_{k^t}^t$
5: Update the posterior ρ^{t+1}
6: **end for**

Weighted TS (WTS), instead of TS, corresponds to a power-spreading strategy where the power profile $\{p_k^t\}$ is chosen as the posterior distribution ρ_k^t, for each round $t = 1, \ldots, T$. This allows WTS to gather information of all frequencies in a single experiment.

The posterior distribution ρ_k^t is updated in two steps. First, we compute a posterior distribution for the real and imaginary parts of the mean of every arm $k = 1, \ldots, K$. This is straightforward to do since the real and imaginary parts of the rewards are independent in rounds and frequency. In fact, since the only source of randomness is the Gaussian additive noise e, we have that

$$\boldsymbol{\mu}_k | (X_k^\ell)_{\ell=1}^{t-1} \sim \mathscr{N}_\mathscr{C}\left(m_k^t, v_k^t I\right), \tag{8.11a}$$

$$m_k^t = \frac{\lambda^2 \sum_{\ell=1}^{t-1} p_k^\ell X_k^\ell}{\sigma^2 + \lambda^2 \sum_{\ell=1}^{t-1} p_k^\ell}, \tag{8.11b}$$

$$v_k^t = \frac{\lambda^2}{1 + \lambda^2/\sigma^2 \sum_{\ell=1}^{t-1} p_k^\ell}, \tag{8.11c}$$

where m_k^t and v_k^t are the complex posterior mean and the complex posterior variance of arm k at round t, respectively.

In a second step, based on this information we compute the probability of each arm being the optimal one ρ_k^t by Monte Carlo simulations. To this end, we generate M samples, for each arm, from the posterior distributions described in (8.11a)–(8.11c). Then, each posterior is updated as the empirical probability, based on M samples, of each frequency attaining the largest ℓ_2-gain.

The pseudo-code for WTS is described in Algorithm 8.4.

Algorithm 8.4: Weighted Thompson sampling

1: Input: M, λ, σ
2: $\rho_k^1 = 1/K$, for all $k = 1, \ldots, K$ (prior distribution)
3: **for** $t = 1$ to T **do**
4: Input design: select the power profile $p^t = \rho^t$
5: Perform the experiment and obtain $(X_k^t)_k$
6: Update the posterior of $\boldsymbol{\mu}_k$ for all k:
7: $m_k^{t+1} = \frac{\lambda^2 \sum_{\ell=1}^{t} p_k^\ell X_k^\ell(\omega_k)}{\sigma^2 + \lambda^2 \sum_{\ell=1}^{t} p_k^\ell}$
8: $v_k^{t+1} = \frac{\lambda^2 \sigma^2}{\sigma^2 + \lambda^2 \sum_{\ell=1}^{t} p_k^\ell}$
9: Update the posterior ρ^{t+1}:
10: Draw $s^l \sim \mathscr{N}_\mathscr{C}(m_k^{t+1}, v_k^{t+1} I)$, $\quad l = 1, \ldots, M$
11: $\rho_k^{t+1} = \frac{1}{M} \sum_{l=1}^{M} 1\{\arg\max_i\{|s_i^l|\} = k\}$
12: $\hat{\beta}^t := \max_{k \in \{1,\ldots,K\}} \frac{\sum_{\ell=1}^{t} p_k^\ell X_k^\ell}{\sum_{\ell=1}^{t} p_k^\ell}$
13: **end for**

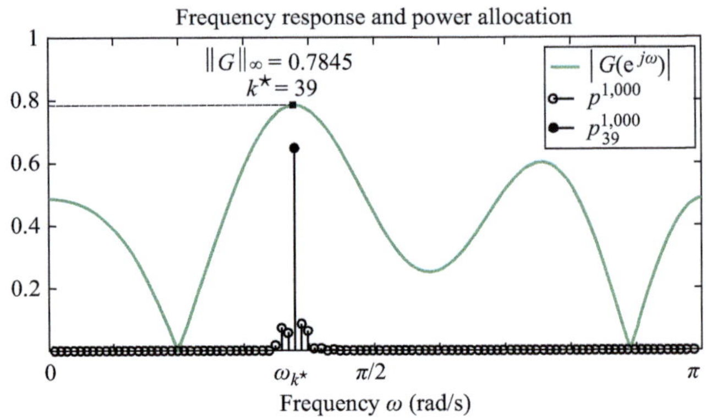

Figure 8.5 Frequency response of the unknown system G, together with the power distribution for the input signal at T = 1,000

The choice of $\hat{\beta}^t$ can be different from the one introduced in Algorithm 8.4. We have chosen the estimation as the maximum averaged weighted reward for each frequency, i.e. it balances the power allocated at some frequency and its observed rewards. Another way to perform the $\mathscr{H}_\infty$-norm estimation from the data collected by Algorithm 8.4 is to consider

$$\hat{\beta}_2^t := \frac{\sum_{\ell=1}^{t} \sum_{k=1}^{K} p_k^\ell X_k^\ell}{T}, \tag{8.12}$$

which is a simple weighted average over all arms and experiments.

8.3.5 An illustrative example

Let G be a tenth order LTI system whose *unknown* frequency response is depicted in Figure 8.5. The noise level is set to $\sigma = 0.1$ and the frequency axis discretized into $K = 100$ different frequencies, with the *unknown* optimal arm (frequency) being $k^\star = 39$, corresponding to $\omega_{39} = 1.225$ (rad/s). To derive the posterior ρ^t, $M = 1{,}000$ Monte Carlo simulations are employed.

Figure 8.5 also depicts the behaviour of WTS when $t = 1{,}000$ in terms of the power allocated to each of the frequencies. We see how WTS is indeed able to identify the best arm, assigning most of the power around the peak frequency.

To illustrate the temporal behaviour of WTS, Figure 8.6(a) compares the regret function between WTS and the well-known TS, which can be easily implemented from Algorithm 8.4, based on 500 iterations. We see that, for large T, both regrets grow logarithmically, but WTS does it in a much smaller scale. In the same line, Figure 8.6(b) depicts the regret difference for the early stages, where WTS incurs quickly in a lower regret due to its ability to collect more information at each round. The 3-σ intervals plotted in the latter figure reveal that the regret of WTS exhibits less variability than that of TS, implying that the performance of WTS is more robust to sample variability.

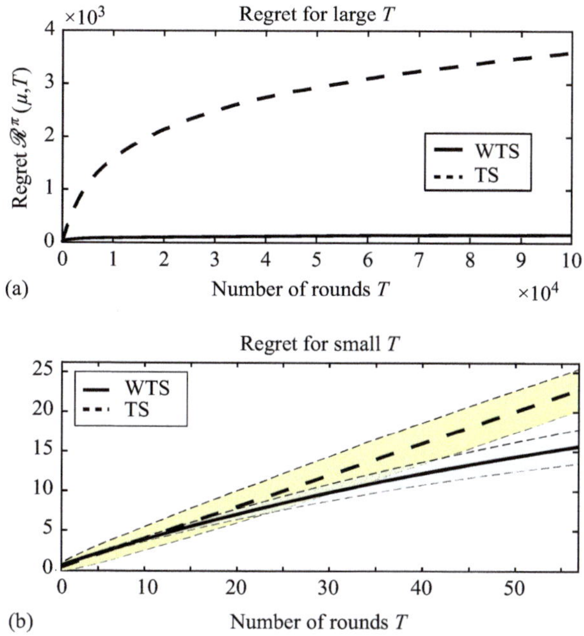

Figure 8.6 Regret performance comparison between WTS and Thompson sampling, when using $\sigma = 0.1$, and the frequency response for G is the one depicted in Figure 8.5. Part (a) shows the performance comparison when T is growing large, while (b) does it when smaller values of T are considered. In (b), light yellow and light blue areas symbolize the confidence bound defined by three times the standard deviation of the regret for TS and WTS, respectively

We finish this section by comparing WTS to the power iterations algorithm in terms of the MSE incurred when estimating $\|G\|_\infty^2$. As Figure 8.7 shows, the convergence rate of the power iterations method outperforms the one of WTS, however, due to the presence of noise, it leads to biased estimations, as explained before. In contrast, WTS performs better when T grows large.

8.4 Extensions to nonlinear systems

The methods presented so far have assumed that the system under study is LTI, so using sinusoidal input signals to excite the system leads to a good estimate of the $\mathcal{H}_\infty$-norm when their lengths are large enough. For example, the MAB method presented in Section 8.3 exploits this property by parametrizing the input signals by amplitude and frequency.

For general nonlinear systems, there is no clear parametrization of the input signal since the shape of the optimal one is system dependent. The approach presented in

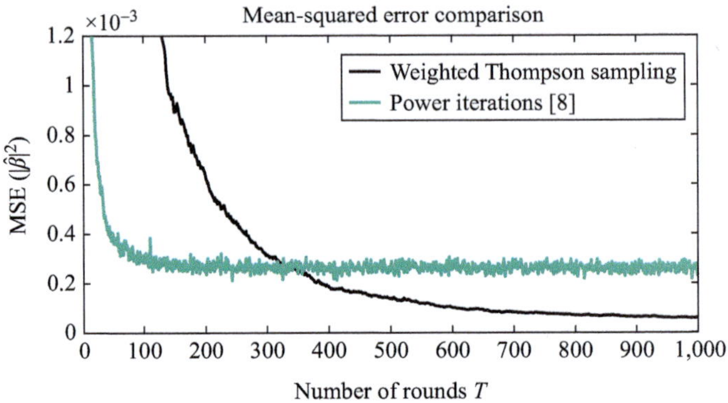

Figure 8.7 Mean-squared error comparison

this section precisely addresses this issue by presenting a novel way of parametrizing input signals within a graph-theoretical framework, firstly introduced in [25]. For other types of input signal parametrization the reader can refer to [7]. The approach is then to find the parameters maximizing the ℓ_2 gain.

The system under study is assumed to be *exponentially forgetting* [26]. Roughly speaking, this means that the current output depends on previous input values in an *exponentially decreasing* fashion.

In the following subsection, we introduce de Bruijn graphs and prime cycles, both being key concepts on which our approach is based.

8.4.1 de Bruijn graphs and prime cycles

An m-de Bruijn graph [27] of symbols $\mathscr{C}$ is a connected graph whose nodes are all the possible combinations of symbols in $\mathscr{C}$ of length m, i.e. each node contains m elements. The edges of de Bruijn graphs connect two nodes when the first $m-1$ elements of a node are equal to the last $m-1$ elements of the other, as shown in Figure 8.8.

The notion of connection allows us to introduce the notion of path, which is a sequence of connected nodes $(n_1, n_2, n_3 \dots)$, with $n_i = (c_{i,1}, c_{i,2}, \dots, c_{i,m}) \in \mathscr{C}^m$ defining a particular input sequence $(u_t, u_{t-1}, u_{t-2}, \dots)$ where $u_j = c_{j,m}$. Thus, path $((1,1,1),(1,1,0),(1,0,1))$ defines the sequence $(u_t, u_{t-1}, u_{t-2}) = (1,0,1)$ in Figure 8.8.

An elementary cycle $(n_1, n_2, \dots, n_1)$ is defined as a sequence of connected nodes (i.e. two adjacent nodes must be connected) where n_1 is the only node appearing twice in the sequence. A prime cycle, instead, is an elementary cycle that does not contain sub-cycles forming elementary sub-cycles. In Figure 8.8, the cycle $((0,1,1),(1,1,0),(1,0,1))$ is prime and elementary, but $((0,1,1),(1,1,0),(1,0,0),(0,0,1),(0,1,0),(1,0,1))$ is not since nodes $(1,0,1)$ and $(0,1,0)$ form an elementary cycle.

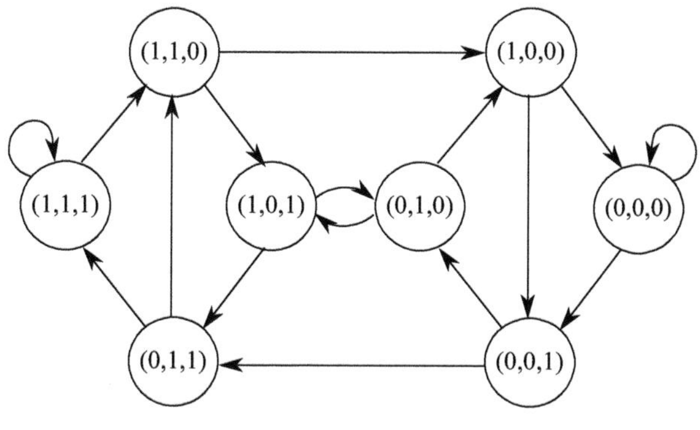

Figure 8.8 Example of a de Bruijn graph with $\mathscr{C} = \{0,1\}$ and $m = 3$

8.4.2 Finding the optimal stationary sequence

A stationary sequence u^π is defined by $\pi \in \mathscr{P}$, where $\mathscr{P}$ is the set of all stationary distributions in the de Bruijn graph. Here, for $\pi = (\pi_1, \pi_2, \ldots, \pi_m)$, $\pi_i \in [0, 1]$ corresponds to the average number of times node i is visited. It is possible to prove that $\mathscr{P}$ is a convex polyhedral set generated by the convex hull of its extreme points $(q_1, \ldots, q_N)$ [28]. That is, any stationary distribution $\pi \in \mathscr{P}$ can be written as the linear combination

$$\pi_i = \sum_{i=1}^{N} \alpha_i q_i, \tag{8.13}$$

with $\alpha_i \geq 0$, $i = 1, \ldots, N$ and $\sum_{i=1}^{N} \alpha_i = 1$.

The importance of the prime cycles relies on the one-to-one relationship existing between each of them and each q_i parameterizing $\mathscr{P}$ [29], implying that N corresponds to the number of prime cycles in the de Bruijn graph. Thus, each stationary sequence can be parametrized by the prime cycles of the corresponding de Bruijn graph.

Observe that all prime cycles are defined by the user choice for $(m, \mathscr{C})$, and then the only parameter we can optimize for is the percentage of time each node is visited for each $\pi \in \mathscr{P}$ by choosing $\alpha = (\alpha_1, \ldots, \alpha_N) \in \mathscr{B}_1 := \{v : \|v\|_1 = 1, v \in \mathbb{R}^m_+\}$ optimally maximizing

$$\frac{\sum_{i=1}^{N} \alpha_i \|Gu^i\|_2}{\sum_{i=1}^{N} \alpha_i \|u^i\|_2}, \tag{8.14}$$

where u^i is the input signal associated with the ith prime cycle.

Under noisy outputs, it is not possible to obtain $\|Gu^i\|_2$ but $\|Gu^i + e\|_2$, where e is the realization of a stochastic process. It may then be possible to tune parameter α within a MAB setup, similar to the one introduced in Section 8.3, where α belongs to the simplex.

8.5 Discussion and extensions

In this chapter, two classes of data-driven algorithms for $\mathcal{H}_\infty$-norm estimation have been presented. Each class has its own merits and disadvantages. For instance, the power iterations method has faster convergence than the MAB algorithms, but it is asymptotically biased.

Most of the research on MAB methods has focused on their asymptotic performance (typically in terms of their regret), while, as seen here, their behaviour for small number of iterations can be relatively slow. Since the performance of power iterations is very good in its initial stages, it would be attractive to combine, if possible, these two methods in order to speed up the transient behaviour of the TS algorithms introduced here.

In the final section of this chapter, a possible extension of ℓ_2-gain estimation methods to nonlinear systems has been discussed, based on the concept of prime cycles of de Bruijn graphs. This idea provides a natural way to extend the decomposition of an input signal into its frequency components, as done for linear systems, to the case of nonlinear systems whose dynamics can be approximated by a Markovian model of finite memory. This would allow the application of TS or other MAB tools to estimate the ℓ_2-gain of such a system.

References

[1] Francis BA. A Course in $\mathcal{H}_\infty$ Control Theory. Springer; 1987.
[2] Zhou K, Doyle J, Glover K. Robust and Optimal Control. Prentice Hall; 1996.
[3] Goodwin GC, Graebe SF, Salgado ME. Control System Design. Prentice Hall; 2000.
[4] Bruinsma NA, Steinbuch M. A fast algorithm to compute the $\mathcal{H}_\infty$-norm of a transfer function matrix. Systems & Control Letters. 1990;14(4):287–293.
[5] Belur MN, Praagman C. An efficient algorithm for computing the $\mathcal{H}_\infty$ norm. IEEE Transactions on Automatic Control. 2011;56(7):1656–1660.
[6] Kailath T. Linear Systems. Prentice Hall; 1980.
[7] Barenthin M. Complexity Issues, Validation and Input Design for Control in System Identification. Stockholm, Sweden: KTH Royal Institute of Technology; 2008.
[8] Rojas CR, Oomen T, Hjalmarsson H, *et al.* Analyzing iterations in identification with application to nonparametric $\mathcal{H}_\infty$-norm estimation. Automatica. 2012;48(11):2776–2790.
[9] Trefethen LN, Bau III D. Numerical Linear Algebra. SIAM; 1997.
[10] Bubeck S, Cesa-Bianchi N. Regret analysis of stochastic and nonstochastic multi-armed bandit problems. Foundations and Trends in Machine Learning. 2012;5(1):1–122.
[11] Lai TL, Robbins H. Asymptotically efficient adaptive allocation rules. Advances in Applied Mathematics. 1985;6:4–22.

[12] Tu S, Boczar R, Recht B. On the Approximation of Toeplitz Operators for Nonparametric $\mathscr{H}_\infty$-norm Estimation. In: 2018 American Control Conference (ACC); 2018. p. 1867–1872.
[13] Albertos P, Sala A. Iterative Identification and Control. Springer; 2002.
[14] Horn RA, Johnson CR. Matrix Analysis. Cambridge: Cambridge University Press; 1985.
[15] Hjalmarsson H. From experiment design to closed-loop control. Automatica. 2005;41:393–438.
[16] Walhberg B, Barenthin M, Hjalmarsson H. Non-parametric methods for $\mathscr{L}_2$-gain estimation using iterative experiments. Automatica. 2010;46(8): 1376–1381.
[17] Oomen T, van der Maas R, Rojas CR, *et al.* Iterative data-driven $\mathscr{H}_\infty$-norm estimation of multivariable systems with application to robust active vibration isolation. IEEE Transactions on Control Systems Technology. 2014;22(6):2247–2260.
[18] Rallo G, Formentin S, Rojas CR, *et al.* Data-Driven $\mathscr{H}_\infty$-Norm Estimation via Expert Advice. In: Proceedings of the 56th IEEE Conference on Decision and Control (CDC 2017). Melbourne, Australia; 2017.
[19] Pintelon R, Schoukens J. System Identification: A Frequency Domain Approach, 2nd ed. New York: IEEE Press; 2012.
[20] Durrett R. Probability: Theory and Examples, 4th ed. Cambridge University Press; 2010.
[21] Robbins H. Some aspects of the sequential design of experiments. Bulletin of the American Mathematical Society. 1952;58(5):527–535.
[22] Müller MI, Valenzuela PE, Proutiere A, *et al.* A Stochastic Multi-Armed Bandit Approach to Nonparametric $\mathscr{H}_\infty$ Estimation. In: Proceedings of the 56th IEEE Conference on Decision and Control (CDC 2017). Melbourne, Australia; 2017.
[23] Thompson WR. On the likelihood that one unknown probability exceeds another in view of the evidence of two samples. Biometrika. 1933;25(3/4): 285–294.
[24] Kaufmann E, Korda N, Munos R. Thompson Sampling: An Asymptotically Optimal Finite-time Analysis. In: Proceedings of the 23rd International Conference on Algorithmic Learning Theory (ALT). Springer-Verlag; 2012. p. 199–213.
[25] Müller MI, Rojas CR. A Markov Chain Approach to Compute the ℓ_2-Gain of Nonlinear Systems. In: Proceedings of the 18th IFAC Symposium on System Identification (SYSID 2018). Stockholm, Sweden; 2018.
[26] Khalil HK. Nonlinear Systems. 3rd ed. Prentice Hall; 2002.
[27] de Bruijn NG, Erdös P. On a combinatorial problem. Koninklijke Nederlandse akademie van Wetenschappen. 1947;4:758–764.
[28] Rockafellar RT. Convex Analysis. Princeton: University Press; 1970.
[29] Zaman A. Stationarity on finite strings and shift register sequences. The Annals of Probability. 1983;11(3):678–684.

Chapter 9

A comparative study of VRFT and set-membership data-driven controller design techniques: active suspension tuning case

Freddy Valderrama[1,2] and Fredy Ruiz[2]

In this chapter, we compare two approaches to the data-driven control (DDC) design problem. In this framework, the controllers are directly identified from data avoiding the plant identification step. The analyzed approaches are virtual reference feedback tuning (VRFT) and set-membership tuning (SMT) controller. They differ in the assumptions about the noise affecting the experimental data and the criteria to select an optimal controller. The former strategy assumes an stochastic description of the unknown signals, while the latter imposes an unknown but bounded (UBB) noise structure. Both methodologies are described and their main theoretical results are reported. The two approaches are evaluated on an experimental case study, consisting of the controller tuning for an active suspension (AS) system. Three Monte Carlo experiments are performed, where 100 controllers are derived from data affected by measurement noise using both methods, and their performance is evaluated on the experimental test-bench. Results show that both approaches offer a similar performance when the size of the dataset is much larger than the dimension of the controller parameters vector. However, for reduced datasets, the SMT approach gives consistent results while the VRFT method is not able to extract useful information. The same behavior is observed when the two approaches are applied to datasets affected by process disturbances. It is observed that the root mean squared error of the resulting loops can be up to 30 times lower using the set membership method for reduced datasets.

9.1 Introduction

In a standard approach to model-based controller design, two optimization problems are to be solved. First, a system model is built (either from first principles, estimated from experimental data, or a mix of both). Second, a controller is selected or tuned, which is possibly optimal in some sense, with respect to the built model. Sometimes

[1]Escuela de Ciencias Básicas, Tecnología e Ingeniería, Universidad Nacional Abierta y a Distancia, Bogotá, Colombia
[2]Departamento de Ingeniería Electrónica, Pontificia Universidad Javeriana, Bogotá, Colombia

an order-reduction step of the resulting controller is required. The performance of the derived controller relies, on great measure, on the quality of the employed model. Note that, obtaining reliable models of industrial processes turns out to be a complex task, and their construction involves excessive costs [1].

DDC approaches do not rely on plant-model identification, since the available input–output data, experimentally collected from the plant, can be directly used to design the controller, reducing the design problem to a single optimization step. For linear systems, different methods of data-driven controller design have been proposed, see, e.g., [2–5]. Those methods can be classified in two groups: sequential methods, where a series of experiments are required to progressively adjust the controller parameters, and batch methods, where a single experiment allows to obtain controller parameters.

VRFT [4] is a well-known method that does not require multiple experiments when there is no noise or the noise model is unavailable. Several satisfactory applications and extensions of this method have been reported, see, e.g., [6–12]. However, dealing with noisy measurements is a challenge in this approach. When a stochastic formulation of noises is employed, it is found that instrumental variables approach solves the problem, but this requires a second experiment or a plant model, see, e.g., [4,13]. Moreover, convergence results are valid for infinitely long datasets only.

A nonstochastic approach to the controller tuning problem for linear systems was proposed in [14], where UBB noise is assumed and a set-membership approach allows to estimate controller parameters in a single step. This framework has been satisfactorily applied in system identification [15,16], filter design from data [17–19] and direct-inversion-based controller design for nonlinear systems [20,21].

In literature, a comprehensive comparative study of DDC techniques can be found in [1], where the authors performed a theoretical and numerical comparison of controller-design approaches assuming stochastic description of noises. However, the performance of the different solutions was evaluated on simulated conditions, and no evaluation on experimental data was provided.

In this chapter, we compare the VRFT and set-membership approaches to the DDC design problem. First, the main characteristics of both methodologies are presented, and their solutions are reported as algorithmic procedures. Then, the two approaches are evaluated on an experimental case study, consisting of the controller tuning for an AS system. Several controller tuning problems, with different loop requirements and controller structures are solved with both methodologies. Monte Carlo experiments are performed, where 100 controllers are derived from data using both methods and their performance is evaluated on the test bench. The experimental results show that the VRFT controllers are strongly affected by the size of the dataset, while the SMT controllers exhibit good performances, even when they are estimated from reduced datasets.

9.2 Problem statement

In this section, the data-driven controller tuning problem is formulated and some assumptions are presented.

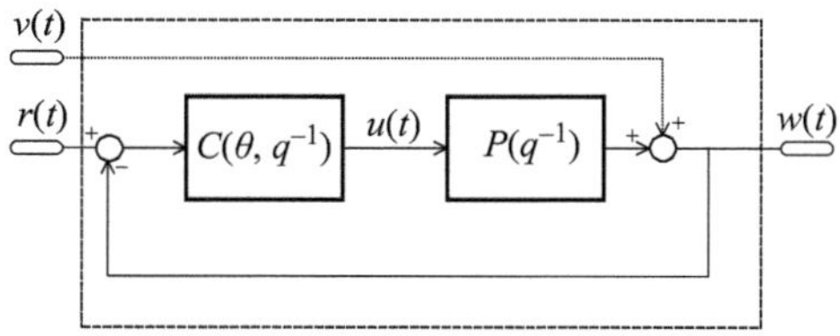

Figure 9.1 Assumed feedback control structure

Consider the discrete-time linear-time invariant single-input single-output feedback control scheme, depicted in Figure 9.1, where q^{-1} denotes backward shift operator, $P(q^{-1})$ is the plant transfer function, $C(\theta, q^{-1})$ is the controller transfer function and θ is vector of controller parameters.

For the system interconnection shown in Figure 9.1, the aim of the controller tuning procedure is to select an optimal controller C^o minimizing some performance criterion. For example, an optimization problem can be stated as

$$C^o(\theta, q^{-1}) = \operatorname{argmin} J(\theta) \tag{9.1}$$

For the cost function

$$J(\theta) = \left\| M(q^{-1}) - \frac{P(q^{-1})C(\theta, q^{-1})}{1 + P(q^{-1})C(\theta, q^{-1})} \right\|_{\mathfrak{s}} \tag{9.2}$$

being $\mathfrak{s}$ a proper system norm and $M(q^{-1})$ a reference model for the closed-loop system, where performance specifications are embedded. If $P(q^{-1})$ is known, problem (9.1) can be seen as a loop-shaping problem and, for $\mathscr{H}_2$, $\mathscr{H}_\infty$ and l_1 norms, known techniques exist to solve it under proper controllability (reachability) and observability (detectability) conditions [22,23].

In the framework analyzed in this chapter, the controller is parametrized as a linear combination of fixed basis functions, that is,

$$C(\theta, q^{-1}) = \sum_{i=1}^{m} \theta_i \beta_i(q^{-1})$$

Then, the controller that solves problem (9.1) is selected from the set:

$$\mathscr{C} = \left\{ C(\theta, q^{-1}) : \theta \in \Theta \subseteq \Re^n \right\}$$

If system $P(q^{-1})$ is unknown, problem (9.1) cannot be solved directly. The common procedure to controller design for unknown plants is to follow a two-step procedure where first a system model $\hat{P}(q^{-1})$ is estimated from data, possibly with some uncertainty measure, and then a controller is obtained solving problem (9.1) for $\hat{P}(q^{-1})$.

The following assumptions define the framework of the data-driven controller tuning problem.

Assumption 9.1. *$P(q^{-1})$ is unknown. The available information on $P(q^{-1})$ is a set of input–output data generated by $P(q^{-1})$, initially at rest,*

$$\mathscr{D} = \{y(t), u(t), t = 1, 2, \ldots, N\} \tag{9.3}$$

where

$$y(t) = \sum_{j=0}^{t} h_j u(t-j) + v(t),$$

h_j are the impulse response coefficients of $P(q^{-1})$ and $v(t)$ is the plant output noise/disturbance.

Assumption 9.2. *An internally stabilizing controller $C(\theta^*, q^{-1})$ exists such that the minimum of* (9.2) *is* 0*, that is,*

$$M(q^{-1}) = \frac{P(q^{-1})C(\theta^*, q^{-1})}{1 + P(q^{-1})C(\theta^*, q^{-1})}$$

Assumption 9.2 is required for the derivation of theoretical results of both controller-tuning methods.

Based on the previous assumption, the DDC tuning problem can be stated as follows:

Problem 9.1. Data-driven controller tuning: *given a dataset $\mathscr{D}$ generated as in Assumption 9.1, a reference model $M(q^{-1})$, a set of basis functions $\beta_i, i = 1, \ldots, m$, and some assumptions about noise/disturbance $v(t)$, find a vector $\hat{\theta}$ that approximately solves problem* (9.1).

9.3 Controller tuning from data

In this section, the two approaches to solve the data-driven controller tuning problem are described.

9.3.1 Set-membership approach

In this method, the noise sequence is modeled as an UBB signal without any statistical assumption about it, as defined in the following assumption:

Assumption 9.3. *Noise $v(t)$ is an (UBB) signal, such that*

$$\|v(t))\|_{\ell_p} \leq \varepsilon_p$$

with $\ell_p \in [2, \infty]$. In this framework, energy and amplitude limited noise sequences can be considered.

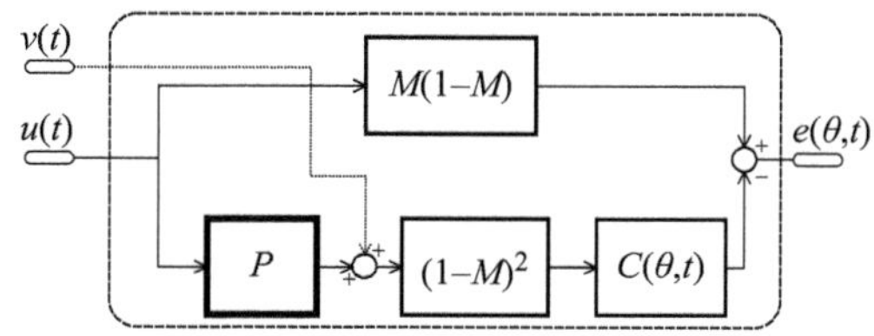

Figure 9.2 Set membership tuning problem block diagram

Recall that the argument of the cost function to be minimized in (9.2) is the model matching error-transfer function,

$$E_m(q^{-1}) = M(q^{-1}) - \frac{P(q^{-1})C(\theta,q^{-1})}{1+P(q^{-1})C(\theta,q^{-1})}. \tag{9.4}$$

Using Assumption 9.2, the model matching transfer function becomes

$$E_m = \frac{C(\theta^*)P}{1+C(\theta^*)P} - \frac{C(\theta)P}{1+C(\theta)P} = \frac{C(\theta^*)P - C(\theta)P}{(1+C(\theta^*)P)(1+C(\theta)P)} \tag{9.5}$$

where the backward shift operator q^{-1} has been removed for simplicity.

Assuming that the sensitivity function for any controller close to the optimum approximates the desired one, i.e.,

$$\frac{1}{1+PC(\theta)} \approx \frac{1}{1+PC^*(\theta)} = (1-M), \tag{9.6}$$

it follows that

$$E_m \approx M(1-M) - C(\theta)\,(1-M)^2\,P. \tag{9.7}$$

Now, in order to avoid using the plant-transfer function in the evaluation of the model-matching error, the input signal $u(t)$ is applied to the system in (9.7), as depicted in the block diagram of Figure 9.2. In this case, the following model-free equation is obtained:

$$e(\theta,t) = M(1-M)u(t) - C(\theta)\,(1-M)^2\,Pu(t) \tag{9.8}$$

$$= M(1-M)u(t) - C(\theta)\,(1-M)^2\,(y(t)-v(t)) \tag{9.9}$$

$$= M(1-M)u(t) - C(\theta)\,(1-M)^2\,y(t) + d(t) \tag{9.10}$$

where $d(t) = -C(\theta)\,(1-M)^2\,v(t)$.

From Assumption 9.3, for any controller $C(\theta)$ guaranteeing an internally stable loop, the noise $d(t)$ can be bounded as

$$\|d(t)\|_{\ell_p} \leq \left\|C(\theta)\,(1-M)^2\right\|_{\ell_p,\ell_p} \varepsilon_p = \delta_p \tag{9.11}$$

where $\left\|C(\theta)\,(1-M)^2\right\|_{\ell_p,\ell_p}$ is the proper induced norm of the system relating $v(t)$ and $d(t)$. Note that, for $p=2$, the induced norm is the H_∞ norm of the system and for $p=\infty$, the induced norm is the ℓ_1 norm of the system.

From the previous analysis, we are in position to define the feasible controller set (FCS), that is, the set of all controllers $C(\theta)$ that are compatible with hypotheses and data.

Definition 9.1. *FCS:*

$$FCS = \left\{ C(\theta) \in \mathscr{C} : \left\| M(1-M)u(t) - C(\theta)\,(1-M)^2 y(t) \right\|_{\ell_p} \leq \delta_p \right\} \tag{9.12}$$

Notice that, in a noise-free case, i.e., $d(t) = 0$, it is possible to find a controller $C(\theta) = C(\theta^*)$, such that $e_m(t) = 0$. However, in any practical setting, we have $\delta_p > 0$, then the set-membership approach is suitable to find a set containing all the controllers $C(\theta, t)$ that are compatible with data, noise bound δ_p and a priori information on the reference model M and the controller structure defined by the basis set. Similarly, the feasible parameters set (FPS) is the set of all parameters $\theta \in \Theta$ that are compatible with hypotheses and data:

$$FPS = \left\{ \theta \in \Theta : \left\| M(1-M)u(t) - C(\theta)\,(1-M)^2 y(t) \right\|_{\ell_p} \leq \delta_p \right\} \tag{9.13}$$

Theorem 9.1. *Given a dataset $\mathscr{D}$, a reference model $M(q^{-1})$ and a set of basis functions $[\beta_1(q^{-1}),\ \beta_2(q^{-1}),\ \ldots,\ \beta_m(q^{-1})]$. If $\delta_p \geq \delta_p^*$, for δ_p^* the solution to the convex optimization problem*

$$\delta_p^* = \min_{\theta \in \Theta, \varepsilon \in \Re} \delta \tag{9.14}$$

$$s.t.$$

$$\left\| M(1-M)u(t) - \sum_{i=1}^{m} \theta_i \beta_i\,(1-M)^2 y(t) \right\|_{\ell_p} \leq \delta$$

$$\delta \geq 0$$

Then, $FPS \neq \emptyset$.

Remark: A non-empty FPS guarantees a non-empty FCS.

The previous theorem gives a tool to determine if a set of a priori hypotheses is compatible with the available dataset. For example, it allows to evaluate if a reference model is achievable with the selected basis functions with acceptable error. More details on this can be found in Section 9.4.

Given a non-empty FPS, the problem of choosing a vector of parameters, and thus a controller, inside this set arises. According to the set membership literature, [24], any element inside is compatible with the available information and is a valid candidate as optimal controller $C(\theta^*, q^{-1})$.

Working within the parameter space, for any vector $\hat{\theta}$ define its estimation error as

$$E_e(\hat{\theta}) = \|\theta^* - \hat{\theta}\|_r \tag{9.15}$$

for any vector norm r. However, it cannot be evaluated because the optimal parameters vector θ^* is unknown. The worst case estimation error of $\hat{\theta}$ is defined as

$$E_{WC}(\hat{\theta}) = \max_{\theta \in FPS} \|\theta - \hat{\theta}\|_r \tag{9.16}$$

where E_{WC} is the maximum distance between the selected solution and any parameters vector compatible with the available information. An estimate $\hat{\theta}^o$ is optimal if it minimizes E_{WC}, i.e.,

$$E_{WC}(\hat{\theta}^o) = \min_{\vartheta \in \Theta} E_{WC}(\vartheta)$$

The following theorem gives a method to find an optimal estimate of θ^* in a worst case setting.

Theorem 9.2. *The vector*

$$\theta^c = \frac{\overline{\theta} + \underline{\theta}}{2}$$

is the Chebyshev center of the FPS and then an optimal estimate of θ^* for $r = \infty$, being $\overline{\theta}$ and $\underline{\theta}$ hard bounds of the parameters feasible intervals, defined as

$$\begin{aligned} \overline{\theta_i} &= \max \theta_i \\ &s.t. \\ \theta &\in FPS \end{aligned} \tag{9.17}$$

and

$$\begin{aligned} \underline{\theta_i} &= \min \theta_i \\ &s.t. \\ \theta &\in FPS \end{aligned} \tag{9.18}$$

for $i = 1, 2, \ldots, m$.

Remark: Note that θ^c belongs to the *FPS*, then

$$C(\theta^c, q^{-1}) \in FCS$$

and therefore $C(\theta^c, q^{-1})$ is an interpolatory estimate for any system norm, i.e.,

$$\|C^*(\theta^*, q^{-1}) - C(\theta^c, q^{-1})\|_{\mathfrak{s}} \leq 2\mathbf{RI}$$

RI being the radius of information, i.e., the lowest achievable uncertainty provided by the Chebyshev center of the *FCS* measured in a given system norm [24].

Based on the results presented above, the following algorithm summarizes the proposed procedure for the tuning of controllers for unknown linear plants. It has been named SMT.

Algorithm 9.1. SMT algorithm

1. *Collect a dataset $\mathscr{D}$ performing an experiment starting with plant at rest.* Note that the assumptions on noise do not require open-loop operation.
2. *Select a proper reference model $M(q^{-1})$ and basis functions $\beta_i(q^{-1})$.*
3. *Solve the feasibility problem in* (9.14). *If the lower bound on noise δ^* is acceptable, go to next step; otherwise, go back to step 2.*
4. *Build the estimate θ^c solving problems* (9.17) *and* (9.18) *for each component of the parameter vector.*
5. *Select as controller the system*

$$C^c(\theta^c, q^{-1}) = \sum_{i=1}^{m} \theta_i^c \beta_i(q^{-1})$$

9.3.2 Tuning via VRFT

As mentioned in the Introduction, VRFT is a well-known method for tuning controllers avoiding to build a model of the plant. Such method is deeply explained in [4], notwithstanding, for the sake of completeness in this section, the main features of the method are explained. In this framework, as in the set membership case, the aim is to modify problem (9.1) avoiding the requirement of the knowledge of the plant model.

The solution is to build a virtual reference signal $r_v(t)$ from the reference model and the available output measurements in $\mathscr{D}$, i.e.,

$$r_v(t) = M^{-1}(q^{-1})y(t),$$

From this signal, the virtual error can be obtained as

$$e_v(t) = r_v(t) - y(t).$$

Figure 9.3 shows the resulting scheme. Note that $e_v(t)$ and $u(t)$ are the input and output of the controller to be identified. The model-based cost function in (9.2), for the case of a $\mathscr{H}_2$ norm, is replaced by the signal-based cost function

$$J_l(\theta) = \frac{1}{N}\sum_{t=1}^{N}\left(u_l(t) - C(\theta, q^{-1})e_l(t)\right)^2, \tag{9.19}$$

following a prediction-error-identification method approach, see, e.g., [25]. The subscript l indicates that signals $e_v(t)$ and $u(t)$ are filtered by the system

$$L(q^{-1}) = (1 - M(q^{-1}))M(q^{-1}), \tag{9.20}$$

i.e., $e_l(t) = L(q^{-1})e_v(t)$ and $u_l(t) = L(q^{-1})u(t)$. It is shown in [4] that this filter guarantees an asymptotic convergence of the argument that minimizes (9.19) to the minimizer of (9.1).

When measurement noise $v(t)$ is considered, the minimizer of (9.19) becomes a biased estimate. In order to get unbiased results, it is necessary to perform a second

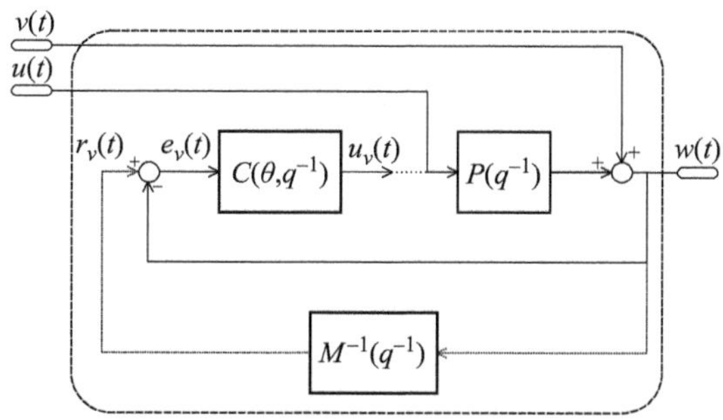

Figure 9.3 VRFT tuning problem block diagram

experiment using the same input (repeated experiment) or to obtain a model of the plant (system identification). These approaches are not evaluated in this chapter because they require more information than the set membership approach.

The following algorithm summarizes the VRFT method.

Algorithm 9.2. VRFT algorithm

1. *Collect a dataset $\mathscr{D}$ performing an experiment starting with plant at rest.* A low noise condition should be guaranteed in order to avoid biased results.
2. *Select a proper reference model $M(q^{-1})$ and basis functions $\beta_i(q^{-1})$.*
3. *Obtain the virtual reference and error signals*

   $$r_v(t) = M^{-1}y(t)$$

   and

   $$e_v(t) = y(t) - r_v(t).$$

4. *Filter the virtual error $e_v(t)$ and plant input $u(t)$ as*

   $$e_l(t) = L(q^{-1})e(t), u_l = L(q^{-1})u(t)$$

   with the system in (9.20).
5. *Solve the least squares problem*

$$\theta^{VR} = \arg\min_{\theta \in \Re^m} \frac{1}{N} \sum_{t=1}^{N} \left(u_l(t) - \sum_{i=1}^{m} \theta_i \beta_i(q^{-1}) e_l(t) \right)^2. \tag{9.21}$$

6. *Select as controller the system*

$$C^{VR}(\theta^{VR}, q^{-1}) = \sum_{i=1}^{m} \theta_i^{VR} \beta_i(q^{-1})$$

9.4 Active suspension tuning case study

The two data-driven controller tuning strategies presented in the previous section are evaluated experimentally using an AS system. The design of controlled suspension systems for road vehicles aims to enhance the vehicle performances with regard to comfort and road handling. AS systems have the ability to store, dissipate and introduce energy to the system. As a result, the trade-offs among conflicting design goals can be better resolved. Comprehensive studies have been published that evaluate the performances of active and semiactive suspensions (see [26,27]).

In this experimental case study, an AS platform from Quanser is employed. The AS is a bench-scale plant to emulate a quarter-car model, controlled by an active mechanism. As is shown in Figure 9.4, the plant consists of three floors/plates on top of each other. The top floor resembles the vehicle body, the middle plate represents the wheel and the bottom plate corresponds to the road. The plates are connected through springs that model a passive suspension and the tire elasticity. A DC motor is also standing between the top and middle plates to emulate the AS mechanism.

The chassis and the wheel are modeled as rigid bodies and static linear characteristics are assumed for the suspension. The spring-mass model is shown in Figure 9.5. The parameters characterizing the model are the sprung mass M_s, representing the mass of the vehicle body while the unsprung mass, M_{us}, represents the tire in the quarter-car model. The spring K_{us} and the damper B_{us} model the stiffness of the tire in contact with the road. The spring K_s and the damper B_s support the body weight over the tire. Values of parameters are shown in Table 9.1.

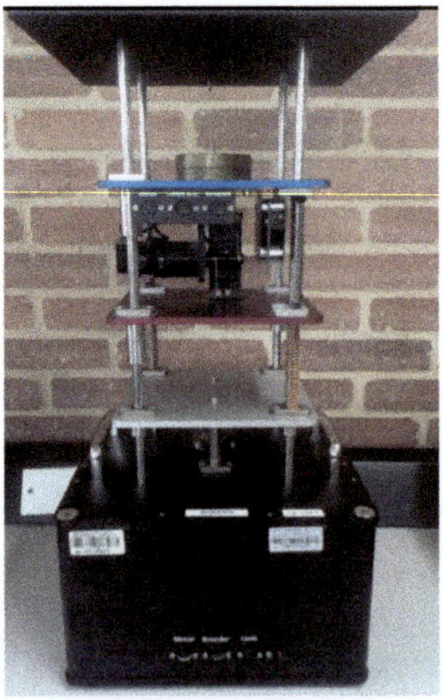

Figure 9.4 Active suspension platform

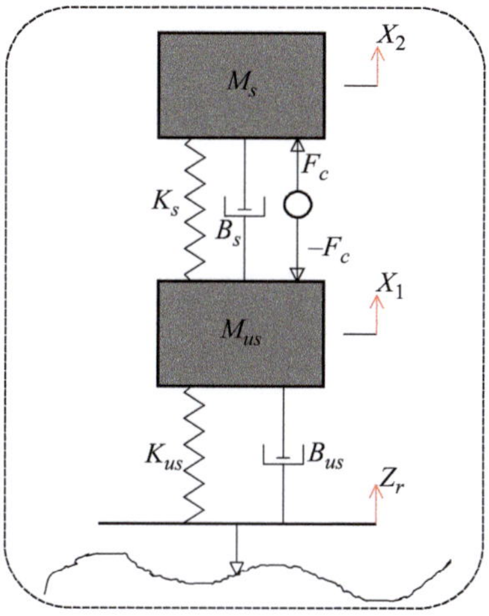

Figure 9.5 Active suspension model

Table 9.1 Active suspension parameters

Parameter	Value	Unit
M_s	2.45	kg
K_s	960	N/m
B_s	7.5	N s/m
M_{us}	1	kg
K_{us}	2,500	N/m
B_{us}	0.01	N s/m

9.4.1 Controller tuning problem

The objectives of controlled suspensions are to enhance comfort and handling. Comfort is related to the sprung mass acceleration, while handling is assured by the contact force between the unsprung mass and the road. In this work, a tracking problem is posed, where the aim is to follow a given suspension displacement. The force F_c (provided by the DC motor) is the manipulated variable (u), measured in Newtons (N), and the output variable is the *suspension travel* ($y = x_2 - x_1$) measured in cm (i.e., separation between the top and middle plates).

The open-loop step response of the actual plant is shown in Figure 9.6. The frequency response of the lumped component model of the system is shown in Figure 9.7, where the nominal parameters in Table 9.1 are employed. The shape of the step and frequency responses show that the AS exhibits an oscillating under-damped behavior.

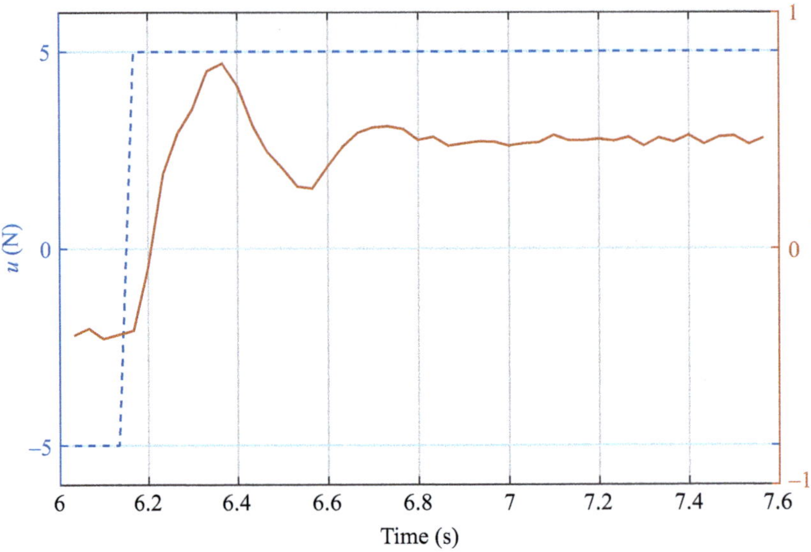

Figure 9.6 Open-loop step response of the active suspension

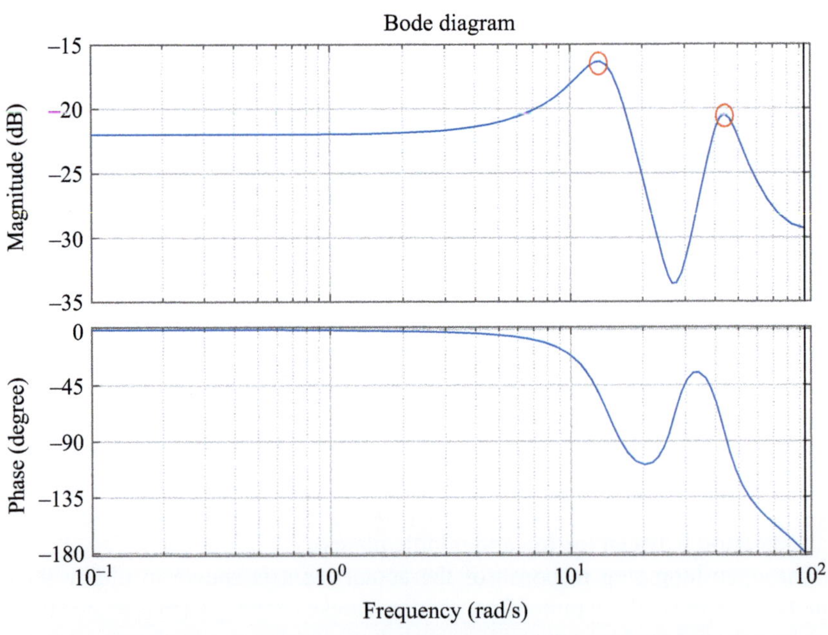

Figure 9.7 Active suspension nominal frequency response

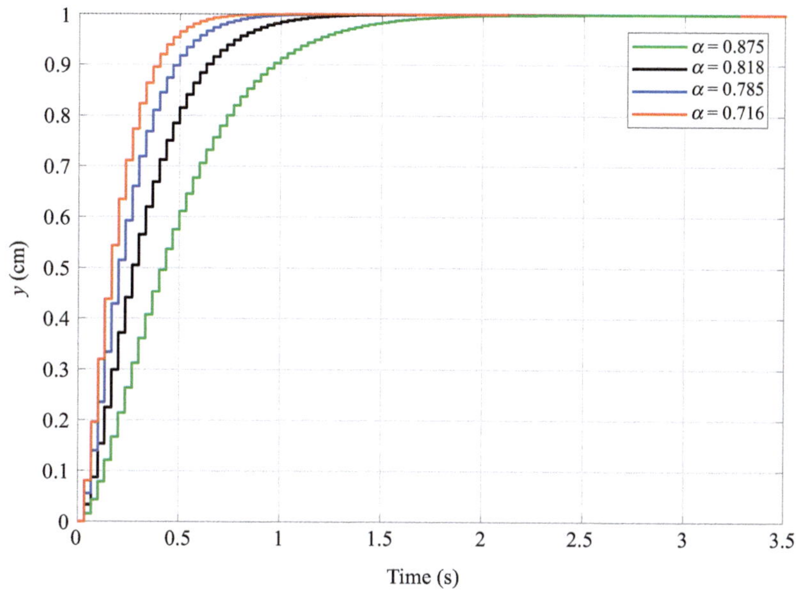

Figure 9.8 Step response of reference models $M(q^{-1})$ for different α values

The loop performance specifications are defined as a critically damped second order reference model M, parametrized by α, as shown in (9.22).

$$M(q^{-1}) = \frac{(1-\alpha)^2 q^{-1}}{1 - 2\alpha q^{-1} + \alpha^2 q^{-2}} \tag{9.22}$$

Note that α indicates the location of the poles defining the desired loop speed and bandwidth. The desired step response for some values of α are shown in Figure 9.8. The sampling frequency is fixed as $F_s = 30$ Hz, that is, one decade after the second resonance peak of the model.

In accordance with the above conditions and requirements, and in order to test the performance of the VRFT and SMT methods, two sets of experiments have been performed. The first one is a Monte Carlo experiment, where the effect of the length of the dataset and the repetitiveness of the methods are evaluated. The second one evaluates the behavior of the approaches in front of process disturbances. In all the tests, $\ell_p = 2$ is fixed for the SMT algorithm (SMT_2), in order to have similar noise treatments in both approaches.

9.4.2 Monte Carlo experiment

In this set of experiments, 100 controllers have been obtained and tested on the AS, for each tuning method. Three cases are addressed, $N = 1{,}000$, $N = 200$ and $N = 100$.

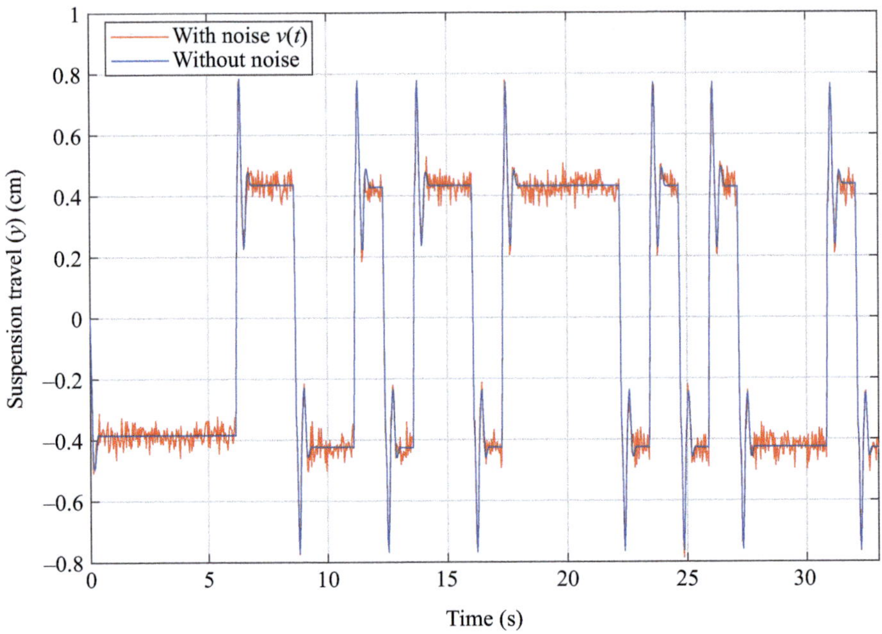

Figure 9.9 Sample of suspension travel with added output noise

Datasets $\mathscr{D}$ required for both algorithms are obtained as follows: first a PRBS signal with N samples is used as input (u) and suspension travel ($x_2 - x_1$) is measured. Output y is obtained adding white noise $v(t)$ with normal distribution. One hundred noise realizations $v(t)$ allow to obtain 100 datasets. Note that, the measured signal $x_2 - x_1$ has very low noise because the sensor resolution is 4.87 μm. For each N, the noise variance is selected such that the signal-to-noise ratio (SNR) is approximately 20 dB. The SNR is calculated as

$$SNR = 10 \log \frac{\sum_{t=1}^{N} y(t)^2}{\sum_{t=1}^{N} v(t)^2}.$$

As an example, one of the realizations with $N = 1{,}000$ is shown in Figure 9.9.

The controllers are selected from an extended PID structure with basis functions β_i of the form

$$\beta_i(q^{-1}) = \frac{q^{1-i}}{1 - q^{-1}}, \quad i = 1, 2, 3, \ldots, m$$

The first step in the tuning procedure is to define the number of basis that should be employed. Using one noise realization $v(t)$, datasets $\mathscr{D}$ for different N values are

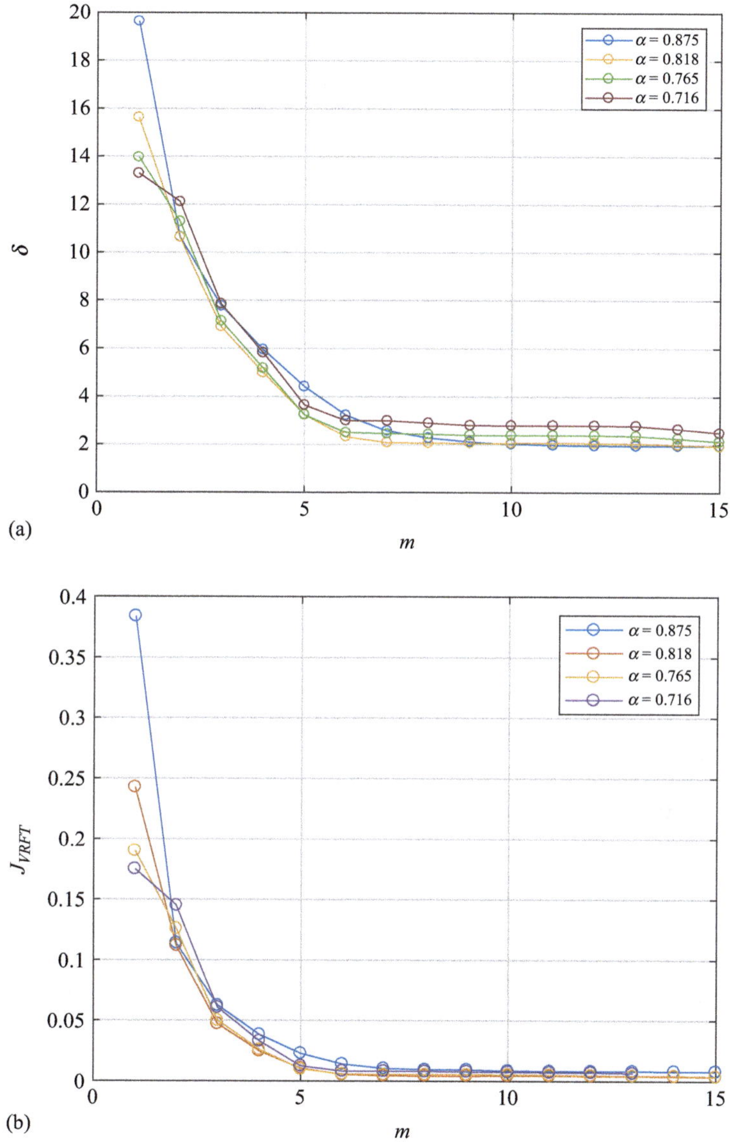

Figure 9.10 Controller structure selection results. Feasibility bounds for SMT and optimal cost for VRFT. $N = 1,000$. (a) SMT_2 and (b) VRFT

build, and for the SMT_2, feasibility problem (9.14) is solved for different reference model parameters α and increasing m values. The same dataset is used for the VRFT method. In this case, for each m value, the cost function in (9.19) is evaluated. Figures 9.10 and 9.11 show the results for $N = 1,000$ and $N = 100$, respectively.

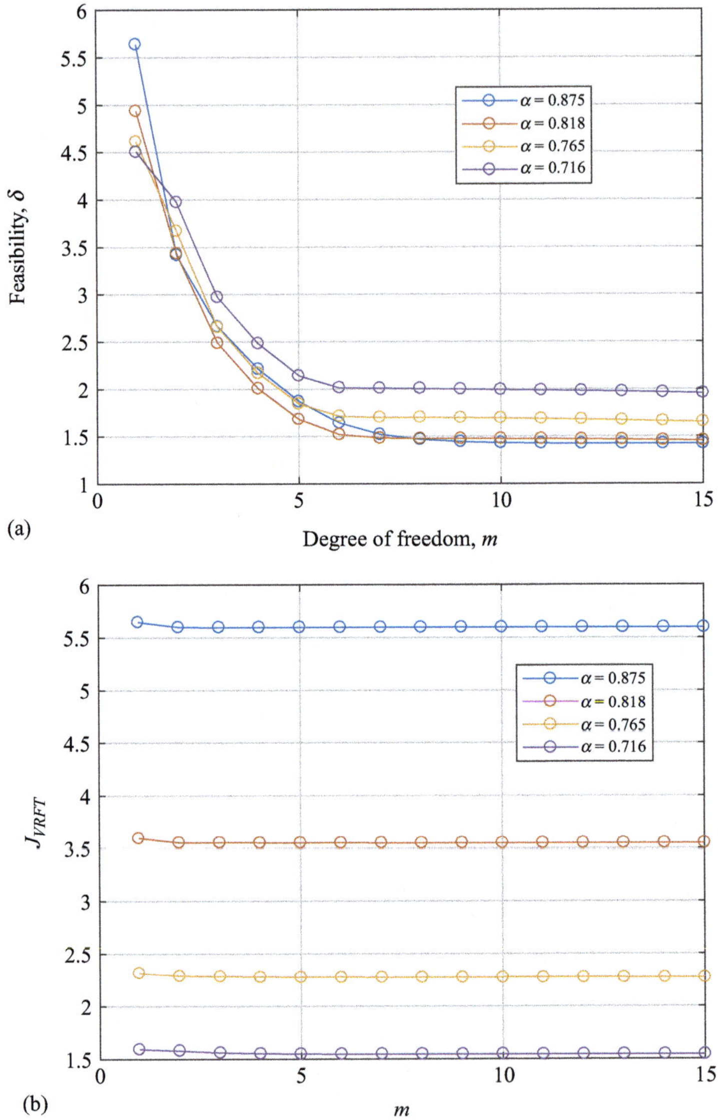

Figure 9.11 Controller structure selection results. Feasibility bounds for SMT and optimal cost for VRFT. N = 100. (a) SMT_2 and (b) VRFT

Note that $m = 7$ is a reasonable number of bases functions when $N = 1{,}000$ for both approaches. Choosing more parameters is unnecessary and would not decrease the error $\delta_2(\theta)$ in (9.11) nor the cost $J_l(\theta)$. On the other hand, choosing fewer parameters would produce an increment of the error in both cases. However, when $N = 100$, in Figure 9.11, it can be noted that the selection of the number of bases is

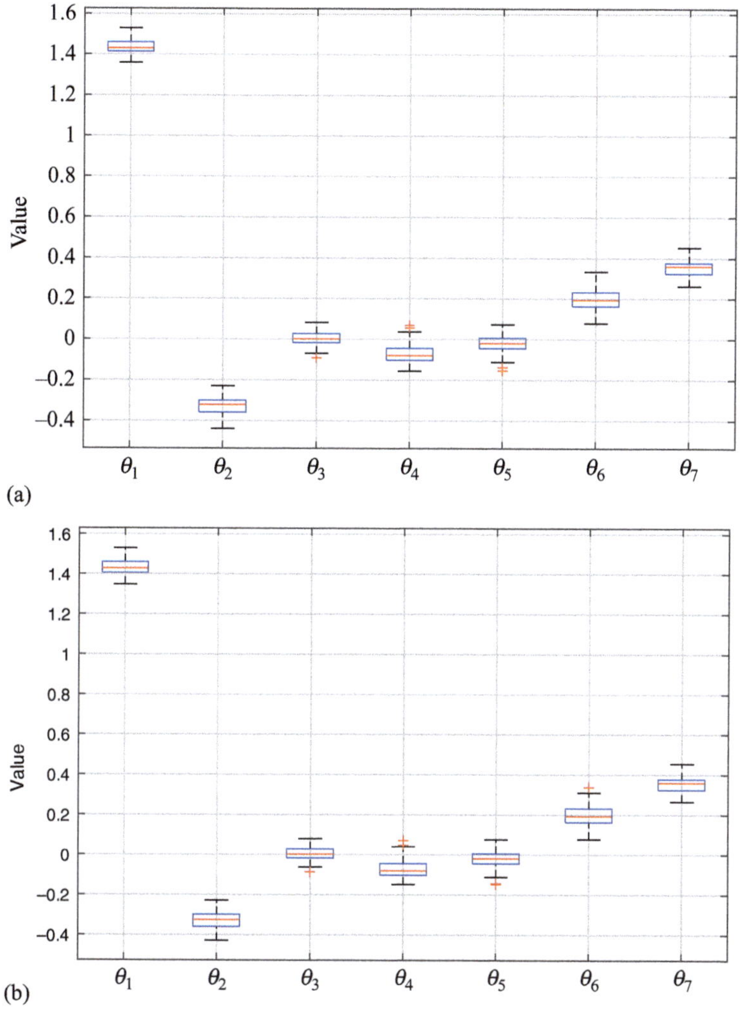

Figure 9.12 Controller parameters dispersion from Monte Carlo experiment. $N = 1{,}000$. (a) SMT_2 and (b) VRFT

not clear for the VRFT approach, while such selection for the case of SMT_2 remains reasonable for all α values of the model reference.

Once the structure of the controller is defined, the estimation of the optimal parameters is performed. For $m = 7$, $N = 1{,}000$ and $\alpha = 0.765$ (i.e., rise time 1 s) a noise bound $\hat{\delta}_2 = 2.7$ is fixed to ensure feasibility (i.e., 10% larger than δ_2^*). Both algorithms are executed for each one of 100 dataset $\mathcal{D}$. The obtained controller parameters dispersion is shown in Figure 9.12. Note that, for this large dataset, both algorithms

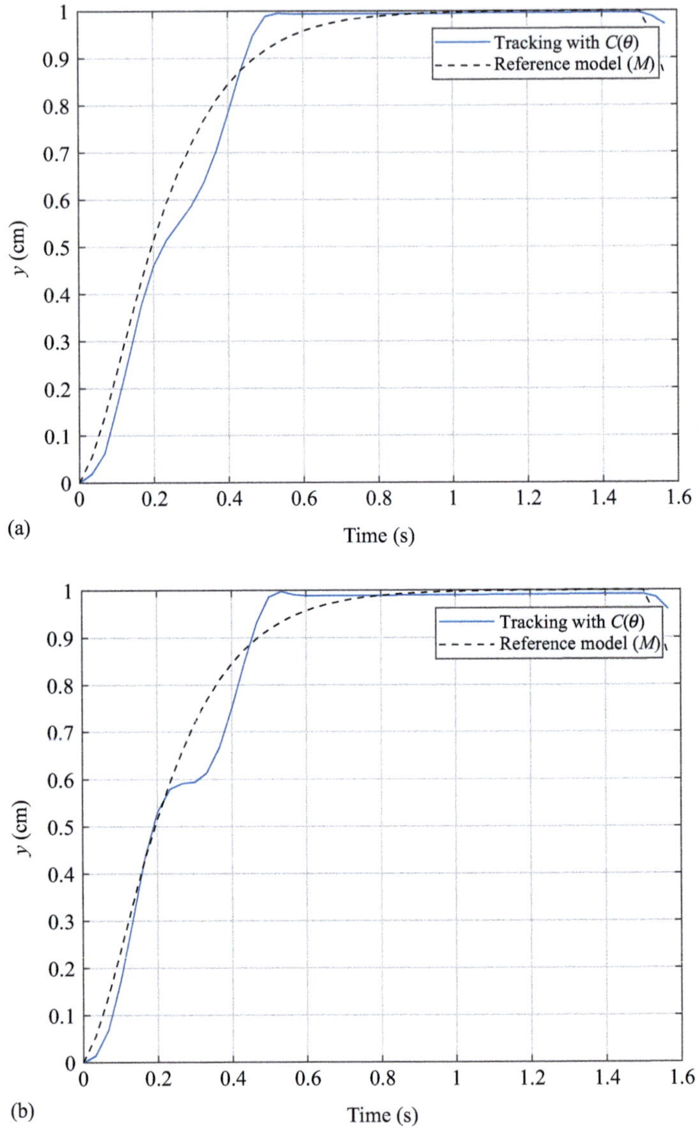

Figure 9.13 Sample closed-loop step response. N=1,000. (a) SMT_2 and (b) VRFT

offer a similar performance with small variations in the obtained parameters. Figure 9.13 shows the experimental step response for one of the obtained controllers in each case.

Finally, for each dataset size N, the performance of the 100 estimated controllers, with each method, is measured experimentally. The quality of the control action is

Table 9.2 Performance results of Monte Carlo experiment

Case	**Measure**	**SMT_2**	**$VRFT$**
$N = 1{,}000$	$\overline{RMSE}, \max(RMSE)$	$0.0031, 0.0039$	$0.0032, 0.0040$
	$\overline{M_e}, \max(M_e)$	$0.1503, 0.1788$	$0.1517, 0.1694$
$N = 200$	$\overline{RMSE}, \max(RMSE)$	$0.0021, 0.0032$	$0.0202, 0.0410$
	$\overline{M_e}, \max(M_e)$	$0.1482, 0.1904$	$0.2889, 0.3581$
$N = 100$	$\overline{RMSE}, \max(RMSE)$	$0.0018, 0.0028$	$0.0566, 0.1100$
	$\overline{M_e}, \max(M_e)$	$0.1051, 0.1555$	$0.3957, 0.5090$

measured as the maximum error (M_e) and the root mean square error ($RMSE$) of the closed-loop step response. The average and worst case results are shown in Table 9.2. As can be seen, for the case of $N = 1{,}000$, the results are comparable, but in the cases $N = 200$ and $N = 100$, both performance measures obtained via SMT_2 algorithm are better than those achieved via $VRFT$ algorithm. Furthermore, it is highlighted that when $N = 100$ the average $RMSE$ of the SMT_2 controllers is about 30 times smaller than the one obtained via $VRFT$. Note also that the behavior of the SMT_2 controllers for small datasets ($N = 100$ and $N = 200$) is superior to the $VRFT$ controllers in a worst case sense. The maximum $RMSE$ for the SMT_2 solutions, among the 100 experiments, is more than ten times lower than those of the $VRFT$ ones, and the maximum peak error (M_e) is almost half.

9.4.3 Process disturbance experiment

In the previous experiment, the tuning methods were evaluated taking into account output measurement noise only. Although, process disturbances are also present in any practical situation. The AS system is able to simulate a road profile by means of a second DC motor driving the lower plate. In this case, the road profile shown in Figure 9.14(a) (filtered Gaussian noise) is applied to the system during the dataset generation experiment, with the same PRBS input $u(t)$ of the first setup. Datasets with lengths $N = 100$, $N = 150$, $N = 200$, $N = 250$ and $N = 300$ are used for testing. As an example, the output signal $y(t)$ obtained when $N = 300$ is depicted in Figure 9.14(b). $m = 7$ is selected for both algorithms, and $\delta_2 = 3.5$ is selected for the SMT_2 solution.

The closed-loop step-response achieved by the controllers are depicted in Figure 9.15. The maximum error (M_e) and $RMSE$ of each controller is shown in Table 9.3. The results indicate that both performance measures obtained via the SMT_2 method are better than those achieved with the $VRFT$ approach, for any N value. Moreover, the step responses show that the SMT_2 method is very robust to the size of the dataset. Similar to the previous experiment, it is highlighted that when $N = 100$, the $RMSE$ is about 16 times smaller using SMT_2 controller than the $VRFT$ one. Also

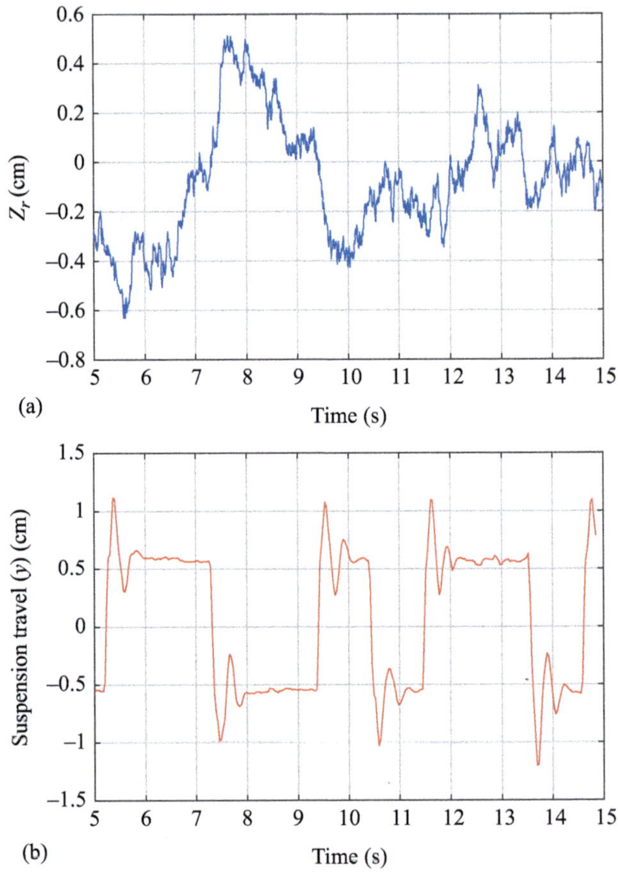

Figure 9.14 Dataset for process disturbance experiment. $N = 300$. (a) Road profile applied as process disturbance and (b) suspension travel for PRBS input and process disturbance

note that, the peak error (M_e) is about two times smaller for the SMT_2 solution than the *VRFT* one, for all N values.

9.5 Conclusions

We have compared the VRFT and SMT control design approaches. They do not rely on a plant model to tune the controllers, but the available input–output data, experimentally collected from the plant, is directly used to design the controller. Although they differ in the considered noise model, it is found that that both approaches offer

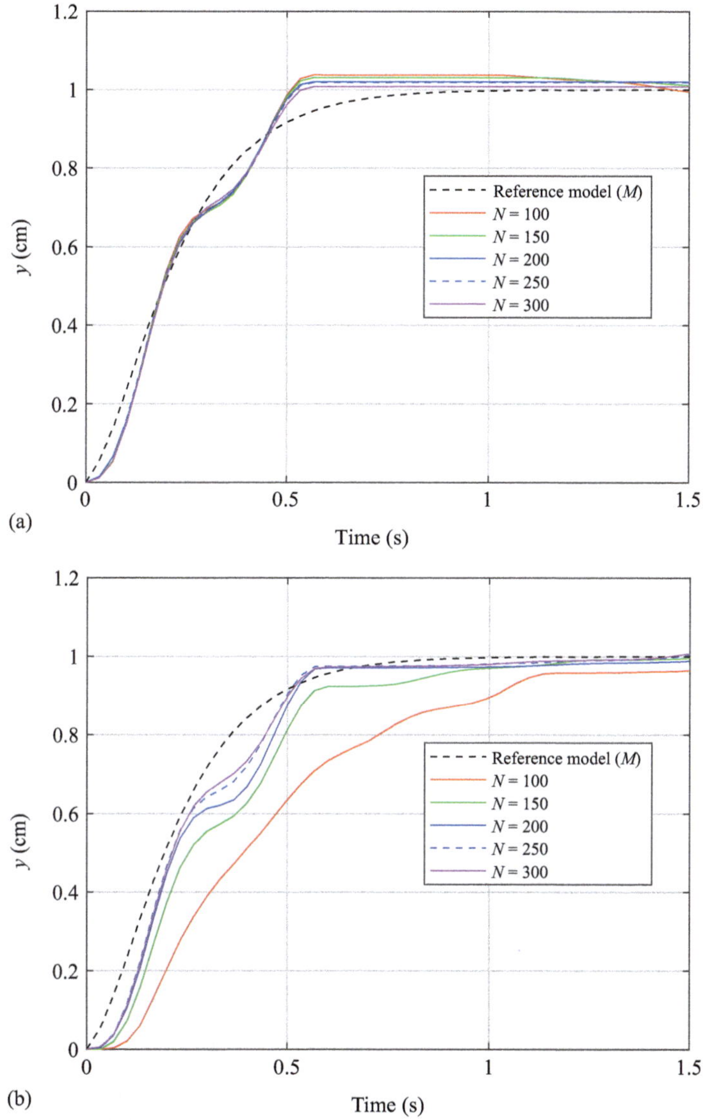

Figure 9.15 Closed-loop step responses: (a) SMT_2 and (b) VRFT

a similar performance when the size of the dataset is much larger than the dimension of the controller parameter vector.

An extended experimental evaluation on an AS system has shown that, using the same information about the plant to be controlled, the VRFT controllers are strongly affected by the size of the dataset, while the SMT controllers exhibit good

Table 9.3 Performance results of process disturbance experiment

Case	Measure	SMT_2	*VRFT*
$N = 100$	*RMSE*	0.0023	0.0380
	M_e	0.0948	0.3385
$N = 150$	*RMSE*	0.0021	0.0091
	M_e	0.0892	0.2180
$N = 200$	*RMSE*	0.0016	0.0045
	M_e	0.0797	0.1765
$N = 250$	*RMSE*	0.0015	0.0026
	M_e	0.0769	0.1274
$N = 300$	*RMSE*	0.0012	0.0023
	M_e	0.0881	0.1230

performances, even when they are estimated from reduced datasets. For a samples to parameters ratio below 15, the VRFT method is not able to detect the required controller structure, and the performance of the derived controllers can be up to 30 times worse than the SMT results.

Acknowledgment

F. Valderrama received a doctoral scholarship from Gobernación de Boyacá, Colombia (call 733 Colciencias).

References

[1] Van Heusden K, Karimi A, Söderström T. On identification methods for direct data-driven controller tuning. International Journal of Adaptive Control and Signal Processing. 2011;25(5):448–465.

[2] Hjalmarsson H. Iterative feedback tuning overview. International Journal of Adaptive Control and Signal Processing. 2002;16(5):373–395.

[3] Karimi A, Miskovic L, Bonvin D. Iterative correlation-based controller tuning: frequency-domain analysis. In: Proceedings of the 41st IEEE Conference on Decision and Control 2002. vol. 4; 2002. p. 4215–4220.

[4] Campi MC, Lecchini A, Savaresi SM. Virtual reference feedback tuning: a direct method for the design of feedback controllers. Automatica. 2002; 38(8):1337–1346.

[5] Van Helvoort J, De Jager B, Steinbuch M. Direct data-driven recursive controller unfalsification with analytic update. Automatica. 2007;43(12): 2034–2046.

[6] Formentin S, Savaresi SM, Re LD. Non-iterative direct data-driven controller tuning for multivariable systems: theory and application. IET Control Theory Applications. 2012;6(9):1250–1257.

[7] Roman R, Radac M, Precup R. Multi-input–multi-output system experimental validation of model-free control and virtual reference feedback tuning techniques. IET Control Theory Applications. 2016;10(12): 1395–1403.

[8] Campi MC, Savaresi SM. Direct nonlinear control design: the virtual reference feedback tuning (VRFT) approach. IEEE Transactions on Automatic Control. 2006;51(1):14–27.

[9] Corleta A, Lorenzini C, Flores JV, *et al.* Data-driven control design applied to uninterruptible power supplies. In: 2016 IEEE Conference on Control Applications (CCA); 2016. p. 1312–1317.

[10] Hedrea E, Radac M, Precup R. Virtual reference feedback tuning for position control of a twin rotor aerodynamic system. In: 2016 IEEE 11th International Symposium on Applied Computational Intelligence and Informatics (SACI); 2016. p. 57–62.

[11] Valderrama F, Hernandez O. Data driven control design applied to single-phase inverters t. In: 2017 IEEE 3rd Colombian Conference on Automatic Control (CCAC); 2017. p. 1–6.

[12] Filho RS, Eckhard D, da Silva GRG, *et al.* Application of virtual reference feedback tuning to a non-minimum phase pilot plant. In: 2016 IEEE Conference on Control Applications (CCA); 2016. p. 1318–1323.

[13] Rojas JD, Tadeo F, Vilanova R. Control of a pH neutralization plant using the VRFT framework. In: 2010 IEEE International Conference on Control Applications; 2010. p. 926–931.

[14] Valderrama F, Ruiz F. Controller design from data under UBB noise. In: Proc Am Control Conf; 2014. p. 5103–5108.

[15] Milanese M, Novara C. Set membership identification of nonlinear systems. Automatica. 2004;40(6):957–975.

[16] Milanese M, Taragna M. $\mathscr{H}_\infty$ set membership identification: a survey. Automatica. 2005;41(12):2019–2032.

[17] Milanese M, Ruiz F, Taragna M. Direct data-driven filter design for uncertain LTI systems with bounded noise. Automatica. 2010;46(11):1773–1784.

[18] Ruiz F, Novara C, Milanese M. Direct design from data of optimal filters for LPV systems. Systems & Control Letters. 2010;59(1):1–8.

[19] Novara C, Ruiz F, Milanese M. Direct filtering: a new approach to optimal filter design for nonlinear systems. IEEE Transactions on Automatic Control. 2013;58(1):86–99.

[20] Formentin S, Novara C, Savaresi SM, *et al.* Data-driven inversion-based control of nonlinear systems. IFAC-PapersOnLine. 2015;48(28):1343–1348. 17th IFAC Symposium on System Identification SYSID 2015.

[21] Tanaskovic M, Fagiano L, Novara C, *et al.* On-line direct control design for nonlinear systems. IFAC-PapersOnLine. 2015;48(28):144–149. 17th IFAC Symposium on System Identification SYSID 2015.

[22] Skogestad S, Postlethwaite I. Multivariable Feedback Control: Analysis and Design. 2nd ed. Chichester, UK: JohnWiley & Sons; 2005.
[23] Zhou K, Doyle JC. Essentials of Robust Control. Upper Saddle River, NJ, USA: Prentice Hall Inc; 1998.
[24] Milanese M, Norton J, Piet-Lahanier H, *et al.* Bounding Approaches to System Identification. New York: Plenum Press; 1996.
[25] Ljung L. System Identification: Theory for the User. 2nd ed. Upper Saddle River, NJ, USA: Prentice Hall Inc; 1999.
[26] Lu J, DePoyster M. Multiobjective optimal suspension control to achieve integrated ride and handling performance. IEEE Transactions on Control Systems Technology. 2002;10(6):807–821.
[27] Savaresi SM, Bittanti S, Silani E, *et al.* On performance evaluation methods and control strategies for semiactive suspension systems. In: Proc. IEEE Conference on Decision and Control. Maui, Hawaii; 2003.

Chapter 10

Relative accuracy of two methods for approximating observed Fisher information

Shenghan Guo[1] and James C. Spall[2,3]

The Fisher information matrix (FIM) has long been of interest in statistics and other areas. It is widely used to measure the amount of information and calculate the lower bound for the variance for maximum likelihood estimation (MLE). In practice, we do not always know the actual FIM. This is often because obtaining the first- or second-order derivative of the log-likelihood function is difficult, or simply because the calculation of FIM is too formidable. In such cases, we need to utilize the approximation of FIM. In general, there are two ways to estimate FIM. One is to use the product of gradient and the transpose of itself, and the other is to calculate the Hessian matrix and then take negative sign. Mostly people use the latter method in practice. However, this is not necessarily the optimal way. To find out which of the two methods is better, we need to conduct a theoretical study to compare their efficiency. In this paper, we mainly focus on the case where the unknown parameter that needs to be estimated by MLE is scalar, and the random variables we have are independent. In this scenario, FIM is virtually Fisher information number (FIN). Using the Central Limit Theorem (CLT), we get asymptotic variances for the two methods, by which we compare their accuracy. Taylor expansion assists in estimating the two asymptotic variances. A numerical study is provided as an illustration of the conclusion. The next is a summary of limitations of this paper. We also enumerate several fields of interest for future study in the end of this paper.

10.1 Introduction

The FIM has great importance and numerous applications in broad areas of mathematical modeling. Nonetheless, calculation of the FIM is not always practical. Sometimes, estimation or approximation makes more sense when feasibility and cost (in all respects) are taken into account. Basically, situations in which we need to

[1]Department of Industrial and Systems Engineering, Rutgers, The State University of New Jersey, Piscataway, NJ, USA
[2]Applied Physics Laboratory, Johns Hopkins University, Laurel, MD, USA
[3]Department of Applied Mathematics and Statistics, Johns Hopkins University, Baltimore, MD, USA

resort to approximation include the following: (1) the analytical forms of the gradient and the Hessian matrix (or the Hessian number when the unknown parameter is scalar) are not accessible nor is any directly usable values and (2) analytical computation of FIN is extremely troublesome that an unacceptably prohibitive cost is resulted.

There are many examples of estimation problems where it is difficult to get FIM. Let us mention several here. In [1], when using a state-space model with nuisance parameters, the FIM is needed to provide a lower bound for the variance of the maximum likelihood estimator. However, the influence of nuisance parameters makes it difficult to get the real value of the FIM. McLachlan and Peel [2, Section 2.15] give an example where we need to use the approximation of the FIM instead of an analytical calculation, which is when finite mixture models are involved. Analytically calculating the FIM (or FIN for a scalar unknown parameter) for mixture density functions could be costly in all respects. In [3], it is also preferable to use the approximation of the FIM since inverse heat condition problems bring challenges to the analytical calculation. A system with binary subsystems is described in [4], for which the approximation of the FIM becomes the optimal option because the complexity of the system makes it almost impossible to provide an analytically calculated result without considerable efforts. Ninness and Henriksen [5] is one of the many examples of using the FIM in the field of system identification.

In these cases, approximation rather than analytical calculation is usually the optimal strategy. This raises the question that what kind of existing methods give the best approximation to the FIM.

According to the definition, when the unknown parameter θ is scalar, the FIM becomes the FIN. We can calculate the FIN in two ways. With a likelihood function $l(\boldsymbol{Z}_n|\theta) = p(\boldsymbol{Z}_n|\theta)$, where $p(\boldsymbol{Z}_n|\theta)$ is the density function of a series of independent random variable $\boldsymbol{Z}_n = [z_1, z_2, \ldots, z_n]^T$, we have the gradient

$$g(\theta|\boldsymbol{Z}_n) = \frac{\partial \log l(\boldsymbol{Z}_n|\theta)}{\partial \theta}$$

and

$$H(\theta|\boldsymbol{Z}_n) = \frac{\partial^2 \log l(\boldsymbol{Z}_n|\theta)}{\partial \theta^2}$$

Denote the FIN by $F_n(\theta)$, then the FIN is defined to be

$$F_n(\theta) = E[g^2(\theta|\boldsymbol{Z}_n)] \tag{10.1}$$

which, under standard conditions (e.g., [6, pp. 352–353]), is equal to

$$F_n(\theta) = -E[H(\theta|\boldsymbol{Z}_n)] \tag{10.2}$$

In practice, people use estimation when the true FIN is inaccessible. Let us define notation

$$g_i(z_i) := g_i(\theta|z_i), H_i(z_i) := H_i(\theta|z_i), \quad i = 1, 2, \ldots, n \tag{10.3}$$

Hereafter, we will use the above notations for simplicity. Then we have

$$\overline{F}_n(\theta^*) = \frac{1}{n}\sum_{i=1}^{n} g_i^2(\theta|z_i) := \frac{1}{n}\sum_{i=1}^{n} g_i^2(z_i) \tag{10.4}$$

$$\overline{H}_n(\theta^*) = \frac{1}{n}\sum_{i=1}^{n} H_i(\theta|z_i) := \frac{1}{n}\sum_{i=1}^{n} H_i(z_i) \tag{10.5}$$

Equations (10.4) and (10.5) are two unbiased estimators for the averaged FIN, i.e., $F_n(\theta^*)/n$, where θ^* represents the true value of the unknown parameter θ.

There are no strict settings for when to use which form. In many settings, people tend to use (10.5), though it appears that there is no theoretical support showing that the Hessian matrix gives a better approximation to the actual FIN. Such reality is the direct motivation for this project. We would like to know which method gives the better estimation. If we have the answer to this inquiry, we will be able to choose the optimal way to approximate FIN in different situations, which will lead to more desirable results.

Studies related to the FIN are common but works that directly tackle our problem are not easily found. Most related works concentrate on the comparison of the expected FIN and the observed FIN. Efron and Hinkley [7] look at the accuracy of these two with regard to approximating variance for the MLE estimator. They come up with an assertion that for the scalar case, the inverse of the observed FIN gives a better estimation with a well-chosen auxiliary statistic. Cao and Spall [8] discuss a similar issue with Efron and Hinkley [7]. For an unknown target parameter θ and its MLE $\hat{\theta}_n$, Cao and Spall [8] adjust the estimation target to

$$\min_{T(\boldsymbol{X})} E\left[\left(n\mathrm{var}\left(\hat{\theta}_n\right) - T(\boldsymbol{X})\right)^2\right]$$

with $T(\boldsymbol{X})$ being an estimate of the variance of normalized $\hat{\theta}_n$ given sample $\boldsymbol{X}$. For this revised problem, Cao and Spall [8] show an opposite conclusion that the inverse of the expected FIN is preferable under certain conditions. Cao and Spall [9] extend this conclusion to multivariate $\boldsymbol{\theta} = [t_1, t_2, \ldots, t_n]^T$ (i.e., to the FIM, not just FIN, case).

The above works are to some extent relevant to our interest, yet they do not lead to a straightforward conclusion for us. Spall [10] comes up with a Monte Carlo method to estimate FIM and gives some theoretical analysis concerning the efficiency of the Hessian matrix form and the product-of-gradient form in approximating the true FIM. It is shown that under certain circumstances, the Hessian matrix form is a better choice. This assertion, as well as the reasoning process, provides some clues for our study. Specifically, we use Taylor expansion to analyze the asymptotic variances just as Spall [10] does, which enables us to simplify the calculation and leads to a relatively succinct result.

10.2 Background

In this section, we provide some background knowledge that we use in this paper in order to help the reader get a clear picture of our study.

10.2.1 The Central Limit Theorem

We can apply the CLT to (10.4) and (10.5) and assert that they are asymptotically normally distributed. Hence, we can analyze their distributions and find out the accuracy of (10.4) and (10.5).

The classic CLT applies to independent and identically distributed (i.i.d.) samples.

10.2.1.1 Lindeberg–Lévy CLT

Suppose $(X_1, X_2, \ldots)$ is a sequence of i.i.d. random samples with finite mean and variance σ^2. Define $S_n := \sum_{i=1}^{n} X_i/n$, where n is the sample size. As n approaches infinity, the random variable $\sqrt{n}(S_n - \mu)$ converges in distribution to a normal random variable with mean 0 and variance σ^2. That is,

$$\sqrt{n}(S_n - \mu) \overset{\text{dist}}{\to} N(0, \sigma^2) \tag{10.6}$$

There are other versions of CLT that apply to independent and nonidentically distributed (i.n.i.d.) samples.

10.2.1.2 Lyapunov CLT

Suppose $(X_1, X_2, \ldots)$ is a sequence of i.n.i.d. random samples with finite mean μ_i and variance σ_i^2. Define $s_n^2 := \sum_{i=1}^{n} \sigma_i^2$, where n is the sample size. If for some $\delta > 0$, the *Lyapunov's condition*

$$\lim_{n\to\infty} \frac{1}{s_n^{2+\delta}} \sum_{i=1}^{n} E[|X_i - \mu_i|^{2+\delta}] = 0$$

is satisfied, then as n approaches infinity, the random variable $s_n^{-1} \sum_{i=1}^{n} (X_i - \mu_i)$ converges in distribution to a standard normal random variable,

$$\frac{1}{s_n} \sum_{i=1}^{n} (X_i - \mu_i) \overset{\text{dist}}{\to} N(0, 1) \tag{10.7}$$

10.2.1.3 Lindeberg CLT

The Lindeberg CLT is built in the same setting with the Lyapunov CLT, with only the *Lyapunov's condition* replaced by the *Lindeberg's condition*, which is much weaker. The *Lindeberg's condition* says that, for every $\epsilon > 0$, if

$$\lim_{n\to\infty} \frac{1}{s_n^2} \sum_{i=1}^{n} E[(X_i - \mu_i)^2 \cdot \mathbf{1}_{\{|X_i - \mu_i| > \epsilon s_n\}}] = 0$$

where $\mathbf{1}_{\{\cdot\}}$ is an indicator, then the result from the Lyapunov CLT holds,

$$\frac{1}{s_n} \sum_{i=1}^{n} (X_i - \mu_i) \overset{\text{dist}}{\to} N(0, 1)$$

10.2.2 Taylor expansion (Taylor series)

The Taylor expansion is very useful for approximating values that are difficult to calculate directly. In this paper, we use the second-order Taylor expansion to estimate the asymptotic variances of (10.4) and (10.5).

According to Rudin [11], the Taylor expansion of a function $f(x)$ that is infinitely differentiable at some point a is

$$f(a) + f'(a)(x - a) + \frac{f''(a)}{2!}(x - a)^2 + \frac{f^{(3)}(a)}{3!}(x - a)^3 + \dots \tag{10.8}$$

where $f'(\cdot)$ denotes the first-order derivative of f with respect to the variable x, $f''(\cdot)$ denotes the second-order derivative of f with respect to the variable x and $f^{(n)}(\cdot)$ denotes the nth derivative of f with respect to the variable x.

Theoretically, we could expand a function till $f^{(n)}(\cdot)(x - a)^n/n!$ with n sufficiently large. However, when using Taylor expansion, people often only expand the function to its second-order derivative and ignore the remaining items that could possibly appear in its Taylor series. This is because the denominator $n!$ in $f^{(n)}(\cdot)(x - a)^n/n!$ increases very fast and items with higher order become insignificant quickly. Based on this rational, we only use the second-order Taylor expansion in the following context. Also, we assume that the density functions we consider are at least twice-differentiable with respect to θ.

10.3 Theoretical analysis

In this section, we present the theoretical analysis on the accuracy of (10.4) and (10.5).

For n independent random variables $\boldsymbol{Z}_n = [z_1, z_2, \dots, z_n]^T$, the likelihood function is

$$l(\theta|\boldsymbol{Z}_n) = p(\boldsymbol{Z}_n|\theta) = \prod_{i=1}^{n} p_i(z_i|\theta) \tag{10.9}$$

where $p_i(\cdot|\theta)$ is the density function for random variable $z_i, i = 1, 2, \dots, n$.

The gradient is the first-order derivative of the log-likelihood function with respect to θ,

$$g(\boldsymbol{Z}_n) = \frac{\partial \log l(\theta|\boldsymbol{Z}_n)}{\partial \theta} = \frac{\partial}{\partial \theta}\left(\sum_{i=1}^{n} \log p_i(z_i|\theta)\right) = \sum_{i=1}^{n} \frac{\partial \log p_i(z_i|\theta)}{\partial \theta} \tag{10.10}$$

The Hessian number turns out to be

$$H(\boldsymbol{Z}_n) = \frac{\partial^2 \log l(\theta|\boldsymbol{Z}_n)}{\partial \theta^2} = \frac{\partial}{\partial \theta}\left(\sum_{i=1}^{n} \frac{\partial \log p_i(z_i|\theta)}{\partial \theta}\right) = \sum_{i=1}^{n} \frac{\partial^2 \log p_i(z_i|\theta)}{\partial \theta^2} \tag{10.11}$$

According to the definition of FIN, we have

$$
\begin{aligned}
F_n(\theta) &= E\left[g^2(\boldsymbol{Z}_n)\right] \\
&= E\left[\left(\sum_{i=1}^{n} \frac{\partial \log p_i(z_i|\theta)}{\partial \theta}\right)^2\right] \\
&= E\left[\sum_{i=1}^{n}\left(\frac{\partial \log p_i(z_i|\theta)}{\partial \theta}\right)^2 + 2\sum_{j<k} \frac{\partial \log p_j(z_j|\theta)}{\partial \theta}\frac{\partial \log p_k(z_k|\theta)}{\partial \theta}\right] \\
&= \sum_{i=1}^{n} E\left[\left(\frac{\partial \log p_i(z_i|\theta)}{\partial \theta}\right)^2\right] + 2\sum_{j<k} \underbrace{E\left[\frac{\partial \log p_j(z_j|\theta)}{\partial \theta}\frac{\partial \log p_k(z_k|\theta)}{\partial \theta}\right]}_{(*)}
\end{aligned} \tag{10.12}
$$

Since z_j and z_k are independent for $j \neq k, j, k = 1, 2, \ldots, n$,

$$
\begin{aligned}
(*) &= E\left[\frac{\partial \log p_j\left(z_j|\theta\right)}{\partial \theta}\frac{\partial \log p_k\left(z_k|\theta\right)}{\partial \theta}\right] \\
&= E\left[\frac{\partial \log p_j\left(z_j|\theta\right)}{\partial \theta}\right] E\left[\frac{\partial \log p_k\left(z_k|\theta\right)}{\partial \theta}\right]
\end{aligned} \tag{10.13}
$$

Assume that the dominated convergence theorem (DCT) applies, so that integration and differentiation can be interchanged. Then

$$
\begin{aligned}
E\left[\frac{\partial \log p(z|\theta)}{\partial \theta}\right] &= \int_{-\infty}^{\infty} \frac{\partial \log p(z|\theta)}{\partial \theta} p(z|\theta)dz = \int_{-\infty}^{\infty} \frac{1}{p(z|\theta)}\frac{\partial p(z|\theta)}{\partial \theta} p(z|\theta)dz \\
&= \int_{-\infty}^{\infty} \frac{\partial p(z|\theta)}{\partial \theta} dz = \frac{\partial}{\partial \theta}\left(\int_{-\infty}^{\infty} p(z|\theta)\, dz\right) = \frac{\partial}{\partial \theta}(\text{constant}) = 0
\end{aligned} \tag{10.14}
$$

Therefore, we have that expression $(*) = 0$. So

$$
F_n(\theta) = \sum_{i=1}^{n} E\left[\left(\frac{\partial \log p_i(z_i|\theta)}{\partial \theta}\right)^2\right] = \sum_{i=1}^{n} E\left[g_i^2(z_i)\right], \tag{10.15}
$$

where we have suppressed the dependence on θ in the summands on the right-hand side. With the alternative definition of FIN,

$$\begin{aligned} F_n(\theta) &= -E[H(z)] = -E\left[\sum_{i=1}^{n} \frac{\partial^2 \log p_i(z_i|\theta)}{\partial \theta^2}\right] \\ &= -\sum_{i=1}^{n} E\left[\frac{\partial^2 \log p_i(z_i|\theta)}{\partial \theta^2}\right] = -\sum_{i=1}^{n} E[H_i(z_i)], \end{aligned} \tag{10.16}$$

with the same suppression of the dependence on θ in the summands.

Scenario A. Independent and identically distributed (i.i.d.) samples

The standard case is where we have i.i.d. samples $\{z_1, z_2, \ldots, z_n\}$, each of which has the (same) density function $p(\cdot|\theta)$. Thus, the averaged FIN, $F_n(\theta^*)/n$, turns out to be either $E[g^2(z_i)]$ or $-E[H(z_i)]$ for any z_i in $\boldsymbol{Z}_n = [z_1, z_2, \ldots, z_n]^T$, with the subscript i in $g_i(\cdot)$ and $H_i(\cdot)$ ignored because of equality in density functions, and where θ^* is the true value of the unknown parameter θ.

According to the CLT, we have the following facts:

$$\sqrt{n}(\overline{F}_n(\theta^*) - F(\theta^*)/n) \overset{\text{dist}}{\rightarrow} N(0, \text{var}[g^2(z_i)]) \tag{10.17}$$

$$\sqrt{n}(\overline{H}_n(\theta^*) - F(\theta^*)/n) \overset{\text{dist}}{\rightarrow} N(0, \text{var}[H(z_i)]) \tag{10.18}$$

Now, what we need to do is to contrast the magnitude of var$[g^2(z_i)]$ and var$[H(z_i)]$. The expression in either (10.4) or (10.5) that has the smaller variance is therefore the better choice for FIN estimation. Since the calculation of var$[g^2(z_i)]$ and var$[H(z_i)]$ is challenging, we turn to the second-order Taylor expansion to give a close approximation to these two. Define $\mu := E(z_i)$. Let

$$g'(\mu) = \left.\frac{\partial g(z_i)}{\partial z_i}\right|_{z_i=\mu}, g''(\mu) = \left.\frac{\partial^2 g(z_i)}{\partial z_i^2}\right|_{z_i=\mu}$$

and

$$H'(\mu) = \left.\frac{\partial H(z_i)}{\partial z_i}\right|_{z_i=\mu}, H''(\mu) = \left.\frac{\partial^2 H(z_i)}{\partial z_i^2}\right|_{z_i=\mu}$$

where $i = 1, 2, \ldots, n$. The second-order expansions for the functions $g^2(\cdot)$ and $H(\cdot)$ are shown in Appendix A. Using the expansions, the variances are

$$\begin{aligned}
\mathrm{var}[g^2(z_i)] \approx\ & 4g^2(\mu)[g'(\mu)]^2\sigma^2 + \left([g'(\mu)]^2 + g(\mu)g''(\mu)\right)^2 \mathrm{var}[(z_i-\mu)^2] \\
& + \frac{1}{16}[g''(\mu)]^4 \mathrm{var}[(z_i-\mu)^4] + [g'(\mu)]^2[g''(\mu)]^2 \mathrm{var}[(z_i-\mu)^3] \\
& + 4g(\mu)g'(\mu)([g'(\mu)]^2 + g(\mu)g''(\mu))E[(z_i-\mu)^3] \\
& + g(\mu)g'(\mu)[g''(\mu)]^2 E[(z_i-\mu)^5] + 4g(\mu)g'(\mu)g''(\mu)E[(z_i-\mu)^4] \\
& + \frac{1}{2}[g''(\mu)]^2([g'(\mu)]^2 + g(\mu)g''(\mu))(E[(z_i-\mu)^6] - \sigma^2 E[(z_i-\mu)^4]) \\
& + 2g(\mu)g''(\mu)([g'(\mu)]^2 + g(\mu)g''(\mu))(E[(z_i-\mu)^5] - \sigma^2 E[(z_i-\mu)^3]) \\
& + \frac{1}{2}g'(\mu)[g''(\mu)]^3(E[(z_i-\mu)^7] - E[(z_i-\mu)^4]E[(z_i-\mu)^3]) \qquad (10.19)
\end{aligned}$$

and

$$\mathrm{var}[H(z_i)] \approx [H'(\mu)]^2\sigma^2 + \frac{1}{4}[H''(\mu)]^2 \mathrm{var}[(z_i-\mu)^2] + H'(\mu)H''(\mu)E[(z_i-\mu)^3] \qquad (10.20)$$

Detailed calculations can be found in Appendix A.

The above results may seem intimidating at first glance, especially that for $\mathrm{var}[g^2(z_i)]$. However, they are not complicated in nature. Let us consider the most common type of density functions, symmetric density. A basic fact about such density functions is that the odd central moments are 0. Applying our results to this kind of density, we are able to drop off a couple of items, giving much more succinct forms:

$$\begin{aligned}
\mathrm{var}[g^2(z_i)] \approx\ & 4g^2(\mu)[g'(\mu)]^2\sigma^2 + ([g'(\mu)]^2 + g(\mu)g''(\mu))^2 \mathrm{var}[(z_i-\mu)^2] \\
& + \frac{1}{16}[g''(\mu)]^4 \mathrm{var}[(z_i-\mu)^4] + [g'(\mu)]^2[g''(\mu)]^2 \mathrm{var}[(z_i-\mu)^3] \\
& + 4g(\mu)[g'(\mu)]^2 g''(\mu)E[(z_i-\mu)^4] + \frac{1}{2}[g''(\mu)]^2 \\
& \times ([g'(\mu)]^2 + g(\mu)g''(\mu))(E[(z_i-\mu)^6] - \sigma^2 E[(z_i-\mu)^4])
\end{aligned} \qquad (10.21)$$

and

$$\mathrm{var}[H(z_i)] \approx [H'(\mu)]^2\sigma^2 + \frac{1}{4}[H''(\mu)]^2 \mathrm{var}[(z_i-\mu)^2] \qquad (10.22)$$

We calculate the difference between $\mathrm{var}[g^2(z_i)]$ and $\mathrm{var}[H(z_i)]$ as follows:

$$\begin{aligned}
\mathrm{var}[g^2(z_i)] - \mathrm{var}[H(z_i)] &\approx (4g^2(\mu)[g'(\mu)]^2 - [H'(\mu)]^2)\sigma^2 \\
&+ \left[\left([g'(\mu)]^2 + g(\mu)g''(\mu)\right) - \frac{1}{4}[H''(\mu)]^2\right]\mathrm{var}[(z_i - \mu)^2] \\
&+ \frac{1}{16}[g''(\mu)]^4\mathrm{var}[(z_i - \mu)^4] + [g'(\mu)]^2[g''(\mu)]^2\mathrm{var}[(z_i - \mu)^3] \\
&+ 4g(\mu)[g'(\mu)]^2 g''(\mu)E[(z_i - \mu)^4] \\
&+ \frac{1}{2}[g''(\mu)]^2([g'(\mu)]^2 + g(\mu)g''(\mu))(E[(z_i - \mu)^6] - \sigma^2 E[(z_i - \mu)^4])
\end{aligned} \tag{10.23}$$

Notice that the third and fourth items on the right-hand side of (10.23) are automatically non-negative. Based on (10.23), we have two different situations:

I. I (i) $4g^2(\mu)[g'(\mu)]^2 - [H'(\mu)]^2 \geq 0$, so the first item in (10.23) is non-negative;
   I (ii) $([g'(\mu)]^2 + g(\mu)g''(\mu)) - \frac{1}{4}[H''(\mu)]^2 \geq 0$, so the second item in (10.23) is non-negative;
   I (iii) $g(\mu)g''(\mu) \geq 0$, so the fifth item in (10.23) is non-negative;
   I (iv) $([g'(\mu)]^2 + g(\mu)g''(\mu))$ and $E[(z_i - \mu)^6] - \sigma^2 E[(z_i - \mu)^4]$ are of the same sign so the last item in (10.23) is non-negative.

II. The items in (10.23) are not all non-negative. Some of them are non-positive while the rest items are non-negative.

In situation I, we have $\mathrm{var}[g^2(z_i)] - \mathrm{var}[H(z_i)] \geq 0$ with certainty given the validity of the Taylor expansion, in which case (10.5) is a better choice than (10.4) to approximate FIN; in situation II, we can have a non-positive or non-negative value for $\mathrm{var}[g^2(z_i)] - \mathrm{var}[H(z_i)]$. The sign of $\mathrm{var}[g^2(z_i)] - \mathrm{var}[H(z_i)]$ cannot be found by checking the four conditions mentioned in situation I. Plugging in individual quantities for further calculation is necessary to figure this out.

We have found the sufficient conditions for $\mathrm{var}[g^2(z_i)] - \mathrm{var}[H(z_i)]$ to be non-negative, i.e. situation I. However, notice that our analysis does not provide necessary conditions for $\mathrm{var}[g^2(z_i)] - \mathrm{var}[H(z_i)]$ to be non-positive or non-negative. If we know that $\mathrm{var}[g^2(z_i)] - \mathrm{var}[H(z_i)]$ is, say, non-negative, we cannot conclude that all the items in (10.23) are all non-negative because the operations among them can give a non-negative value even if some of them are non-positive. As mentioned, further analysis is needed to explore this. In fact, finding the necessary conditions for $\mathrm{var}[g^2(z_i)] - \mathrm{var}[H(z_i)]$ to be non-positive or non-negative is a topic of interest for future study.

Scenario B. Independent and nonidentically distributed (i.n.i.d.) samples

In the more general case, where we have i.n.i.d. samples, the z_i in $\boldsymbol{Z}_n = [z_1, z_2, \ldots, z_n]^T$ have density functions $p_i(\cdot|\theta)$, as well as mean μ_i and variance σ_i^2, $i = 1, 2, \ldots, n$. In order to be precise, we now include the subscript i in $p_i(\cdot|\theta)$, $g_i^2(\cdot)$ and $H_i(\cdot)$, because the individual $p_i(\cdot|\theta)$, $g_i^2(\cdot)$ and $H_i(\cdot)$ are not necessarily identical.

According to the CLT, the following facts are valid:

$$\sqrt{n}(\overline{F}_n(\theta^*) - F(\theta^*)/n) \overset{\text{dist}}{\rightarrow} N(0, s_g^2/n) \tag{10.24}$$

$$\sqrt{n}(\overline{H}_n(\theta^*) - F(\theta^*)/n) \overset{\text{dist}}{\rightarrow} N(0, s_H^2/n) \tag{10.25}$$

where $s_g^2 = \sum_{i=1}^n \text{var}[g_i^2(z_i)]$ and $s_H^2 = \sum_{i=1}^n \text{var}[H_i(z_i)]$.

We calculate s_g^2 and s_H^2 and consider the value of $s_g^2 - s_H^2$ in order to find out which one of (10.4) and (10.5) performs better in the i.n.i.d. samples scenario.

Define $\mu_i := E(z_i)$. For $i = 1, 2, \ldots, n$, let

$$g_i'(\mu_i) = \left.\frac{\partial g_i(z_i)}{\partial z_i}\right|_{z_i=\mu_i}, \quad g_i''(\mu_i) = \left.\frac{\partial^2 g_i(z_i)}{\partial z_i^2}\right|_{z_i=\mu_i}$$

and

$$H_i'(\mu_i) = \left.\frac{\partial H_i(z_i)}{\partial z_i}\right|_{z_i=\mu_i}, \quad H_i''(\mu_i) = \left.\frac{\partial^2 H_i(z_i)}{\partial z_i^2}\right|_{z_i=\mu_i}$$

Then we have

$$\begin{aligned}
s_g^2 &= \sum_{i=1}^n \text{var}[g_i^2(z_i)] \\
&\approx 4\sum_{i=1}^n g_i^2(\mu_i)[g_i'(\mu_i)]^2\sigma_i^2 + \sum_{i=1}^n ([g_i'(\mu_i)]^2 + g_i(\mu_i)g_i''(\mu_i))^2\text{var}[(z_i - \mu_i)^2] \\
&+ \frac{1}{16}\sum_{i=1}^n [g_i''(\mu)]^4\text{var}[(z_i - \mu_i)^4] + \sum_{i=1}^n [g_i'(\mu_i)]^2[g_i''(\mu_i)]^2\text{var}[(z_i - \mu_i)^3] \\
&+ 4\sum_{i-1}^n g_i(\mu_i)g_i'(\mu_i)([g_i'(\mu_i)]^2 + g_i(\mu_i)g_i''(\mu_i))E[(z_i - \mu_i)^3]
\end{aligned}$$

$$+\sum_{i=1}^{n} g_i(\mu_i)g_i'(\mu_i)[g_i''(\mu_i)]^2 E[(z_i-\mu_i)^5]$$

$$+4\sum_{i=1}^{n} g_i(\mu_i)g_i'(\mu_i)g_i''(\mu_i)E[(z_i-\mu_i)^4]$$

$$+\frac{1}{2}\sum_{i=1}^{n}[g_i''(\mu_i)]^2([g_i'(\mu_i)]^2+g_i(\mu_i)g_i''(\mu_i))(E[(z_i-\mu_i)^6]-\sigma^2E[(z_i-\mu_i)^4])$$

$$+2\sum_{i=1}^{n} g_i'(\mu_i)g_i''(\mu_i)([g_i'(\mu_i)]^2$$

$$+g_i(\mu_i)g_i''(\mu_i))(E[(z_i-\mu_i)^5]-\sigma^2E[(z_i-\mu_i)^3])$$

$$+\frac{1}{2}\sum_{i=1}^{n} g_i'(\mu_i)[g_i''(\mu_i)]^3(E[(z_i-\mu_i)^7]-E[(z_i-\mu_i)^4]E[(z_i-\mu_i)^3]) \tag{10.26}$$

and

$$s_H^2=\sum_{i=1}^{n}\mathrm{var}[H_i(z_i)]\approx\sum_{i=1}^{n}[H_i'(\mu_i)]^2\sigma_i^2+\frac{1}{4}\sum_{i=1}^{n}[H_i''(\mu_i)]^2\mathrm{var}[(z_i-\mu_i)^2]$$

$$+\sum_{i=1}^{n} H_i'(\mu_i)H_i''(\mu_i)E[(z_i-\mu_i)^3] \tag{10.27}$$

Recall that we are considering symmetric density functions, which simplifies (10.26) and (10.27) to

$$s_g^2=\sum_{i=1}^{n}\mathrm{var}[g_i^2(z_i)]\approx 4\sum_{i=1}^{n} g_i^2(\mu_i)[g_i'(\mu_i)]^2\sigma_i^2$$

$$+\sum_{i=1}^{n}([g_i'(\mu_i)]^2+g_i(\mu_i)g_i''(\mu_i))^2\mathrm{var}[(z_i-\mu_i)^2]$$

$$+\frac{1}{16}\sum_{i=1}^{n}[g_i''(\mu)]^4\mathrm{var}[(z_i-\mu_i)^4]+\sum_{i=1}^{n}[g_i'(\mu_i)]^2[g_i''(\mu_i)]^2\mathrm{var}[(z_i-\mu_i)^3]$$

$$+4\sum_{i=1}^{n} g_i(\mu_i)g_i'(\mu_i)g_i''(\mu_i)E[(z_i-\mu_i)^4]+\frac{1}{2}\sum_{i=1}^{n}[g_i''(\mu_i)]^2([g_i'(\mu_i)]^2$$

$$+g_i(\mu_i)g_i''(\mu_i))(E[(z_i-\mu_i)^6]-\sigma^2E[(z_i-\mu_i)^4]) \tag{10.28}$$

and

$$s_H^2=\sum_{i=1}^{n}\mathrm{var}[H_i\,(z_i)]\approx\sum_{i=1}^{n}[H_i'\,(\mu_i)]^2\sigma_i^2+\frac{1}{4}\sum_{i=1}^{n}[H_i''\,(\mu_i)]^2\mathrm{var}[(z_i-\mu_i)^2] \tag{10.29}$$

Next, we consider the difference between s_g^2 and s_H^2.

$$
\begin{aligned}
s_g^2 - s_H^2 \approx & \sum_{i=1}^{n} \left(4g_i^2(\mu_i)\left[g_i'(\mu_i)\right]^2 - \left[H_i'(\mu_i)\right]^2\right)\sigma_i^2 \\
& + \sum_{i=1}^{n} \left[\left(\left[g_i'(\mu_i)\right]^2 + g_i(\mu_i)g_i''(\mu_i)\right)^2 - \frac{1}{4}\left[H_i''(\mu_i)\right]^2\right] \\
& \times \mathrm{var}\left[(z_i-\mu_i)^2\right] + \frac{1}{16}\sum_{i=1}^{n}\left[g_i''(\mu)\right]^4 \mathrm{var}\left[(z_i-\mu_i)^4\right] \\
& + \sum_{i=1}^{n}\left[g_i'(\mu_i)\right]^2\left[g_i''(\mu_i)\right]^2 \mathrm{var}\left[(z_i-\mu_i)^3\right] \\
& + 4\sum_{i=1}^{n} g_i(\mu_i)g_i'(\mu_i)g_i''(\mu_i)E\left[(z_i-\mu_i)^4\right] \\
& + \frac{1}{2}\sum_{i=1}^{n}\left[g_i''(\mu_i)\right]^2\left(\left[g_i'(\mu_i)\right]^2 + g_i(\mu_i)g_i''(\mu_i)\right) \\
& \times \left(E\left[(z_i-\mu_i)^6\right] - \sigma^2 E\left[(z_i-\mu_i)^4\right]\right)
\end{aligned}
\tag{10.30}
$$

Notice that the third and fourth items in the above result are automatically non-positive. As with the i.i.d. case, we have two situations based on (10.30):

I′. I′ (i) For all $i = 1, 2, \ldots, n, 4g_i^2(\mu_i)[g_i'(\mu_i)]^2 - [H_i'(\mu_i)]^2 \geq 0$ so that the first item in (30) is non-negative;

I′ (ii) For all $i = 1, 2, \ldots, n, ([g_i'(\mu_i)]^2 + g_i(\mu_i)g_i''(\mu_i))^2 - [H_i''(\mu_i)]^2/4 \geq 0$ so that the second item in (30) is non-negative;

I′ (iii) For all $i = 1, 2, \ldots, n, g_i(\mu_i)g_i''(\mu_i) \geq 0$ so that the fifth item in (10.30) is non-negative.

I′ (iv) $([g_i'(\mu_i)]^2 + g_i(\mu_i)g_i''(\mu_i))$ and $E[(z_i-\mu_i)^6] - \sigma^2 E[(z_i-\mu_i)^4]$ are of the same sign so the last item in (10.30) is non-negative.

II′. The items in (10.34) are not all non-negative. Some of them are non-negative while the remaining items are negative.

In situation I′, we have $s_g^2 - s_H^2 \geq 0$ with certainty, in which case (10.5) is a better choice than (10.4) to approximate FIN; in situation II′, we can have either a non-positive or non-negative value for $s_g^2 - s_H^2$, which cannot be found by checking the four conditions mentioned in I′. Further calculation is necessary to lead to a result indicating the signs of $s_g^2 - s_H^2$.

So far, the analysis has led us to the sufficient conditions for $s_g^2 - s_H^2$ to be non-negative, i.e. situation I′. Nonetheless, our analysis does not provide the necessary conditions. For instance, if we know that $s_g^2 - s_H^2$ is non-negative, we cannot conclude that the items in (30) are all non-negative because the operations among them can give a non-negative value even if some of them are negative. Further analysis is needed to address this issue. One of the proposals for future study is to find out the necessary conditions for $s_g^2 - s_H^2$ to be non-positive or non-negative. Hopefully this issue will be addressed in future work.

Therefore, given the validity of the Taylor expansion-based analysis, we have reached certain results for both the i.i.d. sample case and the i.n.i.d. sample case, though the results appear somewhat unwieldy at first glance, raising concerns about their applicability. It turns out that they fit in well in practice. In effect, for the many density functions that we confront in practice, especially the exponential family, the above results can be substantially simplified, and we are able to apply them to check which one of (10.4) and (10.5) does a better job in estimating the FIN. Illustrations are provided in the section of numerical studies.

10.4 Numerical studies

We have done theoretical analysis. Next, we would like to present some numerical studies as demonstrations.

Example 10.1. Normal distribution $N(\mu, \sigma^2)$ with $\theta = \mu$

We look at a trivial case first. We follow the notations used in the theoretical analysis. Consider an i.i.d. sample set where each component of $\boldsymbol{Z}_n = [z_1, z_2, \ldots, z_n]^T$ is normally distributed with mean μ and variance σ^2. The unknown parameter is $\theta = \mu$. The density function is $p(z|\theta) = (2\pi\sigma^2)^{-1/2} \exp\{-(z-\mu)^2/2\sigma^2\}$. Hence, the log-density function is

$$\log p\,(z|\theta) = -\frac{1}{2}\log 2\pi - \frac{1}{2}\log\sigma^2 - \frac{1}{2\sigma^2}(z-\mu)^2 \tag{10.31}$$

The gradient and the Hessian for an arbitrary random variable in the data set are given by

$$g(z) = \frac{\partial \log p(z|\mu)}{\partial \mu} = \frac{(z-\mu)}{\sigma^2} \tag{10.32}$$

$$H(z) = \frac{\partial^2 \log p(z|\mu)}{\partial \mu^2} = -\frac{1}{\sigma^2} \tag{10.33}$$

We then calculate $g(\mu^*), g'(\mu^*), g''(\mu^*), H(\mu^*), H'(\mu^*)$ and $H''(\mu^*)$, where μ^* denotes the true value of the unknown parameter. Then

$$g(\mu^*) = \frac{1}{\sigma^2}(\mu^* - \mu), g'(\mu^*) = \left.\frac{\partial g(z)}{\partial z}\right|_{z=\mu^*} = \frac{1}{\sigma^2}, g''(\mu^*) = \left.\frac{\partial^2 g(z)}{\partial z^2}\right|_{z=\mu^*} = 0 \tag{10.34}$$

and

$$H(\mu^*) = -\frac{1}{\sigma^2}, H'(\mu^*) = \left.\frac{\partial H(z)}{\partial z}\right|_{z=\mu^*} = 0, H''(\mu^*) = \left.\frac{\partial^2 H(z)}{\partial z^2}\right|_{z=\mu^*} = 0 \tag{10.35}$$

Using the above facts, we check which situations between I and II are fit for $N(\mu, \sigma^2)$.

First, $4g^2(\mu^*)[g'(\mu^*)]^2 - [H'(\mu^*)]^2 = 4[\frac{\mu^*-\mu}{\sigma^2}]^2/\sigma^4 \geq 0$.

Next, $([g'(\mu^*)]^2 + g(\mu^*)g''(\mu^*)) - [H''(\mu^*)]^2/4 = 1/\sigma^8 \geq 0$.

This shows that situation I in the i.i.d. scenario is satisfied, which means that $\text{var}[g^2(z)] - \text{var}[H(z)] > 0$, implying the fact that for normally distributed i.i.d. samples with mean μ and variance σ^2, the average of second-order derivatives in (10.5) should outperform the average of squared derivatives in (10.4) as an approximation for FIN.

In fact, this conclusion could have been made initially. We know that for $N(\mu, \sigma^2)$, the second-order derivative of the log-density function equals the constant $-(\sigma^2)^{-1}$. The variance $\text{var}[H(z)] = \text{var}[(\sigma^2)^{-1}] = 0$. Because $\text{var}[H(z)]$ is automatically no larger than $\text{var}[g^2(z)]$, we should choose (10.5) over (10.4) to be the estimator of FIN. To check if $\text{var}[H(z)] \leq \text{var}[g^2(z)]$ in practice, we use MATLAB® to generate normal random numbers and then calculate (10.4) and (10.5) to see whether (10.5) has a smaller variance than (10.4). Consider three cases where we have (1) $\mu = 1$ and $\sigma^2 = 0.1^2$, (2) $\mu = 0$ and $\sigma^2 = 1$ and (3) $\mu = 5$ and $\sigma^2 = 10^2$.

For a Monte Carlo simulation with a sample size of $n = 100$ and $N = 200$ replicates, the results are given in Table 10.1.

In the table, $\text{var}^{(e)}[X]$ stands for the sample variance of the random variable X, which is defined as $\text{var}^{(e)}[X] = (n-1)^{-1}\sum_{i=1}^{n}(x_i - x'')^2$, $\bar{x} = n^{-1}\sum_{i=1}^{n} x_i$, with x_i denoting the ith sample value, $i = 1, 2, \ldots, n$. It is an unbiased estimator of the theoretical variance, and its value is analytically calculated in order to illustrate the magnitude of the true asymptotic variance. Below, we will use the same notation for the same meaning and purpose.

We do a hypothesis test to see whether $\text{var}^{(e)}[g^2(z)] - \text{var}^{(e)}[H(z)]$ is significantly larger than 0. In particular, we run a t-test for $m = 50$ replicates, the result of which is shown in Table 10.2. As shown, the outcomes for these three cases all coincide with the conclusion of the theoretical analysis. As σ^2 gets larger, $\text{var}^{(e)}[g^2(z)]$ and $\text{var}^{(e)}[H(z)]$ tend to increase, as does $\text{var}^{(e)}[g^2(z)] - \text{var}^{(e)}[H(z)]$.

Table 10.1 Sample variances of (10.5) and (10.6) as well as the difference between them, when the unknown parameter $\theta = \mu$

	$\mu = 1$ **and** $\sigma^2 = 0.1^2$	$\mu = 0$ **and** $\sigma^2 = 1$	$\mu = 5$ **and** $\sigma^2 = 10^2$
$\text{var}^{(e)}[g^2(z)]$	87.35	95.42	95.22
$\text{var}^{(e)}[H(z)]$	0	0	4.9×10^{-35}
$\text{var}^{(e)}[g^2(z)] - \text{var}^{(e)}[H(z)]$	87.35	95.42	95.22

Table 10.2 Hypothesis testing results for $\theta = \mu$ using a one-sided t-test with 50 replicates

	$\mu = 1$ **and** $\sigma^2 = 0.1^2$	$\mu = 0$ **and** $\sigma^2 = 1$	$\mu = 5$ **and** $\sigma^2 = 10^2$
P-Value	3.1×10^{-25}	8.1×10^{-33}	2.0×10^{-19}

Example 10.2. Normal distribution $N(\mu, \sigma^2)$ with $\theta = \sigma^2$
We continue to consider an i.i.d. sample set where each component of $\boldsymbol{Z}_n = [z_1, z_2, \ldots, z_n]^T$ is normally distributed with mean μ and variance σ^2, but this time the unknown parameter is $\theta = \sigma^2$ instead of $\theta = \mu$ and we assume that μ is known. The density function is the same as in Example 10.1. Yet the gradient and Hessian change to

$$g(z) = \frac{\partial \log p(z|\sigma^2)}{\partial \sigma^2} = -\frac{1}{2\sigma^2} + \frac{(z-\mu)^2}{2\sigma^4} \tag{10.36}$$

$$H(z) = \frac{\partial^2 \log p(z|\sigma^2)}{\partial (\sigma^2)^2} = \frac{1}{2\sigma^4} - \frac{(z_i-\mu)^2}{\sigma^6} \tag{10.37}$$

We calculate $g(\mu), g'(\mu), g"(\mu), H(\mu), H'(\mu)$ and $H"(\mu)$:

$$g(\mu) = -\frac{1}{2\sigma^2}, g'(\mu) = \left.\frac{\partial g(z)}{\partial z}\right|_{z=\mu} = 0, g''(\mu) = \left.\frac{\partial^2 g(z)}{\partial z^2}\right|_{z=\mu} = \frac{1}{\sigma^4} \tag{10.38}$$

and

$$H(\mu) = \frac{1}{2\sigma^4}, H'(\mu) = \left.\frac{\partial H(z)}{\partial z}\right|_{z=\mu} = 0, H''(\mu) = \left.\frac{\partial^2 H(z)}{\partial z^2}\right|_{z=\mu} = -\frac{2}{\sigma^6} \tag{10.39}$$

It is easily seen that this example does not fit in situation I, which means that we should plug in $g(\mu), g'(\mu), g''(\mu), H(\mu), H'(\mu)$ and $H''(\mu)$ to get the approximated value for $\mathrm{var}[g^2(z)] - \mathrm{var}[H(z)]$.

$$\begin{aligned}\mathrm{var}[g^2(z)] - \mathrm{var}[H(z)] &\approx 2\sigma^4\left(\frac{1}{4\sigma^{12}} - \frac{1}{\sigma^8}\right) + \frac{1}{16\sigma^{16}}(7!!\sigma^8 - (3\sigma^4)^2) \\ &+ \frac{1}{2\sigma^8}\left(-\frac{1}{2\sigma^6}\right)(5!!\sigma^6 - 3\sigma^6) = \frac{3}{2\sigma^8} > 0\end{aligned} \tag{10.40}$$

where $a!! = a(a-2)(a-4)(a-6)\ldots \times 3 \times 1$ for some odd integer a.

Therefore, we assert that $\mathrm{var}[g^2(z)] - \mathrm{var}[H(z)] > 0$. This implies that for i.i.d. $N(\mu, \sigma^2)$ samples with $\theta = \sigma^2$, (10.5) outperforms (10.4), theoretically. Notice that this example is non-trivial because the Hessian number is no longer a constant (i.e. it is a function of θ). We cannot tell which form between (10.4) and (10.5) is optimal without analysis.

Table 10.3 Sample variances of (10.5) and (10.6) as well as the difference between them, when the unknown parameter $\theta = \sigma^2$

	$\mu = 1$ and $\sigma^2 = 0.1^2$	$\mu = 0$ and $\sigma^2 = 1$	$\mu = 5$ and $\sigma^2 = 10^2$
$\text{var}^{(e)}[g^2(z)]$	3.1×10^8	3.64	3.2×10^{-8}
$\text{var}^{(e)}[H(z)]$	1.9×10^8	2.06	1.9×10^{-8}
$\text{var}^{(e)}[g^2(z)] - \text{var}^{(e)}[H(z)]$	1.4×10^8	1.57	1.2×10^{-8}

Table 10.4 Hypothesis testing results for $\theta = \sigma^2$ using a one-sided t-test with 50 replicates

	$\mu = 1$ and $\sigma^2 = 0.1^2$	$\mu = 0$ and $\sigma^2 = 1$	$\mu = 5$ and $\sigma^2 = 10^2$
P-value	7.8×10^{-5}	3.2×10^{-5}	9.5×10^{-5}

Again, we use MATLAB® to generate normal random numbers and calculate (10.4) and (10.5) to see whether (10.5) has a smaller asymptotic variance than (10.4). Consider the three cases we used in Example 10.1 (1) $\mu = 1$ and $\sigma^2 = 0.1^2$, (2) $\mu = 0$ and $\sigma^2 = 1$ and (3) $\mu = 5$ and $\sigma^2 = 10^2$.

For a program using a single run of $n = 100$ and $N = 200$, the results are given in Table 10.3.

The numerical results show that in practice (10.5) does outperform (10.4) for this case in the sense that (10.5) has a smaller asymptotic variance than (10.4) does. One noteworthy phenomenon is that as the true σ^2 gets larger, the asymptotic variances of the two methods tend to get smaller, which is a little counterintuitive. However, this fact is justified in the sense that σ^2 appears in the denominator of the quantities we estimate (g and H), so the magnitude of the variance is inversely related to σ^2. An extremely large magnitude is resulted for the asymptotic variances for case (1) $\mu = 1$ and $\sigma^2 = 0.1^2$, since the associated denominator σ^2 becomes as small as 0.1^2.

As in the last case study, we do a t-test for $m = 50$ replicates, the result of which is shown in Table 10.4.

The results for our hypothesis test indicate that the negativity of $\text{var}^{(e)}[g^2(z)] - \text{var}^{(e)}[H(z)]$ is significantly different from 0, i.e. larger than 0 in this example. Therefore, (10.5) indeed outperforms (10.4) in the sense that it is more accurate and gives a smaller variance for the estimated FIN. The prevalence of (10.5) as a method to approximate FIN is justified for this individual situation.

Example 10.3. Signal-plus noise

We also present a case study for i.n.i.d. scenario. Consider the signal-plus noise problem where we have i.n.i.d. data with $z_i \sim N(0, \sigma^2 + q_i), q_i = 0.1 \times (i - 10 \times \{i/10\})$, where $\{i/10\}$ denotes the integer part of $i/10, i = 1, 2, \ldots, n$. We would like to estimate the unknown parameter $\theta = \sigma^2$. For an arbitrary i,

$$p(z_i|\sigma^2) = \frac{1}{\sqrt{2\pi(\sigma^2 + q_i)}} \exp\left\{-\frac{z_i^2}{2(\sigma^2 + q_i)}\right\} \tag{10.41}$$

Thus, the log-density function is

$$\log p(z_i|\sigma^2) = -\frac{1}{2}\log 2\pi - \frac{1}{2}\log(\sigma^2 + q_i) - \frac{z_i^2}{2(\sigma^2 + q_i)} \tag{10.42}$$

Next, we compute the gradient and the Hessian:

$$g(z_i) = \frac{\partial \log p(z_i|\sigma^2)}{\partial \sigma^2} = -\frac{1}{2(\sigma^2 + q_i)} + \frac{z_i^2}{2(\sigma^2 + q_i)^2} \tag{10.43}$$

$$H(z_i) = \frac{\partial^2 \log p(z_i|\sigma^2)}{\partial (\sigma^2)^2} = \frac{1}{2(\sigma^2 + q_i)} - \frac{z_i^2}{(\sigma^2 + q_i)^3} \tag{10.44}$$

We calculate $g_i(0), g_i'(0), g_i''(0), H_i(0), H_i'(0)$ and $H_i''(0)$:

$$g_i(0) = -\frac{1}{2\left(\sigma^2 + q_i\right)}, g_i'(0) = \left.\frac{z_i}{2\left(\sigma^2 + q_i\right)^2}\right|_{z_i=0} = 0, g_i''(0) = \frac{1}{\left(\sigma^2 + q_i\right)^2} \tag{10.45}$$

and

$$H_i(0) = \frac{1}{2\left(\sigma^2 + q_i\right)^2}, H_i'(0) = -\left.\frac{2z_i}{\left(\sigma^2 + q_i\right)^3}\right|_{z_i=0} = 0, H_i''(0) = -\frac{2}{\left(\sigma^2 + q_i\right)^3} \tag{10.46}$$

Again, this example does not fit in situation I', which means that we should plug in $g_i(\mu), g_i'(\mu), g_i''(\mu), H_i(\mu), H_i'(\mu)$ and $H_i''(\mu)$ to get the approximated value for $\text{var}[g_i^2(z_i)] - \text{var}[H_i(z_i)]$.

$$\begin{aligned}
&\text{var}[g_i^2(z_i)] - \text{var}[H_i(z_i)] \\
&\approx 2\left(\sigma^2 + q_i\right)^2\left(\frac{1}{4\left(\sigma^2 + q_i\right)^6} - \frac{1}{\left(\sigma^2 + q_i\right)^6}\right) \\
&\quad + \frac{1}{16\left(\sigma^2 + q_i\right)^8}\left(7!!\left(\sigma^2 + q_i\right)^4 - \left(3\left(\sigma^2 + q_i\right)^2\right)^2\right) \\
&\quad + \frac{1}{2\left(\sigma^2 + q_i\right)^4}\left(-\frac{1}{\left(\sigma^2 + q_i\right)^3}\right)\left(5!!\left(\sigma^2 + q_i\right)^3 - 3\left(\sigma^2 + q_i\right)^3\right) \\
&= \frac{3}{2\left(\sigma^2 + q_i\right)^4} > 0
\end{aligned} \tag{10.47}$$

Note that we know whether $s_g^2 - s_H^2$ is non-positive or positive by finding out whether $\text{var}[g_i^2(z_i)] - \text{var}[H_i(z_i)]$ is non-positive or positive. Given the above analysis,

Table 10.5 Sample variances of (10.5) and (10.6) as well as the difference between them, when the unknown parameter $\theta = \sigma^2$

	$\sigma^2 = 0.1^2$	$\sigma^2 = 1$	$\sigma^2 = 10^2$
$\text{var}^{(e)}[g^2(z)]$	6.11×10^7	1.035	3.57×10^{-8}
$\text{var}^{(e)}[H(z)]$	2.52×10^7	0.66	2.00×10^{-8}
$\text{var}^{(e)}[g^2(z)] - \text{var}^{(e)}[H(z)]$	3.59×10^7	0.38	1.57×10^{-8}

Table 10.6 Hypothesis testing results for using a one-sided t-test with 50 replicates

	$\sigma^2 = 0.1^2$	$\sigma^2 = 1$	$\sigma^2 = 10^2$
P-Value	0.0269	0.0025	2.4×10^{-7}

we assert that $\text{var}[g_i^2(z_i)] - \text{var}[H_i(z_i)] \geq 0$. Theoretically, (10.5) should outperform (10.4) in the sense that it has a smaller asymptotic variance. We check this assertion by implementing the example in MATLAB®.

Using three numerical cases where (1) $\sigma^2 = 0.1^2$ (2) $\sigma^2 = 1$ (3) $\sigma^2 = 10^2$, with $n = 100$ and $N = 200$, we get the results as shown in Table 10.5.

One noteworthy phenomenon is the recurrence of the situation confronted in Example 10.2—both variances decrease as σ^2 grows. Clearly, the numerical results coincide with our assertion that (10.5) leads to a smaller asymptotic variance. In order to validate the results, we do a t-test.

As shown in Table 10.6, the t-test demonstrates that the asymptotic variances of (10.4) and (10.5) in the i.n.i.d. scenario appear to be different, i.e., the null hypothesis of no difference is rejected, especially as σ^2gets larger. This fact supports the use of (10.5) as the way to estimate FIN instead of (10.4) for this application.

10.5 Conclusions and future work

This section summarizes the chapter with conclusions and potential future directions.

10.5.1 Conclusion

So far, we have looked at the comparison of accuracy for the two approximation methods in the scalar case and mainly focused our analysis on the family of symmetric density functions. We have considered i.i.d. data and i.n.i.d. data and provided some results for both scenarios.

The example for i.n.i.d. data, signal-plus noise for the scalar case, has a variety of utilizations in practice. According to our theoretical results, when dealing with this sort of problem, the Hessian method is a better choice because it gives a smaller asymptotic variance than the product-of-gradient method does. The difference between their asymptotic variances as estimated by our theory is positive. The numerical study confirms the positiveness of the difference. The same situation arises when the density function is normal with the unknown parameter to be estimated being σ^2. A positive difference between these two variances is well approximated by our theory and then numerically supported in simulation results.

Some interesting phenomena indeed appeared during the research. A salient issue is related to the use of the Taylor expansion. Note that we estimate $\mathrm{var}\left[g^2(z_i)\right]$ (or $\mathrm{var}\left[g_i^2(z_i)\right]$) with the second-order Taylor expansion by first expanding $g(z_i)$ (or $g_i(z_i)$) and then squaring it and proceeding to the rest of the calculation. We do not expand $g^2(z_i)$ (or $g_i^2(z_i)$) directly. This is a tricky part of the study because if we do this in the latter way, we are not able to come up with a valid analysis for the magnitude of the two asymptotic variances. Rather, we find that the magnitude of $\mathrm{var}\left[g^2(z_i)\right]$ (or $\mathrm{var}\left[g_i^2(z_i)\right]$) would be substantially underestimated. Some non-negative items would be neglected because of the incorrect use of the Taylor expansion, thus resulting in an approximation much smaller than the true value.

10.5.2 Future work

In this study, we mainly consider the situation when the explicit form of the density function is accessible, but the true value of Fisher information is not readily available, so we have to estimate it. We compare the accuracy of two methods used to achieve this. The magnitudes of the asymptotic variances of two approximation methods are contrasted and a conclusion is made that the accuracy is case dependent. A couple of conditions are necessary to be looked at to determine which method is to be used. Due to the limitation of time and length of the study, we have only considered the case where the unknown parameter θ is scalar. Two scenarios are looked at, one of which is when we have i.i.d. samples and the other one is when we have i.n.i.d. samples. Examples for both scenarios are provided in the numerical studies, which illustrate our conclusion by numerical results.

Nevertheless, this study has its limitations. It fails to consider the multidimensional case where θ is a vector. In reality, a vector θ of p dimension where p is no less than 2 is often the case we confront. Truly, this brings complexity to our analysis. Also, our study mainly concentrates on symmetric density functions. A broader scope of case study is desired, which can be a potential direction for future study. Furthermore, our theoretical analysis merely gives the sufficient conditions to determine the values for the key quantities $\mathrm{var}\left[g^2(z_i)\right] - \mathrm{var}[H(z_i)]$ and $s_g^2 - s_H^2$. Finding the necessary conditions to decide the magnitudes of these two key quantities is a promising research direction.

Obviously, there is more work to do for our research in future. Three topics to work on in future are listed below, which are representative but not comprehensive:

1. Apply the sufficient conditions we get in this paper to density functions other than the exponential family and see whether the numerical results match the theoretical analysis,
2. Find out the necessary conditions for the two key quantities and $s_g^2 - s_H^2$ to be non-positive or non-negative,
3. Generalize the study to multidimensional situations where θ is a vector.

Appendix A

In this section, we present the detailed calculation of $\mathrm{var}\left[g^2(z_i)\right]$, $\mathrm{var}[H(z_i)]$, $\mathrm{var}\left[g_i^2(z_i)\right]$, $\mathrm{var}[H_i(z_i)]$, s_g^2, and s_H^2. As mentioned above, the calculation of the actual values of $\mathrm{var}\left[g^2(z_i)\right]$, $\mathrm{var}[H(z_i)]$, $\mathrm{var}\left[g_i^2(z_i)\right]$, $\mathrm{var}[H_i(z_i)]$, s_g^2 and s_H^2 is demanding. We utilize the second-order Taylor expansion to give approximations to these values.

For i.i.d. samples, define $\mu := E(z_i)$. In order to derive (10.23) and (10.24), we do a second-order Taylor expansion of $g(z_i)$ at $z_i = \mu, i = 1, 2, \ldots, n$. Using the notation in (10.3), we have

$$g(z_i) \approx g(\mu) + g'(\mu)(z_i - \mu) + \frac{1}{2}g''(\mu)(z_i - \mu)^2 \tag{A.1}$$

Squaring the right-hand side above, we get

$$\begin{aligned} g^2(z_i) &\approx \left(g(\mu) + g'(\mu)(z_i - \mu) + \frac{1}{2}g''(\mu)(z_i - \mu)^2\right)^2 \\ &= g^2(\mu) + [g'(\mu)]^2(z_i - \mu)^2 + \frac{1}{4}[g''(\mu)]^2(z_i - \mu)^4 \\ &\quad + 2g(\mu)g'(\mu)(z_i - \mu) + g(\mu)g''(\mu)(z_i - \mu)^2 + g'(\mu)g''(\mu)(z_i - \mu)^3 \end{aligned} \tag{A.2}$$

Taking expectation of both sides,

$$\begin{aligned} E[g^2(z_i)] \approx\ & g^2(\mu) + ([g'(\mu)]^2 + g(\mu)g''(\mu))\sigma^2 \\ & + \frac{1}{4}[g''(\mu)]^2E[(z_i - \mu)^4] + g'(\mu)g''(\mu)E[(z_i - \mu)^3] \end{aligned} \tag{A.3}$$

Therefore,

$$\begin{aligned} &\mathrm{var}[g^2(z_i)] = E[(g^2(z_i) - E[g^2(z_i)])^2] \\ &\approx E\Bigg[[g'(\mu)]^2(z_i - \mu)^2 + \frac{1}{4}[g''(\mu)]^2(z_i - \mu)^4 + 2g(\mu)g'(\mu)(z_i - \mu) \\ &\quad + g(\mu)g''(\mu)(z_i - \mu)^2 + g'(\mu)g''(\mu)(z_i - \mu)^3 \\ &\quad - \Bigg(([g'(\mu)]^2 + g(\mu)g''(\mu))\sigma^2 - \frac{1}{4}[g''(\mu)]^2E[(z_i - \mu)^4] \\ &\quad - g'(\mu)g''(\mu)E[(z_i - \mu)^3]\Bigg)^2\Bigg] \\ &= E\Bigg[(2g(\mu)g'(\mu)(z_i - \mu) + ([g'(\mu)]^2 + g(\mu)g''(\mu)) \\ &\quad \times ((z_i - \mu)^2 - \sigma^2) + \frac{1}{4}[g'(\mu)]^2((z_i - \mu)^4 - E[(z_i - \mu)^4]) \\ &\quad + g'(\mu)g''(\mu)((z_i - \mu)^3 - E[(z_i - \mu)^3]))^2\Bigg] \\ &= 4E[g^2(\mu)[g'(\mu)]^2(z_i - \mu)^2] + E[([g'(\mu)]^2 + g(\mu)g''(\mu))^2((z_i - \mu)^2 - \sigma^2)^2] \end{aligned}$$

$$+\frac{1}{16}E[[g''(\mu)]^4((z_i-\mu)^4-E[(z_i-\mu)^4])^2]$$
$$+E[[g'(\mu)]^2[g''(\mu)]^2((z_i-\mu)^3-E[(z_i-\mu)^3])^2]$$
$$+4E[g(\mu)g'(\mu)([g'(\mu)]^2+g(\mu)g''(\mu))((z_i-\mu)^3-\sigma^2(z_i-\mu))]$$
$$+E[g(\mu)g'(\mu)[g''(\mu)]^2((z_i-\mu)^5-(z_i-\mu)E[(z_i-\mu)^4])]$$
$$+4E[g(\mu)[g'(\mu)]^2g''(\mu)((z_i-\mu)^4-(z_i-\mu)E[(z_i-\mu)^3])]$$
$$+\frac{1}{2}E\left[[g''(\mu)]^2([g'(\mu)]^2+g(\mu)g''(\mu))\underbrace{((z_i-\mu)^2-\sigma^2)((z_i-\mu)^4-E[(z_i-\mu)^4])}_{(a)}\right]$$
$$+2E\left[g'(\mu)g''(\mu)([g'(\mu)]^2+g(\mu)g''(\mu))\underbrace{((z_i-\mu)^2-\sigma^2)((z_i-\mu)^3-E[(z_i-\mu)^3])}_{(b)}\right]$$
$$+\frac{1}{2}E\left[g'(\mu)[g''(\mu)]^3\underbrace{((z_i-\mu)^3-E[(z_i-\mu)^3])((z_i-\mu)^4-E[(z_i-\mu)^4])}_{(c)}\right]$$
$$=4g^2(\mu)[g'(\mu)]^2\sigma^2+([g'(\mu)]^2+g(\mu)g''(\mu))^2\mathrm{var}[(z_i-\mu)^2]$$
$$+\frac{1}{16}[g''(\mu)]^4\mathrm{var}[(z_i-\mu)^4]+[g'(\mu)]^2[g''(\mu)]^2\mathrm{var}[(z_i-\mu)^3]$$
$$+4(\mu)g'(\mu)([g'(\mu)]^2+g(\mu)g''(\mu))E[(z_i-\mu)^3]$$
$$+g(\mu)g'(\mu)[g''(\mu)]^2E[(z_i-\mu)^5]+4g(\mu)[g'(\mu)]^2g''(\mu)E[(z_i-\mu)^4]$$
$$+\frac{1}{2}[g''(\mu)]^2([g'(\mu)]^2+g(\mu)g''(\mu))(E[(z_i-\mu)^6]-\sigma^2E[(z_i-\mu)^4])$$
$$+2g'(\mu)g''(\mu)([g'(\mu)]^2+g(\mu)g''(\mu))(E[(z_i-\mu)^5]-\sigma^2E[(z_i-\mu)^3])$$
$$+\frac{1}{2}g'(\mu)[g''(\mu)]^3(E[(z_i-\mu)^7]-E[(z_i-\mu)^4]E[(z_i-\mu)^3]) \quad \text{(A.4)}$$

The calculation of the expectations for expressions (a), (b) and (c) are as follows:

$$\begin{aligned}E(a) &= E[(z_i-\mu)^6-\sigma^2(z_i-\mu)^4-(z_i-\mu)^2E[(z_i-\mu)^4]+\sigma^2E[(z_i-\mu)^4]]\\ &= E[(z_i-\mu)^6]-\sigma^2E[(z_i-\mu)^4]-\underbrace{E[(z_i-\mu)^2]}_{=\sigma^2}E[(z_i-\mu)^4]+\sigma^2E[(z_i-\mu)^4]\\ &= E[(z_i-\mu)^6]-\sigma^2E[(z_i-\mu)^4] \end{aligned} \quad \text{(A.5)}$$

$$\begin{aligned}E(b) &= E[(z_i-\mu)^5 - \sigma^2(z_i-\mu)^3 - (z_i-\mu)^2E[(z_i-\mu)^3] + \sigma^2E[(z_i-\mu)^3]] \\ &= E[(z_i-\mu)^5] - \sigma^2E[(z_i-\mu)^3] - \underbrace{E[(z_i-\mu)^2]}_{=\sigma^2}E[(z_i-\mu)^3] + \sigma^2E[(z_i-\mu)^3] \\ &= E[(z_i-\mu)^5] - \sigma^2E[(z_i-\mu)^3] \end{aligned} \tag{A.6}$$

$$\begin{aligned}E(c) &= E[(z_i-\mu)^7 - (z_i-\mu)^4E[(z_i-\mu)^3] - (z_i-\mu)^3E[(z_i-\mu)^4] \\ &\quad + E[(z_i-\mu)^3]E[(z_i-\mu)^4]] \\ &= E[(z_i-\mu)^7] - E[(z_i-\mu)^4]E[(z_i-\mu)^3] - E[(z_i-\mu)^3]E[(z_i-\mu)^4] \\ &\quad + E[(z_i-\mu)^3]E[(z_i-\mu)^4] \\ &= E[(z_i-\mu)^7] - E[(z_i-\mu)^4]E[(z_i-\mu)^3] \end{aligned} \tag{A.7}$$

Next, we do second-order Taylor expansion of $H(z_i)$ at $z_i = \mu, i = 1, 2, \ldots, n$.

$$H(z_i) \approx H(\mu) + H'(\mu)(z_i-\mu) + \frac{1}{2}H''(\mu)(z_i-\mu)^2 \tag{A.8}$$

Taking the expectation on both sides,

$$E[H(z_i)] \approx H(\mu) + H'(\mu)\underbrace{E[z_i-\mu]}_{=0} + \frac{1}{2}H''(\mu)\underbrace{E[(z_i-\mu)^2]}_{=\sigma^2} = H(\mu) + \frac{1}{2}\sigma^2H''(\mu) \tag{A.9}$$

Therefore,

$$\begin{aligned}\mathrm{var}(H(z_i)) &= E[(H(z_i) - E[H(z_i)])^2] \\ &\approx E\left[\left(H'(\mu)(z_i-\mu) + \frac{1}{2}H''(\mu)(z_i-\mu)^2 - \frac{1}{2}\sigma^2H''(\mu)\right)^2\right] \\ &= E\left[\left(H'(\mu)(z_i-\mu) + \frac{1}{2}H''(\mu)[(z_i-\mu)^2 - \sigma^2]\right)^2\right] \\ &= [H'(\mu)]^2\underbrace{E[(z_i-\mu)^2]}_{=\sigma^2} + \frac{1}{4}[H''(\mu)]^2E[((z_i-\mu)^3 - \sigma^2)^2] \\ &\quad + H'(\mu)H''(\mu)(E[(z_i-\mu)^3] - \sigma^2\underbrace{E[z_i-\mu]}_{0}) \\ &= [H'(\mu)]^2\sigma^2 + \frac{1}{4}[H''(\mu)]^2(E[(z_i-\mu)^4] - \sigma^4) \\ &\quad + H'(\mu)H''(\mu)E[(z_i-\mu)^3] \end{aligned} \tag{A.10}$$

For i.n.i.d. samples, let $\mu_i := E(z_i)$. The calculation procedure is similar to that for the i.i.d. case. The only two modifications are that $E(z_i) = \mu_i$ instead of an identical mean value and individual var $\left(g_i^2(z_i)\right)$ and var $(H_i(z_i))$ are summed to get s_g^2 and s_H^2.

References

[1] Spall, J. C. and Garner, J. P. (1990), "Parameter Identification for State-Space Models with Nuisance Parameters," *IEEE Transactions on Aerospace and Electronic Systems*, vol. 26(6), pp. 992–998.

[2] McLachlan, G. J. and Peel, D. (2000), *Finite Mixture Models*, Wiley, New York.

[3] Favennec, Y. (2007), "Hessian and Fisher Matrices for Error Analysis in Inverse Heat Conduction Problems," *Numerical Heat Transfer, Part B: Fundamentals*, vol. 52(4), pp. 323–340.

[4] Spall, J. C. (2014), "Identification for Systems with Binary Subsystems," *IEEE Transactions on Automatic Control*, vol. 59(1), pp. 3–17.

[5] Ninness, B. and Henriksen, S. (2010), "Bayesian System Identification via Markov Chain Monte Carlo Techniques," *Automatica*, vol. 46, pp. 40–51.

[6] Spall, J. C. (2003), Introduction to Stochastic Search and Optimization: Estimation, Simulation, and Control, Wiley, Hoboken, NJ.

[7] Efron, B. and Hinkley, D. V. (1978), "Assessing the Accuracy of the Maximum Likelihood Estimator: Observed Versus Expected Fisher Information," *Biometrika*, vol. 65, no. 3, pp. 457–482.

[8] Cao, X. and Spall, J. C. (2009), "Preliminary Results on Relative Performance of Expected and Observed Fisher Information," *Proceedings of the IEEE Conference on Decision and Control*, Shanghai, P.R. China, 16–18 December 2009, pp. 1538–1543.

[9] Cao, X. and Spall, J. C. (2012), "Relative Performance of Expected and Observed Fisher Information in Covariance Estimation for Maximum Likelihood Estimates," *Proceedings of the American Control Conference*, Montréal, Canada, 27–29 June 2012, pp. 1871–1876.

[10] Spall, J. C. (2005), "Monte Carlo Computation of the Fisher Information Matrix in Nonstandard Settings," *Journal of Computational and Graphical Statistics*, December 2005, vol. 14(4), pp. 889–909.

[11] Rudin, W. (1976), Principles of Mathematical Analysis (3rd ed.), McGraw-Hill, New York.

Chapter 11

A hierarchical approach to data-driven LPV control design of constrained systems

Dario Piga[1], Simone Formentin[2], Roland Tóth[3], Alberto Bemporad[4], and Sergio Matteo Savaresi[2]

Modeling is recognized to be one of the toughest and most time-consuming tasks in modern nonlinear control engineering applications. Linear parameter-varying (LPV) models deal with such complex problems in an effective way, by exploiting well-established tools for linear systems while, at the same time, being able to accurately describe highly nonlinear and time-varying plants. When LPV models are derived from experimental data, it is difficult to estimate a priori how modeling errors will affect the closed-loop performance. In this work, a method is proposed to directly map data onto LPV controllers. Specifically, a hierarchical structure is proposed both to maximize the system performance and to handle signal constraints. The effectiveness of the approach is illustrated via suitable simulation tests.

11.1 Introduction

LPV modeling represents an effective tool to describe many nonlinear time-varying systems using linear *input–output* (IO) maps, wherein changes of an exogenous measurable variable, called scheduling signal, accounts for nonlinear behavior and time dependency [1]. Using standard robust and gain-scheduling techniques for linear systems, it has been shown that simple and effective controllers can be devised for such complex systems [2].

However, most LPV control design techniques rely on the availability of an accurate physical model of the plant. The effect of modeling errors between the LPV model and the physical plant is often unpredictable on the controller; hence, the resulting closed-loop performance might be severely jeopardized. Even in those applications where gathering data to identify and validate a model of the plant is not costly nor time-consuming, estimating a compact LPV description of the plant which

[1]IDSIA Dalle Molle Institute for Artificial Intelligence SUPSI-USI, Manno, Switzerland
[2]Dipartimento di Elettronica, Informazione e Bioingegneria, Politecnico di Milano, Milano, Italy
[3]Department of Electrical Engineering, Eindhoven University of Technology, Eindhoven, The Netherlands
[4]IMT School for Advanced Studies Lucca, Lucca, Italy

accurately captures the control-relevant aspects of the dynamics is a cumbersome procedure due to the absence of efficient tools that support identification for control in the LPV context. In fact, when deriving a model of the plant, one always trades off accuracy and complexity, and, most of the times, with the current identification tools it is not possible to decide a priori how accurate the model should be to achieve the desired closed-loop performance.

Furthermore, low complexity models of LPV systems are more efficiently derived using IO model structures [1,3,4], which allow to extend linear time-invariant (LTI) prediction-error methods to the LPV framework avoiding the curse of dimensionality present in the identification of state-space LPV models [5,6]. On the other hand, most of the control design methods are based on a state-space representation of the system (except some recent works, e.g., [7,8]) and minimal state-space realization of complex IO models is difficult to accomplish [1]. The above problems show that LPV control of nonlinear time-varying models has a great potential (see, e.g., some successful applications in [9–12]) but also suffers from some substantial practical limitations, mostly related to modeling issues.

In this chapter, a *method is proposed to design fixed-order LPV controllers in an IO form using directly experimental data. In fact, this corresponds of designing controllers without deriving a model of the system*. This approach permits to avoid the critical (and time-consuming) approximation steps related to modeling, identification and state-space realization and it results in an automatic procedure in which only the desired closed-loop behavior has to be specified by the user. Recent results in data-driven LPV model structure selection [13,14] employing *least squares support vector machines* (LS-SVMs) [15] are exploited, and the controller synthesis problem is formulated as an optimization problem. *Instrumental-variable* (IV)-based identification techniques are used to efficiently cope with the noise affecting the signal measurements so that the resulting optimization problem is convex and the bias error is guaranteed to converge to zero as the number of data samples increases. The basic idea behind this method was first presented in [16].

An outer *model predictive controller* (MPC) is then added to the above loop, acting as a *reference governor*, aiming at enhancing the closed-loop performance in case the desired performance is not fully achieved and ensuring that signal constraints are not violated. We will show that also the whole control design procedure does not depend on the plant knowledge, according to the direct data-driven philosophy of the method. To the best of the authors' knowledge, this is the only method dealing with direct data-driven control design for constrained LPV systems and was first presented in [17].

The remainder of this chapter is organized as follows. First of all, the proposed methodology is positioned within the existing literature in Section 11.2. The mathematical formulation of the design problem is provided in Section 11.3, whereas Section 11.4 outlines the main idea behind our suggested method. Then, Section 11.5 illustrates the technical derivation of the data-driven design method for LPV control, and Section 11.6 describes the rationale behind the model-free reference governor for performance boosting and constraints handling. The effectiveness of the developed control design method is shown by means of a simulation case study in Section 11.7. The chapter is ended by some concluding remarks.

11.2 Related works

Direct controller tuning approaches using a single set of IO data, also known as non-iterative data-driven control, have been already studied in the *LTI* framework [18]. Well-established approaches, like *virtual reference feedback tuning* (VRFT) [19,20] and *non-iterative correlation-based tuning* (CbT) [21], have been widely discussed in the literature, see, e.g., the recent results in [22–27]. Other recently introduced approaches are, e.g., [28,29].

The first attempt to extend a data-driven method, namely, the VRFT method, to LPV systems has been presented in [30], where data-driven gain-scheduled controller design has been proposed to realize a user-defined LTI closed-loop behavior. Although satisfactory performance has been shown for slowly varying scheduling trajectories, this methodology is far from being generally applicable to LPV systems. As a matter of fact, in the method presented in [30], the controller must be linearly parameterized and the reference behavior must be LTI. The latter requirement represents a strict limitation, since an LTI behavior might be difficult to realize in practice, as it may require too demanding input signals and dynamic dependence of the controller on the scheduling signal. On the other hand, the LPV extension of non-iterative CbT has been found to be unfeasible, as the derivation of this approach is based on the commutation of the plant and the controller in the tuning scheme [31]. Unfortunately, such a commutation does not generally hold for parameter-varying transfer operators [32]. The recent work in [33] also deals with LPV direct data-driven control, but the framework is completely different from the one proposed herein. Specifically, in [33], the system is given in state-space form with measurable state vector, the optimal controller is assumed to be Lipschitz continuous and the whole method is developed in a deterministic set-membership setting. A direct data-driven LPV solution has been presented in a stochastic framework for feed-forward precompensator tuning in [34]. However, also in this case, no dynamic dependance is accounted for, and the final objective is constrained to be LTI.

11.3 Problem formulation

Let $\mathscr{G}$ be an unknown *single-input single-output* LPV system, which generates data according to the relation[1]

$$y_o(t) = G(p, t, q^{-1})u(t), \tag{11.1}$$

[1]This formulation is adopted to represent a general situation and can be greatly simplified in specific cases.

where $G(p, t, q^{-1})$ is a time-domain operator describing the behavior of $\mathscr{G}$, q^{-1} denotes the backward shift operator, $u(t) \in \mathbb{R}$ is the input signal at the time sample t, $y_o(t) \in \mathbb{R}$ is the noise-free output and $p(t) \in \mathbb{P} \subseteq \mathbb{R}^{n_p}$ is a set of n_p (exogenous) measurable scheduling variables. Without loss of generality, the case of $n_p = 1$ will be discussed in this work in order to keep the notation simple. The measured output of the system is supposed to be corrupted by an additive, zero mean, stationary colored noise $w(t)$, i.e.,

$$y(t) = y_o(t) + w(t). \tag{11.2}$$

The system $\mathscr{G}$ is assumed to be *bounded-input bounded-output* (BIBO) *stable* in $\mathscr{L}_\infty$ norm [1,35] and, according to this notion, stability is defined as follows:

Definition 11.1 (BIBO stability). *The system $\mathscr{G}$ is BIBO stable if the following condition holds for all $p(t) \in \mathbb{P}$ and all trajectories $\{u(t), y(t), p(t)\}$ satisfying* (11.1)*:*

$$\sup_{t\geq 0} |u(t)| < \infty \implies \sup_{t\geq 0} |y(t)| < \infty.$$

The first objective of the considered control problem is to achieve a desired closed-loop behavior $\mathscr{M}_p$, given by a state-space representation[2]

$$x_M(t+1) = A_M(p,t)x_M(t) + B_M(p,t)g(t), \tag{11.3a}$$

$$y_d(t) = C_M(p,t)x_M(t) + D_M(p,t)g(t), \tag{11.3b}$$

where y_d denotes the desired closed-loop output for a given reference signal g. In the sequel, the operator $M(p, t, q^{-1})$ is used as a shorthand form to indicate the mapping of g to y_d in terms of (11.3a) and (11.3b), i.e., $y_d(t) = M(p, t, q^{-1})g(t)$ for all trajectories of p and g. In case the reference model is given in an IO form, this can be realized in a state-space representation using the so-called maximally augmented realization form [37] or the approaches presented in [38].

The class of controllers $\mathscr{K}_p(\theta)$ is described in terms of the IO representation

$$A_K(p, t, q^{-1}, \theta)u(t) = B_K(p, t, q^{-1}, \theta)(g(t) - y(t)), \tag{11.4}$$

where A_K and B_K are polynomials in the backward shift operator q^{-1} (i.e., $q^{-1}u(t) = u(t-1)$) whose coefficients, parameterized with the design parameters collected in θ, can be any (possibly dynamic) bounded function of the scheduling variable p.

Remark 11.1 (Dynamic dependence on the scheduling signal). *The controller should be assumed to be dynamically dependent on p in order to have enough flexibility to achieve the user-defined behavior $\mathscr{M}_p$. As a matter of fact, it is well-known that a static dependence would be a rather strong assumption for most real-world systems [1,32].*

[2]In this work, we assume $\mathscr{M}_p$ is given as a part of the problem specification, e.g., the system matrices could be affine functions of the scheduling signal with certain dynamics. Notice that the choice of the reference model within a model-free setting is otherwise an open research problem. An interesting approach for setting $\mathscr{M}_p$ in the LTI framework can be found in [36].

Assume that a collection of open-loop data $\mathscr{D}_N = \{u(t), y(t), p(t)\}$, $t \in \mathscr{I}_1^N = \{1, \ldots, N\}$, from (11.1) is available. Assume also that the following statements hold the following:

A1. There exists a value θ° such that the controller $\mathscr{K}_\mathrm{p}(\theta^\circ)$ realizes $\mathscr{M}_\mathrm{p}$ in closed loop.
A2. Equation (11.4) is globally identifiable, i.e., for any two instances of the parameter vector θ, namely, $\theta^{(1)}$ and $\theta^{(2)}$, there exists a trajectory of u and p such that the response of the feedback interconnection of (11.1) and (11.4) is different if and only if $\theta^{(1)} \neq \theta^{(2)}$. This implies that θ° is unique.
A3. The dataset $\mathscr{D}_N$ is persistently exciting with respect to the used parameterization, i.e., based on $\mathscr{D}_N$, θ° can be uniquely determined.

Therefore, the direct data-driven model-reference control problem addressed in this work can be formally stated as follows.

Problem 11.1 (Data-driven model-reference control). *Given a reference model* (11.3a) *and* (11.3b) *and a controller structure $\mathscr{K}_\mathrm{p}(\theta)$ as defined in* (11.4)*, determine the parameter vector $\hat{\theta}$ defining the controller $\mathscr{K}_\mathrm{p}(\hat{\theta})$ from a noisy dataset $\mathscr{D}_N = \{u(t), y(t), p(t)\}_{t=1}^N$, so that $\hat{\theta}$ asymptotically converges to θ° as $N \to \infty$.*

In order to handle actuator limitations, safety requirements and performance specifications, constraints on the IO signal are also considered. Specifically, the following constrained control problem is addressed along with Problem 11.1.

Problem 11.2 (Control of constrained systems). *When the loop is closed with the synthesized controller $\mathscr{K}_\mathrm{p}(\theta)$, the following constraints on the IO signals should not be violated:*

$$u_{\min} \leq u(t) \leq u_{\max}, \quad \Delta u_{\min} \leq u(t) - u(t-1) \leq \Delta u_{\max}, \tag{11.5a}$$

$$y_{\min} \leq y(t) \leq y_{\max}, \quad \forall t \in \mathbb{Z},\ t \geq 0. \tag{11.5b}$$

11.4 A hierarchical approach

The *constrained data-driven model-reference control problem* (namely, Problems 11.1 and 11.2) is tackled through the hierarchical architecture illustrated in Figure 11.1, which integrates

- An *inner linear controller* $\mathscr{K}_\mathrm{p}(\theta)$, described by (11.4) and designed from the available dataset $\mathscr{D}_N$ according to the model-reference control specifications formalized in Problem 11.1.
- An *outer linear MPC*, designed based on the desired closed-loop model $\mathscr{M}_\mathrm{p}$ according the control requirements formalized in Problem 11.2. The MPC controller, according to a *receding horizon* strategy, designs online the optimal reference to the inner loop in order to fulfill the constraints (11.5a) and (11.5b),

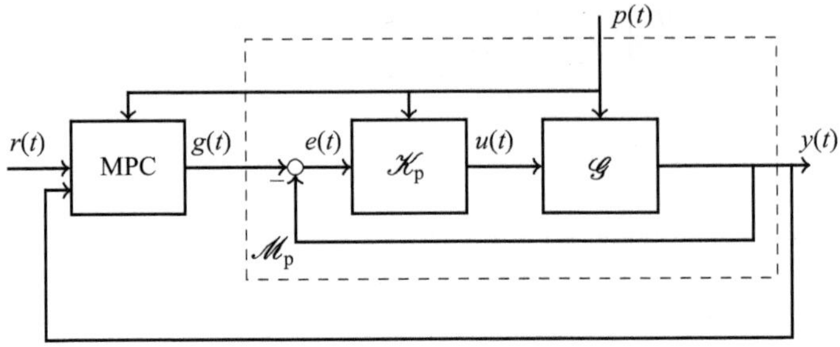

Figure 11.1 Hierarchical control architecture: the inner controller $\mathscr{K}_\mathrm{p}$ is designed from data to provide tracking capabilities for the unconstrained system $\mathscr{G}$ and to match $\mathscr{M}_\mathrm{p}$ by the inner loop as much as possible, whereas the outer predictive controller enhances the performance and guarantees that the constraints are not violated

acting as a reference governor [39]. Besides constraint fulfillment, the outer MPC may be also used to enhance the performance of the inner closed-loop system.

By merging the two controllers together in a hierarchical fashion, one can choose a low-demanding (e.g., with slow dynamics and low damping factor) inner closed-loop behavior $\mathscr{M}_\mathrm{p}$, which is known to be easily achievable by the inner controller $\mathscr{K}_\mathrm{p}(\theta)$ (for this, only a rough knowledge of the process dynamics is required). The tasks of optimizing the closed-loop performance and fulfilling the IO constraints are then left to the outer MPC, which is designed based on the closed-loop dynamics (modeled by $\mathscr{M}_\mathrm{p}$) and based on the model of the inner controller $\mathscr{K}_\mathrm{p}$.

The design of the inner controller $\mathscr{K}_\mathrm{p}(\theta)$ and of the outer MPC is discussed in Sections 11.5 and 11.6, respectively.

11.5 Data-driven inner controller design

In this section, we address the design of the inner controller $\mathscr{K}_\mathrm{p}$ according the data-driven model-reference control problem (namely, Problem 11.1). To start, the following additional assumption is taken:

A4. The model reference $M(p,t,q^{-1})$ is invertible.

The inverse of a LPV mapping is defined as follows:

Definition 11.2 (Left inverse). *Consider a causal LPV map $M(p,t,q^{-1})$ with input g, scheduling signal p and output y_d. The LPV mapping $M^\dagger(p,t,q^{-1})$ that gives g as output when fed by y_d, for any trajectory of p, is called the* left inverse *of $M(p)$, i.e., $M^\dagger(p,t,q^{-1})M(p,t,q^{-1}) = 1$.*

Computation of the *left inverse* of an LPV map is discussed in Section 11.5.1.

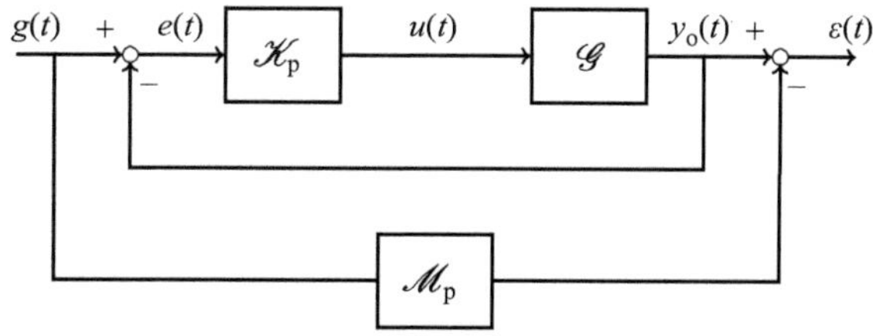

Figure 11.2 Closed-loop behavior matching scheme. Both the plant $\mathscr{G}$ the controller $\mathscr{K}_\mathrm{p}$ and the reference model $\mathscr{M}_\mathrm{p}$ depend on the scheduling variable p

Due to assumption **A1**, notice that Problem 11.1 can be reformulated as the optimization task

$$\min_{\theta,\varepsilon} \|\varepsilon\|_{\ell_2}^2 \tag{11.6a}$$

$$\text{s.t.} \quad \varepsilon(t) = M(p,t)g(t) - y_\mathrm{o}(t), \qquad \forall t \in \mathscr{I}_1^N, \tag{11.6b}$$

$$y_\mathrm{o}(t) = G(p,t)u(t), \qquad \forall t \in \mathscr{I}_1^N, \tag{11.6c}$$

$$A_K(p,t,\theta)u(t) = B_K(p,t,\theta)(g(t) - y_\mathrm{o}(t)), \qquad \forall t \in \mathscr{I}_1^N. \tag{11.6d}$$

over a generic time interval $\mathscr{I}_1^N$, denoting the set of instants from $t = 1$ until $t = N$, where the time-domain operator $G(p,t)$ of the open-loop process $\mathscr{G}$ is, in principle, identified from $\mathscr{D}_N$. The argument q^{-1} has been dropped for the sake of readability.

Notice that this corresponds to matching the desired dynamical behavior with the actual closed-loop system, as graphically illustrated in Figure 11.2.

If a consistent method is used for the identification of $G(p,t)$ (e.g., the PEM method in [1] in the case $G(p,t)$ is an LPV model), and its dependence on the scheduling variable p is correctly parameterized, the estimate of the system asymptotically converges to the real $G(p,t)$ and the controller resulting as the solution of (11.6a)–(11.6d) makes the closed-loop system asymptotically converge to (11.3a) and (11.3b), thus solving Problem 11.1. However, such a model-reference problem is very hard to solve in the given form. In what follows, it will be shown that the problem can be further reformulated in a different fashion, which requires neither to parameterize nor to identify the system $\mathscr{G}$.

The proposed approach is based on two-key ideas. The first one is that, *under Assumption* **A4**, *the dependence on the choice of g can be annihilated.* As a matter of fact, by rewriting the constraint (11.6b) as

$$g(t) = M^\dagger(p,t,q^{-1})\varepsilon(t) + M^\dagger(p,t,q^{-1})y_\mathrm{o}(t), \tag{11.7}$$

where $M^\dagger$ denotes the left inverse of M, the optimization (11.6a)–(11.6d) can be reformulated as

$$\min_{\theta,\varepsilon} \|\varepsilon\|_{\ell_2}^2 \tag{11.8a}$$

$$\text{s.t. } y_o(t) = G(p,t)u(t) \qquad \forall t \in \mathcal{I}_1^N \tag{11.8b}$$

$$A_K(p,t,\theta)u(t) = B_K(p,t,\theta)(M^\dagger(p,t)\varepsilon(t) + M^\dagger(p,t)y_o(t) - y_o(t)), \qquad \forall t \in \mathcal{I}_1^N \tag{11.8c}$$

Notice that, unlike in (11.6a)–(11.6d), the reference signal (11.7) is expressed as a projection of y_o and ε. Specifically, such a projection corresponds to the sum of two terms:

- The reference trajectory that would produce the data y_o as an output, in case the closed-loop system is equal to $\mathcal{M}_p$.
- A term compensating the mismatch between $\mathcal{M}_p$ and the actual closed-loop system. The term depends on the designed controller $\mathcal{K}_p(\theta)$ and thus parameterized by θ.

In this way, an explicit trajectory of g is not required, as (11.7) corresponds to a *virtual reference signal* satisfying the above given conditions. In other words, problem (11.8a)–(11.8c) still corresponds to a closed-loop model matching task, but now it can be solved based on the open-loop data $\mathcal{D}_N$.

Now we take another observation. *Since the available dataset $\mathcal{D}_N$ is generated according to the system equation* (11.8b)*, $\mathcal{D}_N$ can be used—under Assumption **A3** and $w = 0$—as a substitute to represent the dynamics of the open-loop process.*

Consider then the problem

$$\min_{\theta,\varepsilon} \|\varepsilon\|_{\ell_2}^2 \tag{11.9a}$$

$$\begin{aligned} \text{s.t.} \quad & A_K(p,t,\theta)u(t) = B_K(p,t,\theta)(M^\dagger(p,t)\varepsilon(t) + M^\dagger(p,t)y(t) - y(t)), \\ & \{u(t), y(t), p(t)\} \in \mathcal{D}_N, \end{aligned} \tag{11.9b}$$

Notice that such a problem is independent of the analytical description of $G(p,t)$ and therefore no model identification is needed to solve it. If the data are noiseless, *the global minimizer of* (11.9a) *and* (11.9b) *coincides with that of* (11.6a)–(11.6d), *providing the optimal controller parameters θ° achieving $\mathcal{M}_p$ in closed-loop and yielding $\varepsilon = 0$.*

Let us now consider the practical situation, where the output is noisy (i.e., $w \neq 0$)). In this scenario, (11.9a) and (11.9b) leads to a bias estimate of the optimal controller θ°. Indeed, the optimization problem (11.9a) and (11.9b) pushes for $\varepsilon = 0$, whereas the minimizer of (11.8a)–(11.8c) would yield ε such that

$$M^\dagger(p,t)\varepsilon(t) = M^\dagger(p,t)w(t) - w(t). \tag{11.10}$$

Proper stochastic treatment of the noise will be discussed to handle the discrepancy between (11.8a)–(11.8c) and (11.9a) and (11.9b), in order to obtain a solution of (11.9a) and (11.9b) that still asymptotically converges to the solution of (11.6a)–(11.6d).

11.5.1 Inversion of the reference model

Before proceeding, let us analyze the problem of computing $M^\dagger$. For reference maps given in the state-space form (11.3a) and (11.3b), the following proposition can be employed:

Proposition 11.1 (Computing the left inverse). *Consider* (11.3a) *and* (11.3b) *and assume that there exists a* $D_M^{-1}(p,t)$ *which satisfies* $D_M^{-1}(p,t)D_M(p,t) = 1$ *and is bounded* $\forall p \in \mathbb{P}^{\mathbb{Z}}$. *Then,*

$$x_{M^\dagger}(t+1) = A_{M^\dagger}(p,t)x_{M^\dagger}(t) + B_{M^\dagger}(p,t)y_{\mathrm{d}}(t) \tag{11.11a}$$

$$g(t) = C_{M^\dagger}(p,t)x_{M^\dagger}(t) + D_{M^\dagger}(p,t)y_{\mathrm{d}}(t) \tag{11.11b}$$

defines the left-inverse of (11.3a) *and* (11.3b), *where the matrix functions are given as*

$$\begin{aligned} A_{M^\dagger}(p,t) &= A_M(p,t) - B_M(p,t)D_M^{-1}(p,t)C_M(p,t),\\ B_{M^\dagger}(p,t) &= B_M(p,t)D_M^{-1}(p,t),\\ C_{M^\dagger}(p,t) &= -D_M^{-1}(p,t)C_M(p,t),\\ D_{M^\dagger}(p,t) &= D_M^{-1}(p,t). \end{aligned} \tag{11.12}$$

Proof. Consider the generic form of the output equation in (11.11a) and (11.11b). Substituting (11.3b) gives

$$g(t) = C_{M^\dagger}(p,t)x_{M^\dagger}(t) + D_{M^\dagger}(p,t)C_M(p,t)x_M(t) + D_{M^\dagger}(p,t)D_M(p,t)g(t). \tag{11.13}$$

Choose $x_{M^\dagger} = x_M$. Since D_M is assumed to be invertible, it follows that

$$D_{M^\dagger}(p,t) = D_M^{-1}(p,t), \quad C_{M^\dagger}(p,t) = -D_M^{-1}(p,t)C_M(p,t).$$

Moreover, since $x_{M^\dagger} = x_M$, the state equations of $M^\dagger$ and M can be combined such as

$$A_{M^\dagger}(p,t)x_M(t) + B_{M^\dagger}(p,t)y_{\mathrm{d}}(t) = A_M(p,t)x_M(t) + B_M(p,t)g(t), \tag{11.14}$$

where the left-hand term can be rewritten as

$$A_{M^\dagger}(p,t)x_M(t) + B_{M^\dagger}(p,t)C_M(p,t)x_M(t) + B_{M^\dagger}(p,t)D_M(p,t)g(t). \tag{11.15}$$

Based on the above given equations, the expressions of $A_{M^\dagger}$ and $B_{M^\dagger}$ follow as

$$\begin{aligned} A_{M^\dagger}(p,t) &= A_M(p,t) - B_{M^\dagger}(p)C_M(p,t), \\ B_{M^\dagger}(p,t) &= B_M(p,t)D_M^{-1}(p,t), \end{aligned}$$

which completes the proof. □

In the case of $D_M = 0$, D_M can be approximated by $\hat{D}_M = \varepsilon_D$ where $0 < \varepsilon_D \ll 1$ (as it is common in robust control [40]) to compute the inverse mapping. Another practical way to overcome the inverse in this case with a relative degree delayed inversion is discussed in [16].

11.5.2 Data-driven controller design

Let us set the following structure for the controller in (11.4):

$$A_K(p,t,q^{-1}) = 1 + \sum_{i=1}^{n_{\mathrm{a}_K}} a_i^K(p,t)q^{-i}, \tag{11.16a}$$

$$B_K(p,t,q^{-1}) = \sum_{i=0}^{n_{\mathrm{b}_K}} b_i^K(p,t)q^{-i}, \tag{11.16b}$$

with orders $n_{\mathrm{a}_K} \in \mathbb{N}$ and $n_{\mathrm{b}_K} \in \mathbb{N}$ specified by the user.

To select a controller structure such that the reference model is achievable is a hard task when the model of the system is unknown. Therefore, the p-dependent coefficient functions $a_i^K(p,t)$ and $b_i^K(p,t)$ characterizing the LPV controller (11.16a) and (11.16b) are not a priori parameterized, i.e., the dependence of $a_i^K(p,t)$ and $b_i^K(p,t)$ on p is not a priori specified. The data-driven synthesis procedure for the controller coefficient functions $a_i^K(p,t)$ and $b_i^K(p,t)$ is now discussed.

Let us write the p-dependent functions $a_i^K(p,t)$ and $b_i^K(p,t)$ in (11.16a) and (11.16b) as

$$a_i^K(p,t) = \theta_i^\top \psi_i(p,t) \tag{11.17a}$$

$$b_i^K(p,t) = \theta_{i+n_{\mathrm{a}_K}+1}^\top \psi_{i+n_{\mathrm{a}_K}+1}(p,t), \tag{11.17b}$$

where $i \in \mathcal{I}_1^{n_{\mathrm{a}_K}}, j \in \mathcal{I}_1^{n_{\mathrm{b}_K}}$ (defining the sets of indexes from 1 to n_{a_K} and from 1 to n_{b_K}, respectively) and $\theta_i \in \mathbb{R}^{n_\mathrm{H}}$ is a vector of unknown parameters and $\psi_i(p,t)$ (with $i \in \mathcal{I}_1^{n_{\mathrm{a}_K}+n_{\mathrm{b}_K}+1}$) is a set of unknown nonlinear maps from the original scheduling space $\mathbb{P}$ to an n_H-dimensional space, commonly referred to as the *feature space*. Note that neither the maps ψ_i nor the dimension n_H of the vectors θ_i and ψ_i are specified. Potentially, θ_i and $\psi_i(p,t)$ can be infinite-dimensional vectors (i.e., $n_\mathrm{H} = \infty$).

Let us define the signals

$$\xi(t) = M^\dagger(p,t)y(t) - y(t), \quad \xi_\mathrm{o}(t) = M^\dagger(p,t)y_\mathrm{o}(t) - y_\mathrm{o}(t) \tag{11.18}$$

and the vector $x \in \mathbb{R}^{n_\mathrm{f}}$ (with $n_\mathrm{f} = n_{\mathrm{a}_K} + n_{\mathrm{b}_K} + 1$) as

$$x(\xi,t) = \left[-u(t-1) \ \cdots \ -u(t-n_{\mathrm{a}K}) \ \ \xi(t) \ \cdots \ \xi(t-n_{\mathrm{b}_K})\right]^\top. \tag{11.19}$$

with $x_i(\xi, t)$ denoting the ith component of the vector $x(\xi, t)$. Based on the introduced notation, the constraint in (11.9a) and (11.9b) defining the controller behavior can be rewritten in the regression form:

$$u(t) = \sum_{i=1}^{n_\mathrm{f}} \theta_i^\top \psi_i(p,t) x_i(\xi, t) + \underbrace{B_K(p,t,q^{-1}) M^\dagger(p,t)\varepsilon(t)}_{\varepsilon_u(t)}. \tag{11.20}$$

To avoid the bias problem discussed in (11.10), synthesis of the controller based on a data record $\mathscr{D}_N = \{u(t), y(t), p(t)\}_{t=1}^N$ is formulated as an *IV*-based estimation problem [14]:

$$\min_{\theta_i, \varepsilon_u} \frac{1}{2} \sum_{i=1}^{n_\mathrm{f}} \theta_i^\top \theta_i + \frac{\gamma}{2N^2} \sum_{i=1}^{n_\mathrm{f}} \left\| \sum_{t=1}^{N} z_i(t)\varepsilon_u(t) \right\|_2^2 \tag{11.21a}$$

$$\text{s.t. } \varepsilon_u(t) = u(t) - \sum_{i=1}^{n_\mathrm{f}} \theta_i^\top \psi_i(p,t) x_i(\xi, t), \quad \forall t \in \mathscr{I}_1^N, \tag{11.21b}$$

where $z_i(t) \in \mathbb{R}^{n_\mathrm{H}}$ is the so-called instrument, a vector with the same dimension of $\psi_i(p,t)$ (thus, $z_i(t)$ can be an infinite-dimensional vector) which is chosen by the user so that $z_i(t)$ is uncorrelated with the noise term $\xi_\mathrm{o}(t) - \xi(t) = (M^\dagger(p,t) - 1)w(t)$, i.e.,

$$\mathbb{E}\{z_i(t)(M^\dagger(p,t) - 1)w(t)\} = 0, \ \forall t \in \mathscr{I}_1^N, \ i = 1, \ldots, n_\mathrm{f}. \tag{11.22}$$

Note that, since the map $\psi_i(p,t)$ does not depend on the noise $w(t)$, an instrument $z_i(t)$ satisfying condition (11.22) can be defined as follows:

$$z_i(t) = \psi_i(p,t) x_i(\hat{\xi}, t), \tag{11.23}$$

where

$$x(\hat{\xi}, t) = \begin{bmatrix} -u(t-1) & \cdots & -u(t-n_{\mathrm{a}_K}) & \hat{\xi}(t) & \cdots & \hat{\xi}(t-n_{\mathrm{b}_K}) \end{bmatrix}^\top, \tag{11.24}$$

with $\hat{\xi}(t)$ being an approximation (independent of the noise $w(t)$) of the noise-free signal $\xi_\mathrm{o}(t)$. This choice of the instrument is inspired by the LTI framework. A possible way to approximate $\xi_\mathrm{o}(t)$, preserving independence of the noise realization $w(t)$, is to take a second measurement of the output $y_\mathrm{o}(t)$, which would be uncorrelated with the realization of noise characterizing the original dataset.

Note that the second term in the objective function of (11.21a) and (11.21b) aims at minimizing the projection (under the instrument $z(t)$) of the colored residual $\varepsilon_u(t) = B_K(p,t,\theta) M^\dagger(p,t)\varepsilon(t)$, while the regularization term $\sum_{i=1}^{n_\mathrm{f}} \theta_i^\top \theta_i$ is included to prevent overfitting. In fact, since the dimension n_H of the parameter vector θ_i is not specified, penalizing the two norms of θ_i is essential to achieve an accurate estimate of the functions $a_i^K(p,t)$ and $b_i^K(p,t)$ in terms of bias/variance trade-off. The regularization parameter γ should be tuned by, e.g., cross validation, taking into account this trade-off.

Before showing how to solve (11.21a) and (11.21b) exactly, let us study the asymptotic properties of the solution of (11.21a) and (11.21b). To this aim, let us introduce the matrix notation:

$$\Psi_i = \begin{bmatrix} \psi_i^\top(p,1) \\ \vdots \\ \psi_i^\top(p,N), \end{bmatrix}, \quad E = \begin{bmatrix} \varepsilon_u(1) \\ \vdots \\ \varepsilon_u(N) \end{bmatrix}, \tag{11.25a}$$

$$X_i(\xi) = \begin{bmatrix} x_i(\xi,1) & 0 & \cdots & 0 \\ 0 & x_i(\xi,2) & & \vdots \\ \vdots & & \ddots & 0 \\ 0 & \cdots & 0 & x_i(\xi,N) \end{bmatrix}. \tag{11.25b}$$

Then, with the instrument $z_i(t)$ as in (11.23), problem (11.21a) and (11.21b) can be written in the compact form:

$$\min_{\theta_i,E} \frac{1}{2}\sum_{i=1}^{n_f} \theta_i^\top\theta_i + \frac{\gamma}{2N^2}\sum_{i=1}^{n_f} \left\| \Psi_i^\top X_i(\hat{\xi})E \right\|_2^2 \tag{11.26a}$$

$$\text{s.t. } E = U - \sum_{i=1}^{n_f} X_i(\xi)\Psi_i\theta_i. \tag{11.26b}$$

Proposition 11.2 (Consistency of controller parameters). *Given θ° as the optimal parameter minimizing* (11.6a)–(11.6d)*, the controller parameters $\hat{\theta}_{\mathrm{NP,IV}}$, obtained by minimizing problem* (11.26a) *and* (11.26b)*, are such that, with probability 1 (w.p. 1),*

$$\lim_{N\to\infty} \hat{\theta}_{\mathrm{NP,IV}} = \theta^\circ - R^{-1}\gamma^{-1}\theta^\circ, \tag{11.27}$$

where

$$R = \lim_{N\to\infty} \gamma^{-1}I + \frac{1}{N^2}\Psi^\top ZZ^\top\Psi.$$

and

$$\Psi = \begin{bmatrix} \psi_1^\top(p,1)x_1(\tilde{\xi},1) & \cdots & \psi_{n_f}^\top(p,1)x_{n_f}(\tilde{\xi},1) \\ \cdots & \cdots & \cdots \\ \psi_1^\top(p,N)x_1(\tilde{\xi},N) & \cdots & \psi_{n_f}^\top(p,N)x_{n_f}(\tilde{\xi},N) \end{bmatrix}, \tag{11.28a}$$

$$Z = \begin{bmatrix} z_1^\top(1) & \cdots & z_{n_f}^\top(1) \\ \cdots & \cdots & \cdots \\ z_1^\top(N) & \cdots & z_{n_f}^\top(N) \end{bmatrix}. \tag{11.28b}$$

Proof. Let us rewrite problem (11.21a) and (11.21b) as:

$$\min_{\theta} \frac{1}{2}\theta^\top\theta + \frac{\gamma}{2N^2}\left\|Z^\top(U-\Psi\theta)\right\|_2^2, \tag{11.29}$$

with $\theta = \left[\theta_1^\top \ \cdots \ \theta_{n_f}^\top\right]^\top$, whose solution is achieved for

$$\hat{\theta}_{\mathrm{NP,IV}} = \left(\gamma^{-1}I + \frac{1}{N^2}\Psi^\top ZZ^\top \Psi\right)^{-1} \frac{1}{N^2}\left(\Psi^\top ZZ^\top U\right). \tag{11.30}$$

The solution θ° of Problem (11.8a)–(11.8c)—or equivalently the solution of Problem (11.6a)–(11.6d)—satisfies the following condition (since, under the given assumptions, $\varepsilon = 0$ if and only if the optimum is achieved):

$$U = \Psi_o\theta^\circ = \Psi\theta^\circ + \underbrace{(\Psi_o - \Psi)}_{\Delta\Psi}\theta^\circ, \tag{11.31}$$

where Ψ_o is defined as the noise-free matrix:

$$\Psi_o = \begin{bmatrix} \psi_1^\top(p,1)x_1(\xi_o,1) & \cdots & \psi_{n_f}^\top(p,1)x_{n_f}(\xi_o,1) \\ \cdots & \cdots & \cdots \\ \psi_1^\top(p,N)x_1(\xi_o,N) & \cdots & \psi_{n_f}^\top(p,N)x_{n_f}(\xi_o,N) \end{bmatrix}. \tag{11.32}$$

Substitution of (11.31) in (11.30) leads to

$$\hat{\theta}_{\mathrm{NP,IV}} = \theta^\circ + \left(\gamma^{-1}I + \frac{1}{N^2}\Psi^\top ZZ^\top \Psi\right)^{-1}\left(\frac{1}{N^2}\Psi^\top ZZ^\top \Delta\Psi\theta^\circ - \gamma^{-1}\theta^\circ\right).$$

Since the instruments $z_i(t)$ are, by construction, not correlated with the noise w, the term $(1/N)Z^\top\Delta\Psi$ converges to 0 w.p. 1 as N goes to infinity. Thus, w.p. 1:

$$\lim_{N\to\infty} \hat{\theta}_{\mathrm{NP,IV}} = \theta^\circ - R^{-1}\gamma^{-1}\theta^\circ. \tag{11.33}$$

□

Note that the bias term $R^{-1}\gamma^{-1}\theta^\circ$ in the estimate of the controller parameters $\hat{\theta}_{\mathrm{NP,IV}}$ does not depend on the realization of the noise $w(t)$, and it is only due to the regularization term $(1/2)\sum_{i=1}^{n_f}\theta_i^\top\theta_i$ used in Problem (11.26a) and (11.26b).

However, the controller parameter vector $\hat{\theta}_{\mathrm{NP,IV}}$ minimizing (11.26a) and (11.26b) cannot be computed directly since an explicit representation of the feature maps $\psi_i(p,t)$ and of the instruments $z_i(t)$ would be needed. In order to overcome this problem, a dual formulation of (11.26a) and (11.26b) is considered, which allows us to solve problem (11.26a) and (11.26b) without the need to specify the feature maps $\psi_i(p,t)$ and without the need to compute the optimal parameters θ_i.

11.5.3 Dual problem

Let us define the *Lagrangian* $\mathscr{L}(\alpha,\theta,e)$ associated with Problem (11.26a) and (11.26b):

$$\mathscr{L}(\alpha,\theta,E) = \frac{1}{2}\sum_{i=1}^{n_f}\theta_i^\top\theta_i + \frac{\gamma}{2N^2}\sum_{i=1}^{n_f}\left\|\Psi_i^\top X_i(\hat{\xi})E\right\|_2^2 - \alpha^\top\left(E - U + \sum_{i=1}^{n_f}X_i(\xi)\Psi_i\theta_i\right),$$

where $\alpha \in \mathbb{R}^N$ is the collection of *Lagrange multipliers*. Due to the affine nature of the constraints and convexity of the cost function, the global optimum of (11.26a) and (11.26b) is obtained when the following *Karush–Kuhn–Tucker* conditions are fulfilled for all $i = 1, \ldots, n_f$:

$$\frac{\partial \mathscr{L}}{\partial \theta_i} = 0 \rightarrow \theta_i = \Psi_i^\top X_i(\xi)\alpha, \tag{11.34a}$$

$$\frac{\partial \mathscr{L}}{\partial E} = 0 \rightarrow \alpha = \frac{\gamma}{N^2} \sum_{i=1}^{n_f} X_i(\hat{\xi})\Psi_i \Psi_i^\top X_i(\hat{\xi})E, \tag{11.34b}$$

$$\frac{\partial \mathscr{L}}{\partial \alpha} = 0 \rightarrow E = U - \sum_{i=1}^{n_f} X_i(\xi)\Psi_i\theta_i. \tag{11.34c}$$

By substituting (11.34a) into (11.34c), we obtain

$$E = U - \sum_{i=1}^{n_f} X_i(\xi)\Psi_i \Psi_i^\top X_i(\xi)\alpha. \tag{11.35}$$

Then, substitution of (11.35) into (11.34b) leads to

$$\alpha = \frac{\gamma}{N^2} \sum_{i=1}^{n_f} X_i(\hat{\xi})\Psi_i \Psi_i^\top X_i(\hat{\xi})U - \frac{\gamma}{N^2} \sum_{i=1}^{n_f} X_i(\hat{\xi})\Psi_i \Psi_i^\top X_i(\hat{\xi}) \sum_{j=1}^{n_f} X_j(\xi)\Psi_j \Psi_j^\top X_j(\xi)\alpha,$$

which has the solution

$$\alpha = R_D^{-1}(\Psi_i)\frac{1}{N^2} \sum_{i=1}^{n_f} X_i(\hat{\xi})\Psi_i \Psi_i^\top X_i(\hat{\xi})U, \tag{11.36}$$

with

$$R_D(\Psi_i) = \gamma^{-1}I + \frac{1}{N^2} \sum_{i=1}^{n_f} X_i(\hat{\xi})\Psi_i \Psi_i^\top X_i(\hat{\xi}) \sum_{j=1}^{n_f} X_j(\xi)\Psi_j \Psi_j^\top X_j(\xi). \tag{11.37}$$

The importance of the LS-SVM approach lies in the fact that the Lagrangian multipliers α can be computed without the proper knowledge of the feature maps $\psi_i(p, t)$ characterizing the matrix Ψ_i, as discussed in the following.

Define the so-called *Grammian* matrix as $\Omega_i = \Psi_i \Psi_i^\top$. According to the Mercer's theorem [41,42], we can define our Grammian matrices Ω_i (with $i = 1, \ldots, n_f$) in terms of kernel functions. More specifically, the (t, k)th entry of each Ω_i given by the inner product,

$$[\Omega_i]_{t,k} = \langle \psi_i(p, t), \psi_i(p, k) \rangle, \tag{11.38}$$

can be described by a positive definite *kernel* function $\kappa_i(p, t, k)$, i.e.,

$$[\Omega_i]_{t,k} = \langle \psi_i(p, t), \psi_i(p, k) \rangle = \kappa_i(p, t, k). \tag{11.39}$$

Specification of the kernels instead of the maps ψ_i is called the *kernel trick* [43], and it provides the solution for (11.36) in terms of

$$\alpha = R_{\mathrm{D}}^{-1}(\Omega_i)\frac{1}{N^2}\sum_{i=1}^{n_{\mathrm{f}}} X_i(\hat{\xi})\Omega_i X_i(\hat{\xi})U \tag{11.40}$$

with

$$R_{\mathrm{D}}(\Omega_i) = \gamma^{-1}I + \frac{1}{N^2}\sum_{i=1}^{n_{\mathrm{f}}} X_i(\hat{\xi})\Omega_i X_i(\hat{\xi})\sum_{j=1}^{n_{\mathrm{f}}} X_j(\xi)\Omega_i X_j(\xi).$$

A typical choice of kernel, which provides uniformly effective representation of a large class of smooth functions, is the *radial basis function* (RBF) kernel:

$$\kappa_i(p(t), p(k)) = \exp\left(-\frac{\|p(t)-p(k)\|_2^2}{\sigma_i^2}\right), \tag{11.41}$$

where $\sigma_i > 0$ is a so-called hyper-parameter characterizing the width of the RBF and it is tuned by the user (e.g., through cross validation).

Once the Lagrangian multipliers α are computed through (11.40), the p-dependent coefficient functions $a_i^K(p,t)$ and $b_i^K(p,t)$ characterizing the LPV controller (11.16a) and (11.16b) are obtained from (11.17a) and (11.17b) and (11.34a), i.e.,

$$a_i^K(\centerdot) = \psi_i^\top(\centerdot)\theta_i = \psi_i^\top(\centerdot)\Psi_i^\top X_i(\xi)\alpha = \tag{11.42a}$$

$$= \sum_{t=1}^{N} \underbrace{\psi_i^\top(\centerdot)\psi_i(p,t)x_i(\xi,t)}_{\kappa_i(p,t,\centerdot)}\alpha_t, \tag{11.42b}$$

$$b_i^K(\centerdot) = \psi_{i+n_{\mathrm{a}K}+1}^\top(\centerdot)\theta_{i+n_{\mathrm{a}K}+1} = \psi_{i+n_{\mathrm{a}K}+1}^\top(\centerdot)\Psi_{i+n_{\mathrm{a}K}+1}^\top X_{i+n_{\mathrm{a}K}+1}(\xi)\alpha =$$

$$= \sum_{t=1}^{N} \underbrace{\psi_{i+n_{\mathrm{a}K}+1}^\top(\centerdot)\psi_{i+n_{\mathrm{a}K}+1}(p,t)x_{i+n_{\mathrm{a}K}+1}(\xi,t)}_{\kappa_{i+n_{\mathrm{a}K}+1}(p,t,\centerdot)}\alpha_t. \tag{11.42c}$$

Note that the resulting controller coefficient functions only depend on the available observations in $\mathscr{D}_N$ and the specified kernel functions $\kappa_i(p,t,\centerdot)$. On the other hand, no prior knowledge of the system dynamics and feature maps $\psi_i(p,t)$ is required.

Finally, notice that, since the overall procedure above turns out to be an IVs version of an SVM, it is referred to as IV-SVM in the literature.

11.6 Outer controller design

Once the inner controller $\mathscr{K}_{\mathrm{p}}(\theta)$ is designed, the outer MPC controller, acting as a reference governor, can designed based for the interconnected reference plant $\mathscr{M}_{\mathrm{p}}'$ of Figure 11.3 under the assumption of perfect model matching (the dynamics of the inner closed-loop system are now described by the known reference model $\mathscr{M}_{\mathrm{p}}$). The interconnected LPV reference model $\mathscr{M}_{\mathrm{p}}'$ describes the relationship between $g(t)$ and $u(t)$, $y(t)$. In this formulation, the role of the inner LPV controller $\mathscr{K}_{\mathrm{p}}(\theta)$ is to

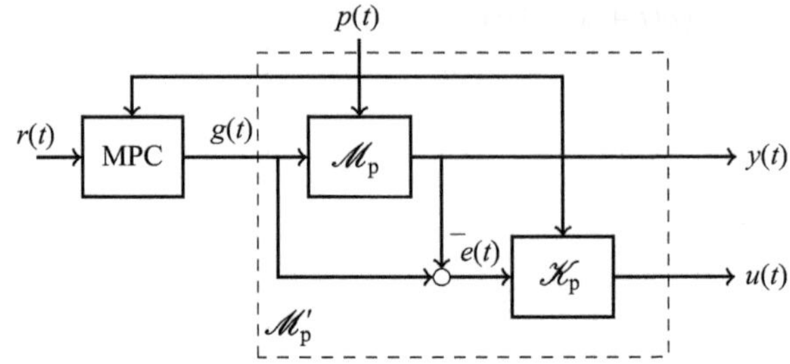

Figure 11.3 Equivalent single-input two-output LPV model describing the relationship between the MPC designed reference signal g(t), the plant input and the achieved output signal

transform the behavior of the unknown plant $\mathscr{G}_p$ into that of a known, usually simpler, and a priori specified LPV model $\mathscr{M}_p$.

Consider the following, not necessarily minimal, state-space realization of $\mathscr{M}'_p$

$$\begin{cases} \xi(t+1) = \bar{A}_M(p(t))\xi(t) + \bar{B}_M(p(t))g(t) \\ \begin{bmatrix} y(t) \\ u(t) \end{bmatrix} = \bar{C}_M(p(t))\xi(t) + \begin{bmatrix} 0 \\ \bar{D}_M(p(t)) \end{bmatrix} g(t), \end{cases} \tag{11.43}$$

where the matrices $\bar{A}_M(p(t)), \bar{B}_M(p(t)), \bar{C}_M(p(t)), \bar{D}_M(p(t))$ can be derived from the description of the reference model $\mathscr{M}_p$ (Equation (11.3a) and (11.3b)) and the inner controller $\mathscr{K}_p$ (Equation (11.4)).

Based on the prediction model (11.43), the outer MPC controller is designed both to impose IO constraints and to possibly improve the tracking quality of the reference signal r. As shown in the equivalent scheme of Figure 11.3, only the reference model $\mathscr{M}_p$ and the model of the controller $\mathscr{K}_p(\theta)$ are needed to predict the behavior of $u(t)$ and $y(t)$. Then, also in this second step, a model of $\mathscr{G}_p$ is not required.

The design method is as follows: by assuming that the state vector $\xi(t)$ of the inner-loop model $\mathscr{M}_p$ is fully accessible or, alternatively, estimated from measurements of u, y and p, for example, by means of a linear time-varying (LTV) Kalman filter, the reference tracking MPC problem can be formulated, at each time instant t, as the following optimization problem (provided the future trajectory of $p(t)$ is given for the following k steps)

$$\min_{\{g(t+k|t)\}_{k=1}^{N_u}} Q_y \sum_{k=1}^{N_p} (y(t+k|t) - r(t+k))^2 + Q_u \sum_{k=1}^{N_p} (u(t+k|t) - u_{\text{ref}}(t+k))^2$$
$$+ Q_{\Delta u} \sum_{k=1}^{N_p} (u(t+k|k) - u(t+k-1|t))^2 + Q_g \sum_{k=1}^{N_u} (r(t+k) - g(t+k|t))^2 + Q_\varepsilon \varepsilon^2$$

$$
\begin{aligned}
\text{s.t. } & \xi(t+k|t) = \bar{A}_M(p(t+k-1|t))\xi(t+k-1|t) + \bar{B}_M(p(t+k-1|t))g(t+k-1|t) \\
& \begin{bmatrix} y(t+k|t) \\ u(t+k|t) \end{bmatrix} = \bar{C}_M(p(t+k|t))\xi(t+k|t) + \begin{bmatrix} 0 \\ \bar{D}_M(p(t+k|t)) \end{bmatrix} g(t+k|t) \\
& -V_y\varepsilon + y_{\min} \leq y(t+k|t) \leq y_{\max} + V_y\varepsilon, \quad -V_u\varepsilon + u_{\min} \leq u(t+k|t) \leq u_{\max} + V_u\varepsilon \\
& -V_{\Delta u}\varepsilon + \Delta u_{\min} \leq u(t+k|t) - u(t+k-1|t) \leq \Delta u_{\max} + V_{\Delta u}\varepsilon \\
& k = 1, \dots, N_{\mathrm{p}}, \\
& g(t+N_{\mathrm{u}}+j|t) = g(t+N_{\mathrm{u}}|t), \quad \xi(t|t) = \xi(t), \quad g(t) = g(t|t), \\
& j = 1, \dots, N_{\mathrm{p}} - N_{\mathrm{u}}
\end{aligned}
\tag{11.44}
$$

where N_{p} and N_{u} denote the prediction and control horizon, respectively; Q_y, Q_u, $Q_{\Delta u}$, Q_g, Q_ϵ are nonnegative weights; u_{ref} is a desired input reference; and V_y, V_u, $V_{\Delta u}$ are positive vectors that are used to soften the constraints, so that (11.44) always admits a solution, that can be computed via *quadratic programming* (QP).

In the MPC formulation (11.44), the following terms are penalized: (i) the tracking error between the reference signal r and the output y; (ii) the tracking error between the input reference signal u_{ref} and the manipulated variable u; (iii) the increments of the plant input u; (iv) the error between the reference signal r and the MPC output g and (v) the violation of the constraints. From a practical point of view, the goal of the penalty on $r-g$ is to guarantee that the reference signal g for the inner loop does not differ too much from the global reference signal r, so as to avoid to excite unmodeled dynamics. In case $p(t+k)$ is known at time t for the future N_p steps, we set $p(t+k|t) = p(t+k)$ and call the MPC formulation (11.44) *LTV-MPC*. In case future values of p are not known, we set $p(t+k|t) \equiv p(t)$ and call the formulation *gain-scheduled MPC*, in which the prediction model is LTI but depends on $p(t)$, and therefore the controller itself is a simplified form of LPV MPC. Note that stability is not theoretically guaranteed here (it is an important theme that is currently under investigation).

When both the nominal reference model $\mathcal{M}_{\mathrm{p}}$ and the inner controller $\mathcal{K}_{\mathrm{p}}$ are LTI, problem (11.44) is a more standard LTI-MPC problem, which has computational advantages over LTV-MPC and LPV-MPC. In such a case, the matrices involved in the QP problem can be precomputed off-line. An explicit MPC approach [44,45] may further reduce the upper control layer to a piecewise affine function. However, if this happens, the plant itself is likely to be LTI, as both the closed-loop dynamics and the controller can be well captured by an LTI structure. In that case, simpler data-driven controller tuning methods could be applied, see, e.g., [19]. If the plant is LPV, even when the selected reference model $\mathcal{M}_{\mathrm{p}}$ is LTI, a parameter-varying controller $\mathcal{K}_{\mathrm{p}}$ is needed to achieve the desired behavior.

11.7 Case study: servo-positioning system

The effectiveness of the proposed hierarchical data-driven control design approach is shown in this section on a realistic case study regarding the LPV control of a

Figure 11.4 Servo positioning system with an unbalanced disk

servo positioning system shown in Figure 11.4, representing a voltage-controlled DC motor. The system has viscous friction and an additional mass is mounted on the disc connected to the rotor to make the mass distribution inhomogeneous. Note that in practical utilization of this application, an important control objective is to attenuate the load disturbance caused by the unbalanced disc and realize an LTI behavior of the closed-loop system, i.e., hide the position dependent/nonlinear nature of the system from the user.

11.7.1 System description

The mathematical model of the considered servo-positioning system, used to simulate its dynamics behavior, is represented by the continuous-time state-space equation (where τ denotes the continuous time index, which is different from its discrete time counterpart t):

$$\begin{bmatrix} \dot{\alpha}(\tau) \\ \dot{\omega}(\tau) \\ \dot{I}(\tau) \end{bmatrix} = \left(\begin{bmatrix} 0 & 1 & 0 \\ 0 & -\frac{b}{J} & \frac{K}{J} \\ 0 & -\frac{K}{L} & -\frac{R}{L} \end{bmatrix} + \begin{bmatrix} 0 & 1 & 0 \\ \frac{mgl}{J} & 0 & 0 \\ 0 & 0 & 0 \end{bmatrix} \mathrm{sinc}(\alpha(\tau)) \right) \begin{bmatrix} \alpha(\tau) \\ \omega(\tau) \\ I(\tau) \end{bmatrix} + \begin{bmatrix} 0 \\ 0 \\ \frac{1}{L} \end{bmatrix} V(\tau) \quad (11.45)$$

with an output equation $y(\tau) = \alpha(\tau)$. $V(\tau)$ (V) is the control input voltage over the armature, $I(\tau)$ (mA) is the current, $\alpha(\tau)$ (rad) is the angular position and $\omega(\tau)$ (rad/s) is the angular velocity of the disc. The parameters characterizing the DC motor are taken from [46] and are reported in Table 11.1. In terms of data acquisition, an ideal *zero-order-hold* (ZOH) setting is assumed where the actuation and output sampling is synchronized with a sampling period $T_s = 10$ ms.

Table 11.1 Physical parameters of the DC motor [46]

Symbol	Description	Value	Unit
R	Motor resistance	9.5	Ω
L	Motor inductance	0.84×10^{-3}	H
K	Motor torque constant	53.6×10^{-3}	N m/A
J	Complete disk inertia	2.2×10^{-4}	N m^2
b	Friction coefficient	6.6×10^{-5}	N m s/rad
M	Additional mass	0.07	kg
l	Mass distance from the center	0.042	m

Note that, by choosing the output $\alpha(\tau)$ as scheduling variable (i.e., $p(\tau) = \alpha(\tau)$), the DC motor model (11.45) becomes LPV.

11.7.2 Desired inner closed-loop behavior

The desired closed-loop behavior $\mathscr{M}_\text{p}$ for the inner loop is described by the state-space representation

$$\begin{aligned} x_M(t+1) &= 0.99x_M(t) + 0.01g(t), \\ \alpha_M(t) &= x_M(t). \end{aligned} \tag{11.46}$$

Namely, $\mathscr{M}_\text{p}$ is selected as a discrete-time LTI first-order plant, under the sampling time T_s and a continuous pole at 0.15 Hz. From now on, we denote by $\alpha_M(t)$ the desired closed-loop trajectory for the angular position of the disc $\alpha(t)$.

11.7.3 Inner controller design

To gather data for the synthesis of the inner controller $\mathscr{K}_\text{p}$, we apply to the motor a discrete-time filtered zero-mean white noise voltage (followed by a ZOH block) with uniform distribution in the interval $[-10 \;\; 10]$ V. The input filter is a first order digital filter with a cutoff frequency of 2 Hz. The output measurements are corrupted by an additive colored noise $w(t)$ with zero mean and variance such that the signal-to-noise ratio, representing the ratio between the signal and noise variances, is 18 dB. A second measurement of the output is taken to build the instruments $z_i(t)$ used in (11.21a) and (11.21b). A training dataset $\mathscr{D}_N$ with $N = 1{,}500$ IO pairs is constructed.

The chosen structure for the inner controller $\mathscr{K}_\text{p}$ is given by a fourth-order LPV system with dynamic dependence on the scheduling signal and integral action, described by

$$\begin{aligned} u(t) &= \sum_{i=1}^{4} a_i^K(p(t-1))u(t-i) + \sum_{j=0}^{4} b_j^K(p(t-1))e_{int}(t-j) \\ e_{int}(t) &= e_{int}(t-1) + (g(t) - y(t)). \end{aligned}$$

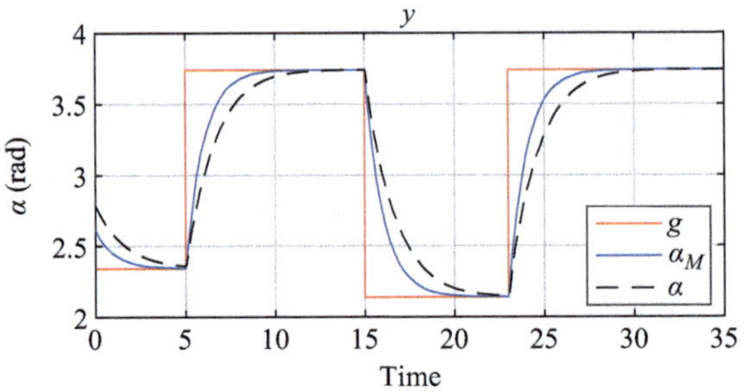

Figure 11.5 Inner loop behavior: reference signal $g(\tau)$ (red), desired step response of the shaft angle $y_d(\tau)$ (solid blue) and actual controlled output $y(\tau)$ (dashed black)

Gaussian kernels (Equation (11.41)) are used to estimate the coefficient functions a_i^K and b_j^K. The width σ of the kernels and the value of the hyper-parameter γ in (11.21a) and (11.21b) are chosen through cross-validation. More specifically, a two-dimensional grid search over the hyper-parameters σ and γ is performed. For each value of σ and γ, a controller $\mathscr{K}_\text{p}$ is estimated and the one providing the best performance, in terms of mean square error, is chosen. The controller performance is measured on a calibration set $\mathscr{D}_{N_\text{C}} = \{u(t), y(t)\}_{t=1}^{N_\text{C}}$ of $N_\text{C} = 500$ samples generated from an open-loop experiment and evaluated in terms of its capabilities to reconstruct the (open-loop) input sequence $u(t)$ given the *virtual tracking error* $e(t) = M^\dagger y(t) - y(t)$. The computed values for the hyper-parameters are $\sigma = 2$ and $\gamma = 160$.

11.7.4 Achieved inner closed-loop behavior

A closed-loop simulation, with a piecewise linear reference signal g, is performed to assess the quality of the synthesized controller $\mathscr{K}_\text{p}$. The response of the inner closed-loop system is shown in Figure 11.5 and compared with the output $y_d = \alpha_M$ of the desired reference-model $\mathscr{M}_\text{p}$. The input voltage $u(t) = V(t)$ calculated by the controller $\mathscr{K}_\text{p}$ and applied to the motor is plotted in Figure 11.6. Although an LTI closed-loop system would be expected to be barely achievable, results in Figure 11.5 show a good matching between the actual output y and the output y_d of the desired reference model $\mathscr{M}_\text{p}$.

However, the closed-loop system exhibits slow dynamics, with a 10%–90% rise time of about 4 s and a 2%-settling time (defined as the time required by the output to remain within a 2% error band) of about 6 s. Due to the limited degrees of the freedom in the controller structure, it has not been possible to achieve reference models $\mathscr{M}_\text{p}$ with faster dynamics.

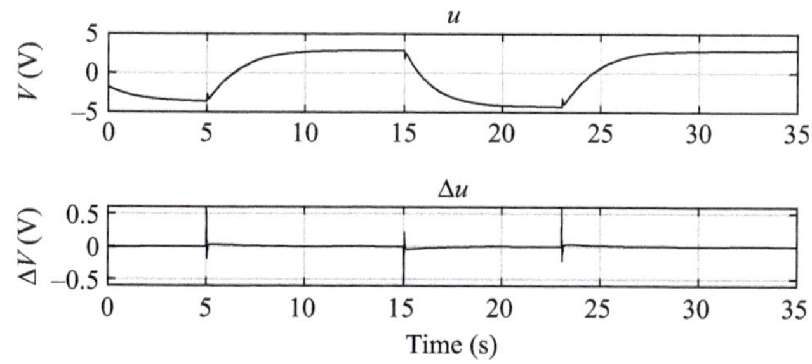

Figure 11.6 Inner loop behavior: plant input $V(\tau)$ and input increments $\Delta V(tT_s) = V(tT_s) - V((t-1)T_s)$

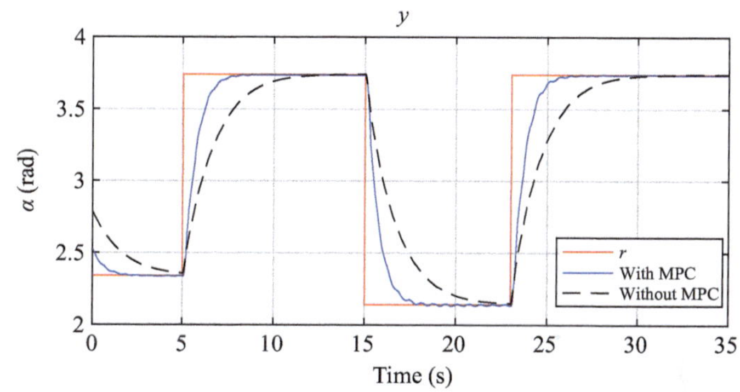

Figure 11.7 Closed-loop behavior: reference signal $r(\tau)$ (red) and controlled output $y(\tau)$ (solid blue). The inner loop output achieved without outer MPC is illustrated by the dashed black signal

11.7.5 Outer controller design

An outer MPC is designed based on the reference model $\mathscr{M}_\mathrm{p}$ (used to describe the behavior of the inner loop) and the LPV controller $\mathscr{K}_\mathrm{p}$, in order to achieve the following objectives: (i) enhance the performance of the inner loop, in terms of rise and settling time and (ii) enforce the following constraint on the input rate: $V(tT_s) - V((t-1)T_s) \leq 0.2V, t = 1, 2, \ldots$.

The horizons and the weights defining the MPC with cost function (11.44) are tuned through closed-loop simulation, by using the model $\mathscr{M}_\mathrm{p}$ to simulate the behavior of the inner closed-loop system. The chosen values for weights and parameters are $N_\mathrm{p} = 10$, $N_\mathrm{u} = 10$, $Q_y = 7$, $Q_u = 0$, $Q_{\Delta u} = 0.2$ and $Q_g = 1.7$.

The response of the closed-loop system for the same reference signal used in the previous analysis (Section 11.7.4) is plotted in Figure 11.7, while the input applied to the motor is plotted in Figure 11.8. For the sake of comparison, the output of the

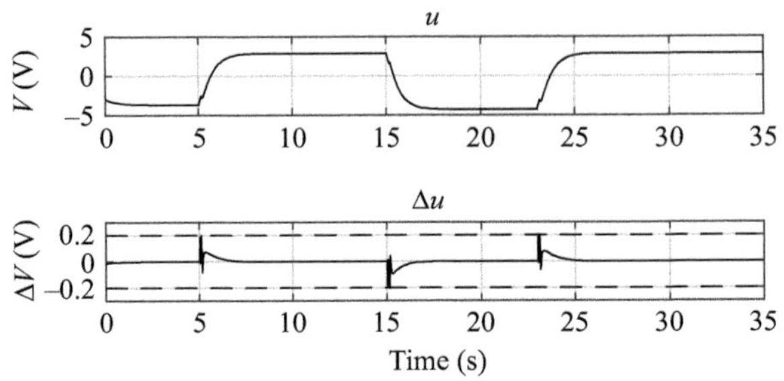

Figure 11.8 *Example 1: closed-loop behavior. Input voltage $V(\tau)$ and input increments $\Delta V(tT_s) = V(tT_s) - V((t-1)T_s)$, constrained between ± 0.2 V (dashed lines)*

inner loop achieved without the proposed hierarchical structure is also displayed in Figure 11.7. The obtained results show that, although constraints on the variation of the input voltage are enforced, the hierarchical MPC structure achieves a faster reference tracking than the inner-loop system, with a 10%–90% rise time of about 1.3 s (about 3× smaller than the inner loop rise time) and a 2%-settling time of about 2.2 s (about 2.7× smaller than the inner loop settling time). The obtained results show that the proposed hierarchical architecture makes the choice of the reference model $\mathscr{M}_\text{p}$ less critical than with standard data-driven control design methods (see, e.g., [16]). Indeed, with the chosen structure of the controller $\mathscr{K}$ and the reference model $\mathscr{M}_\text{p}$, it was not possible to achieve closed-loop systems with a cutoff frequency larger than 0.15 Hz. The task of enhancing the performance is then left to the outer MPC loop, whose design is based on $\mathscr{M}_\text{p}$.

All computations have been carried out on an i7 2.40-GHz Intel core processor with 4 GB of RAM running MATLAB® R2016b. The computation time of the MPC layer is 18 ms (including various MATLAB overheads), based on the code generated by the *model predictive control toolbox* [47], which is already in the order of magnitude of the sampling time $T_s = 10$ ms. Although computational feasibility is not the main aim of this case study, it is realistic to assume that the controller could be implemented in real-time to control the motor by adopting a fast C implementation of the QP constructor of problem (11.44) and QP solver (see, e.g., [48]).

11.8 Conclusions

In this chapter, a data-driven approach to directly map experimental data onto LPV controller parameters guaranteeing a desired performance specification has been discussed. The method only requires one set of measurements and the specification of a desired closed-loop behavior. In order to compensate for conservative choices of the reference model as well as to handle IO signal constraints, a predictive controller has been shown to be computable from the same set of data, so as to be used as an outer reference governor. The effectiveness of the proposed approach has been illustrated with a benchmark simulation case study.

Future research will be dedicated to the analysis of the properties of the hierarchical loop as well as to extensive experimental tests on challenging control engineering applications.

References

[1] Tóth R. Modeling and identification of linear parameter-varying systems. vol. 403. Berlin: Springer; 2010.

[2] Mohammadpour J, Scherer CW. Control of linear parameter varying systems with applications. New York: Springer Science & Business Media; 2012.

[3] Bamieh B, Giarre L. Identification of linear parameter varying models. International Journal of Robust and Nonlinear Control. 2002;12(9):841–853.

[4] Piga D, Cox P, Tóth R, *et al.* LPV system identification under noise corrupted scheduling and output signal observations. Automatica. 2015;53:329–338.

[5] Cox PB, Tóth R, Petreczky M. Towards efficient maximum likelihood estimation of LPV-SS models. Automatica. 2018;97:392–403.

[6] van Wingerden J, Verhaegen M. Subspace identification of bilinear and LPV systems for open-and closed-loop data. Automatica. 2009;45(2):372–381.

[7] Wollnack S, Abbas HS, Tóth R, *et al.* Fixed-structure LPV-IO controllers: An implicit representation based approach. Automatica. 2017;83:282–289.

[8] Wollnack S, Werner H. LPV-IO controller design: An LMI approach. In: American Control Conference (ACC), 2016. IEEE; 2016. p. 4617–4622.

[9] Lu B, Wu F, Kim SW. Switching LPV control of an F-16 aircraft via controller state reset. IEEE Transactions on Control Systems Technology. 2006;14(2):267–277.

[10] Cerone V, Piga D, Regruto D. Set-membership LPV model identification of vehicle lateral dynamics. Automatica. 2011;47(8):1794–1799.

[11] Hashemi SM, Abbas HS, Werner H. Low-complexity linear parameter-varying modeling and control of a robotic manipulator. Control Engineering Practice. 2012;20(3):248–257.

[12] Prempain E, Postlethwaite I, Benchaib A. A linear parameter variant $\mathscr{H}_\infty$ control design for an induction motor. Control Engineering Practice. 2002;10(6):633–644.

[13] Tóth R, Laurain V, Zheng WX, *et al.* Model structure learning: A support vector machine approach for LPV linear-regression models. In: Proc. of the 50th IEEE Conference on Decision and Control and European Control Conference, Orlando, Florida, USA; 2011. p. 3192–3197.

[14] Laurain V, Tóth R, Piga D, *et al.* An instrumental least squares support vector machine for nonlinear system identification. Automatica. 2015;54:340–347.

[15] Suykens J, Van Gestel T, De Brabanter J, *et al.* Least squares support vector machines. Singapore: World Scientific Publishing; 2002.

[16] Formentin S, Piga D, Tóth R, *et al.* Direct learning of LPV controllers from data. Automatica. 2016;65:98–110.
[17] Piga D, Formentin S, Bemporad A. Direct data-driven control of constrained systems. IEEE Transactions on Control Systems Technology. 2018;26(4):1422–1429.
[18] Bazanella AS, Campestrini L, Eckhard D. Data-driven controller design: The $\mathscr{H}_2$ approach. Netherlands: Springer; 2011.
[19] Campi MC, Lecchini A, Savaresi SM. Virtual reference feedback tuning: a direct method for the design of feedback controllers. Automatica. 2002;38(8):1337–1346.
[20] Formentin S, Savaresi SM, Del Re L. Non-iterative direct data-driven controller tuning for multivariable systems: theory and application. IET Control Theory & Applications. 2012;6(9):1250–1257.
[21] van Heusden K, Karimi A, Bonvin D. Data-driven model reference control with asymptotically guaranteed stability. International Journal of Adaptive Control and Signal Processing. 2011;25(4):331–351.
[22] Formentin S, van Heusden K, Karimi A. A comparison of model-based and data-driven controller tuning. International Journal of Adaptive Control and Signal Processing. 2013, DOI:101002/acs2415.
[23] Formentin S, De Filippi P, Corno M, *et al.* Data-driven design of braking control systems. IEEE Transactions on Control Systems Technology. 2013;21(1): 186–193.
[24] Formentin S, Cologni AL, Previdi F, *et al.* A data-driven approach to control of batch processes with an application to a gravimetric blender. IEEE Transactions on Industrial Electronics. 2014;61(11):6383–6390.
[25] Formentin S, Corno M, Savaresi SM, *et al.* Direct data-driven control of linear time-delay systems. Asian Journal of Control. 2012;14(3):652–663.
[26] Formentin S, Karimi A, Savaresi SM. Optimal input design for direct data-driven tuning of model-reference controllers. Automatica. 2013;49(6): 1874–1882.
[27] Formentin S, Karimi A. Enhancing statistical performance of data-driven controller tuning via L2-regularization. Automatica. 2014;50(5):1514–1520.
[28] Radac MB, Precup RE, Petriu EM, *et al.* Data-driven reference trajectory tracking algorithm and experimental validation. IEEE Transactions on Industrial Informatics. 2013;9(4):2327–2336.
[29] Formentin S, Karimi A. A data-driven approach to mixed-sensitivity control with application to an active suspension system. IEEE Transactions on Industrial Informatics. 2013;9(4):2293–2300.
[30] Formentin S, Savaresi S. Virtual reference feedback tuning for linear parameter-varying systems. In: Proc. of the 18th IFAC World Congress, Milan, Italy; 2011. p. 10219–10224.
[31] Karimi A, Van Heusden K, Bonvin D. Noniterative data-driven controller tuning using the correlation approach. In: European Control Conference; 2007. p. 5189–5195.

[32] Tóth R, Willems JC, Heuberger PSC, *et al.* The behavioral approach to linear parameter-varying systems. IEEE Transactions on Automatic Control. 2011;56(11):2499–2514.
[33] Novara C. Direct design from data of LPV feedback controllers. In: Proc. of the 52nd IEEE Conference on Decision and Control, Florence, Italy; 2013. p. 4098–4103.
[34] Butcher M, Karimi A. Data-driven tuning of linear parameter-varying precompensators. International Journal of Adaptive Control and Signal Processing. 2009;24(7):592–609.
[35] Desoer C, Vidyasagar M. Feedback systems: Input-output properties. vol. 55. New York: Academic Press, Inc.; 1975.
[36] Selvi D, Piga D, Bemporad A. Towards direct data-driven control design of optimal controllers. In: European Control Conference; 2018.
[37] Tóth R. Maximum LPV-SS realization in a static form. Technical report TUE-CS-2013-003, Department of Electrical Engineering, TU Eindhoven; 2013.
[38] Tóth R, Abbas HS, Werner H. On the state-space realization of LPV input-output models: Practical approaches. IEEE Transactions on Control Systems Technology. 2012;20(1):139–153.
[39] Bemporad A. Reference governor for constrained nonlinear systems. IEEE Transactions on Automatic Control. 1998;43(3):415–419.
[40] Zhou K, Doyle J. Essentials of robust control. vol. 180. Upper Saddle River, NJ: Prentice Hall; 1998.
[41] Mercer J. Functions of positive and negative type, and their connection with the theory of integral equations. Philosophical Transactions of the Royal Society of London, Series A. 1909;209:415–446.
[42] Cucker F, Smale S. On the mathematical foundations of learning. Bulletin of the American Mathematical Society. 2001;39(1):1–49.
[43] Vapnik V. Statistical learning theory. New York: Wiley-Interscience; 1998.
[44] Bemporad A, Morari M, Dua V, *et al.* The explicit linear quadratic regulator for constrained systems. Automatica. 2002;38(1):3–20.
[45] Bemporad A. A multiparametric quadratic programming algorithm with polyhedral computations based on nonnegative least squares. IEEE Transactions on Automatic Control. 2015;60(11):2892–2903.
[46] Kulcsar B, Dong J, van Wingerden J, *et al.* LPV subspace identification of a DC motor with unbalanced disc. In: IFAC Symposium on System Identification. vol. 15; 2009. p. 856–861.
[47] Bemporad A, Morari M, Ricker NL. Model predictive control toolbox for MATLAB 5.0. The Mathworks, Inc.; 2015. http://www.mathworks.com/help/mpc/ug/adaptive-mpc.html.
[48] Cimini G, Bernardini D, Bemporad A, *et al.* Online model predictive torque control for permanent magnet synchronous motors. In: 2015 IEEE Int. Conf. on Industrial Technology. Seville, Spain; 2015.

Chapter 12
Set membership fault detection for nonlinear dynamic systems

Milad Karimshoushtari[1], Luigi Spagnolo[2], and Carlo Novara[1]

In this chapter, an innovative approach to fault detection for nonlinear dynamic systems is proposed, based on the recently introduced quasi-local set membership-identification method, overcoming some relevant issues proper of the "classical" techniques. The approach is based on the direct identification from experimental data of a suitable filter and related uncertainty bounds. These bounds are used to detect when a change (e.g., a fault) has occurred in the dynamics of the system of interest. The main advantage of the approach compared to the existing methods is that it avoids the utilization of complex modeling and filter design procedures, since the filter/observer is directly designed from data. Other advantages are that the approach does not require to choose any threshold (as typically done in many "classical" techniques), and it is not affected by under-modeling problems. An experimental study regarding fault detection for a drone actuator is finally presented to demonstrate the effectiveness of the proposed approach.

12.1 Introduction

Consider a discrete-time nonlinear system in state-space form:

$$\begin{aligned} z_k &= f^o(z_{k-1}, u_{k-1}) + d_k \\ y_k &= z_{i,k} \end{aligned} \tag{12.1}$$

where $z_k \in \mathbb{R}^{n_z}$ is the state, $y_k \in \mathbb{R}$ is the output, $z_{i,k}$ is a component of z_k, $u_k \in \mathbb{R}^{n_u}$ is the input, $d_k \in \mathbb{R}^{n_d}$ is a bounded disturbance and $k = 0, 1, 2, \ldots$ is the discrete time index. Assume that the input u_k and the state z_k are measured. Note that the assumption of measuring the state is not strictly necessary: the fault-detection approach proposed in the following can be applied with minor modifications using an input–output system representation, see Remark 12.1.

[1]Department of Electronics and Telecommunications, Politecnico di Torino, Turin, Italy
[2]LADISPE Laboratory, Politecnico di Torino, Turin, Italy

A “classical” approach to fault detection is to identify a model of the system (12.1) and to design a filter/observer on the basis of the identified model. The designed filter/observer is then used to generate online a suitable residual signal. The fault is detected when the residual exceeds a given threshold, see, e.g., [1–10]. However, the design of the filter/observer may be hard in the presence of nonlinear and/or uncertain dynamics. Indeed, designing an optimal filter from a nonlinear model is in general not possible, and approximate filters only, such as the extended Kalman filters, can be actually obtained. These kinds of filters may often be inaccurate and not even guarantee the estimation error stability. Moreover, the choice of the threshold may be critical, especially when poor prior information on the system is available. Another relevant issue is that, in real-world applications, the system (12.1) is unknown and only approximate models can be identified from finite data; evaluating the effects of the modeling error on the estimation error of the filter designed from the approximated model is a largely open problem.

Set membership fault-detection methods have been introduced to efficiently deal with modeling errors [11–17]. These methods have been mainly developed for linear systems, while only few of them deal with nonlinear systems [15–17]. Typically, in set membership methods, a suitable estimation interval is computed online and the fault is detected when one or more measured variables fall outside this interval.

In this chapter, following this set membership philosophy, an innovative approach to fault detection is considered, allowing us to overcome the above issues. The main advantage of this approach compared to the existing methods (“classical” and set membership) is that it avoids the utilization of difficult filter design procedures, since the filter/observer is directly designed from data. Other advantages with respect to the “classical” methods are that the approach does not require to choose any threshold, and it is not affected by modeling errors since no model is used. A further interesting feature is that the approach is computationally simple in both the design and implementation phases.

The method proposed in this chapter represents an improvement with respect to the one of [18]. Indeed, in [18], filter design is performed by means of the so-called local nonlinear set membership-identification method, where the filter is obtained in the form of a linear combination of given basis functions. In this chapter, a so-called quasi-local method is presented, where no filter parametric form needs to be assumed. The filter is obtained directly from experimental data in a nonparametric closed form, thus not requiring the choice of a suitable set of basis functions. Such a quasi-local approach is similar to the so-called global approach of [19] but leads to the derivation of significantly less conservative uncertainty bounds.

The chapter is organized as follows. In Section 12.2, the fault-detection problem is formulated in the set membership framework, defining the types of assumptions and optimality concepts considered. In Sections 12.3 and 12.4, the nonlinear set membership-identification method of [18,19] is summarized. Two versions of this method are discussed: the global one, where the filter is obtained in closed-form, and the local one, where the filter is obtained in the form of a linear combination of given basis functions, solving a convex optimization problem. In Section 12.5, the quasi-local approach is presented. In Section 12.6, two algorithms are presented to find the model parameters and an algorithm to build an adaptive set membership model. In Section 12.7 a summary of the fault-detection procedure is given. Finally, in Section 12.8, the method is tested on a drone propeller in a real experimental setting.

12.2 Nonlinear set membership fault detection

Suppose that the function f^o in (12.1) is not known but a set of noise corrupted data is available, given by

$$\mathcal{D} = \{\widetilde{x}_k, \widetilde{y}_k\}_{k=1}^{L} \tag{12.2}$$

where $\widetilde{x}_k \doteq (\tilde{z}_{k-1}, \tilde{u}_{k-1})$. The proposed fault-detection approach consists of the following main steps. First, a set membership filter for the system (12.1) is defined from the dataset (12.2) which gives us tight bounding functions $\overline{f}$ and $\underline{f}$, such that

$$\underline{f}(\widetilde{x}_k) \leqslant f_i^o(\widetilde{x}_k) \leqslant \overline{f}(\widetilde{x}_k), \ \forall k.$$

Then, a fault-detection system F is defined, on the basis of the bounding functions $\overline{f}$ and $\underline{f}$. The inputs of F is $\widetilde{x}_k \doteq (\tilde{z}_{k-1}, \tilde{u}_{k-1})$, the outputs are the following:

$$\left.\begin{aligned} \overline{y}_k &\doteq \overline{f}(\widetilde{x}_k) + \varepsilon_k \\ \underline{y}_k &\doteq \underline{f}(\widetilde{x}_k) - \varepsilon_k \end{aligned}\right\} \ k > L. \tag{12.3}$$

where ε_k is a bound on the noise d_k affecting the system. It will be shown in the following sections how to construct the functions $\overline{f}$ and $\underline{f}$ and to properly choose the involved parameters (e.g., ε_k). The rationale behind this fault-detection scheme can be explained as follows:

Since $\tilde{y}_k = f_i^o(\widetilde{x}_k) + d_k$, we have that $\tilde{y}_k \leqslant \overline{y}_k$, $\forall k$. Similarly, it holds that $\tilde{y}_k \geqslant \underline{y}_k$, $\forall k$. It follows that $\tilde{y}_k > \overline{y}_k$ or $\tilde{y}_k < \underline{y}_k$ only if the function f^o has changed, i.e., only if some structural change has occurred in the system (12.1). On the basis of this result, fault detection is performed by checking online if $\tilde{y}_k > \overline{y}_k$ or $\tilde{y}_k < \underline{y}_k$: a fault is detected as soon as one of these two inequalities is satisfied.

Remark 12.1. *If the system state is not measured, the following input–output representation can be considered:*

$$\begin{aligned} y_k &= f^o(x_k) + d_k \\ x_k &= (y_{k-1}, \ldots, y_{k-n_y}, u_{k-1}, \ldots, u_{k-n_u}) \end{aligned} \tag{12.4}$$

where $y_k \in \mathbb{R}^{n_y}$ *is the measured output,* $u_k \in \mathbb{R}^{n_u}$ *is the measured input and* $d_k \in \mathbb{R}^{n_d}$ *is an unknown bounded disturbance. Note that the function* f^o *in (12.4) is different from the one in (12.1). The proposed fault-detection approach can be applied considering this representation without significant modifications.*

Remark 12.2. *The proposed fault-detection approach can be applied to each state component (or output component, in the case where the input–output representation (12.4) is used), in order to obtain a multidimensional fault-detection system, allowing us to improve the detection performance with respect to a mono-dimensional case.*

12.2.1 Problem formulation

Consider a nonlinear function f^o defined by

$$y = f^o(x) \tag{12.5}$$

where $x \in \mathcal{X} \subset \mathbb{R}^{n_x}$, $\mathcal{X}$ is a compact connected set and $y \in \mathbb{R}$. Suppose that f^o is not known but a set of noise-corrupted data $\mathcal{D} = \{\widetilde{x}_k, \widetilde{y}_k\}_{k=1}^{L}$ is available, described by

$$\widetilde{y}_k = f^o(\widetilde{x}_k) + d_k, \quad k = 1, 2, \ldots, L \tag{12.6}$$

where d_k is a noise. Assume that the noise sequence $d = (d_1, d_2, \ldots, d_L)$ is unknown but bounded:

$$\|d\|_q \leqslant \mu \tag{12.7}$$

where $\|\cdot\|_q$ is the vector ℓ_q norm. A general formulation is developed in the following, allowing us to deal with the most important cases (i.e., $q = 2, \infty$) in a unified framework. As well known, the ℓ_2 norm is related to the energy of the considered signal, while the ℓ_∞ norm to its amplitude. The choice of this norm can be carried out on the basis of the prior knowledge on the energy or amplitude of the involved noise (if available) or by means of a trial and error procedure.

In the following sections, the problem of deriving from the available data an approximation $\hat{f}$ of f^o and evaluating tight estimate bounds on f^o is considered. The approximation is required to be accurate on the whole domain $\mathcal{X}$. The accuracy is measured by means of the following approximation error:

$$e\left(\hat{f}\right) \doteq \|f^o - \hat{f}\|_p \tag{12.8}$$

where $\|\cdot\|_p$ is a functional L_p norm, defined as

$$\|f\|_p \equiv \|f(\cdot)\|_p \doteq \begin{cases} [\int_{\mathcal{X}} |f(x)|^p \, dx]^{1/p}, p \in [1, \infty) \\ \text{ess sup}_{x\in\mathcal{X}} |f(x)|, p = \infty. \end{cases} \tag{12.9}$$

For a given estimate $\widehat{f}$, the related L_p error is given by (12.8). This error cannot be exactly computed since f^o is not known. It is only known that $f^o \in FFS$, where FFS is called the *feasible function set*, that is, the set of all possible functions consistent with the available prior information and measured data. The formal definition of FFS will be given in the next sections, for three relevant specific cases. This motivates the following definition of identification error, often indicated as worst case or guaranteed error.

Definition 12.1. Worst case approximation error *of* $\widehat{f}$:

$$EN(\widehat{f}) \doteq \sup_{f \in FFS} \|f - \widehat{f}\|_p. \quad \square \tag{12.10}$$

An optimal approximation is defined as a function f^{op} which minimizes the worst case approximation error.

Definition 12.2. *An approximation* f^{op} *is* optimal *if*

$$EN(f^{op}) = \inf_{\widehat{f}} EN(\widehat{f}) \doteq \mathcal{R}_{\mathcal{I}}.$$

$\mathcal{R}_{\mathcal{I}}$ *is called the* radius of information *and is the minimum worst case error that can be achieved on the basis of the available prior and experimental information.* □

Finding optimal approximations may be hard or not convenient, and suboptimal solutions can be looked for. In particular, approximations called interpolatory are often considered in the literature, see, e.g., [20,21].

Definition 12.3. *An approximation* f^I *is* interpolatory *if*

$$f^I \in FFS. \quad \square$$

A fundamental property of an interpolatory approximation is that it guarantees a worst case error degradation of at most 2 [20,21]. An approximation with this property is called almost optimal.

Definition 12.4. *An approximation* f^{ao} *is* almost optimal *if*

$$EN\,(f^{ao}) \leqslant 2 \inf_{\widehat{f}} EN(\widehat{f}). \quad \square$$

In this chapter, the following problem is considered.

Problem 12.1. *From the data set* (12.2)*, find an approximation* $\hat{f}$ *of* f^o

1. *optimal or almost optimal and*
2. *equipped with tight interval estimates* $\overline{f}, \underline{f}$ *for* f^o. □

Remark 12.3. *In the set membership (SM) and approximation theory literature, two optimality concepts are typically considered: local and global optimality [19,20]. The worst case error (12.10) is a local error, since it depends also on the function* f^o *and the data* $\mathcal{D}$*, i.e.,* $EN(\widehat{f}) = EN(\widehat{f}, f^o, \mathcal{D})$*. A global identification error is also often considered, defined as*

$$EN^g(\widehat{f}) \doteq \sup_{\substack{f^o \in \mathcal{F}(\Gamma) \\ \mathcal{D} \in \{d : \|d\|_q \leqslant \mu\}}} EN(\widehat{f}, f^o, \mathcal{D}).$$

An approximation f^g *is called globally optimal if* $EN^g(f^g) = \inf_{\widehat{f}} EN^g(\widehat{f})$*. This is the optimality concept usually investigated in the SM context and approximation theory literature [20]. Note that a locally optimal algorithm* f^{op} *is globally optimal, but* f^g *is not in general locally optimal. Therefore, the local optimality concept investigated in this chapter is stronger and thus less conservative than the global optimality concept investigated in the abovementioned literature.*

12.3 Nonlinear set membership identification: global approach

In order to ensure a bound on the approximation error (12.8), some assumptions have to be made on the noise affecting the data and on the unknown function f^o. In this chapter, the noise sequence $d = (d_1, d_2, \ldots, d_L)$ in (12.6) is assumed to be bounded according to (12.7). Differently from [22], where f^o is assumed to be parametrized by a finite set of basis functions, a mild regularity assumption is made here on f^o, not requiring any knowledge on its parametric form. In particular, we assume that the function f^o is Lipschitz continuous on $\mathcal{X}$:

$$f^o \in \mathcal{F}(\Gamma) \tag{12.11}$$

for some $\Gamma < \infty$, where

$$\mathcal{F}(\Gamma) \doteq \{f : |f(x) - f(\widehat{x})| \leqslant \Gamma \|x - \widehat{x}\|_2, \forall x, \widehat{x} \in \mathcal{X}\}.$$

This allows us to introduce the *FFS*.

Definition 12.5. *The FFS is*

$$FFS \doteq \left\{f \in \mathcal{F} : \|\widetilde{y} - f(\widetilde{x})\|_q \leqslant \mu\right\} \tag{12.12}$$

where $\widetilde{y} = (\widetilde{y}_1, \ldots, \widetilde{y}_L)$ *and* $f(\widetilde{x}) \doteq (f(\widetilde{x}_1), \ldots, f(\widetilde{x}_L))$. □

According to this definition, *FFS* is the set of all functions consistent with the prior assumptions and data. If the prior assumptions hold, then $f^o \in FFS$, that is an important property for evaluating the accuracy of any estimate.

In the set membership framework, the validation of prior assumptions is a fundamental step. It is usual to introduce the concept of prior assumption validation as consistency with the available data: the prior assumptions are considered validated if at least one estimate consistent with these assumptions and the data exists, i.e., if *FFS* is not empty, see, e.g., [21,23].

Definition 12.6. *Prior assumptions are validated if* $FFS \neq \varnothing$ □

The following theorem gives a necessary and a sufficient condition for the validation of prior assumptions. Let us define the following functions:

$$\begin{aligned} \overline{f}(x) &\doteq \min_{k=1,\ldots,L} (\overline{h}_k + \Gamma \|x - \widetilde{x}_k\|_2) \\ \underline{f}(x) &\doteq \max_{k=1,\ldots,L} (\underline{h}_k - \Gamma \|x - \widetilde{x}_k\|_2) \end{aligned} \tag{12.13}$$

where $\overline{h}_k \doteq \widetilde{y}_k + \varepsilon_k$ and $\underline{h}_k \doteq \widetilde{y}_k - \varepsilon_k$.

Necessary and sufficient conditions for checking the assumptions validity are now given.

Theorem 12.1.

1. *A necessary condition for prior assumptions to be validated is* $\overline{f}(\widetilde{x}_k) \geqslant \underline{h}_k, \underline{f}(\widetilde{x}_k) \leqslant \overline{h}_k$, $k = 1, \ldots, L$.
2. *A sufficient condition for prior assumptions to be validated is* $\overline{f}(\widetilde{x}_k) > \underline{h}_k, \underline{f}(\widetilde{x}_k) < \overline{h}_k$, $k = 1, \ldots, L$.

Proof. See [19]. □

Note that the fact that prior assumptions are validated, i.e., that they are consistent with the present data, does not exclude that they may be invalidated by future data.

In the reminder of chapter, it is assumed that the sufficient condition holds. If not, values of the constants appearing in the assumptions on function f^o and on the noise d_k have to be suitably modified. The above validation Theorem can be used for assessing the values of such constants so that sufficient conditions hold.

Now let the function f^c be defined as

$$f^c(x) \doteq \frac{1}{2}[\underline{f}(x) + \overline{f}(x)] \tag{12.14}$$

where $\underline{f}(x)$ and $\overline{f}(x)$ are given in (12.13). The next result shows that the estimate f^c is optimal according to the Definition 12.2 for any L_p norm.

Theorem 12.2. *Assume the following:*

1. *The noise affecting the measurements* $\{\widetilde{x}_k, \widetilde{y}_k\}_{k=1}^{L}$ *is bounded according to (12.7).*
2. *The function* f^o *is Lipschitz continuous according to (12.11).*

Then, for $q = \infty$ *and for any* $p \in [1, \infty]$*:*

1. *The approximation* f^c *defined in* (12.14) *is optimal.*
2. *The worst case approximation error of* f^c *is given by*

$$E(f^c) = \frac{1}{2}\|\overline{f} - \underline{f}\|_p = \inf_{\widehat{f}} EN(\widehat{f}) = \mathcal{R}_{\mathcal{I}}. \tag{12.15}$$

Proof. See [19]. □

12.3.1 Interval estimates

The interval estimates $\overline{f}, \underline{f}$ of the unknown function f^o are now given for the cases where the noise is bounded in ℓ_∞ norm (i.e., $q = \infty$ in (12.7)).

Theorem 12.3. *Assume the following:*

1. *The noise affecting the measurements* $\{\widetilde{x}_k, \widetilde{y}_k\}_{k=1}^{L}$ *is bounded according to (12.7).*
2. *The function* f^o *is Lipschitz continuous according to (12.11).*

Then, for $q = \infty$ *and for any* $x \in \mathcal{X}$, $f^o(x)$ *is tightly bounded as*

$$\underline{f}(x) \leqslant f^o(x) \leqslant \overline{f}(x) \tag{12.16}$$

Proof. See [19]. □

$\overline{f}$ and $\underline{f}$ are also called *optimal bounds* since they are tightest upper and lower bounds of $\overline{f}^o$.

12.4 Nonlinear set membership identification: local approach

Suppose that a preliminary approximation f^* of the function f^o has been obtained using any method. This approximation is of the form

$$f^*(x) = \sum_{i=1}^{N} a_i \phi_i (x) \tag{12.17}$$

where $\phi_i : \mathcal{X} \to \mathbb{R}$ are Lipschitz continuous basis functions and $a_i \in \mathbb{R}$ are parameters identified by means of some suitable algorithm (two algorithms will be presented in Section 12.4.2). The choice of the basis functions ϕ_i is clearly an important step of the identification process, see, e.g., [24–26]. In several cases of practical interest, the basis functions are known a priori to belong to some "large" set of functions, see, e.g., the example presented in [22]. The sparse approximation algorithms presented below can be applied in these cases to select within this "large" set the functions which are important for providing an accurate description of the system under investigation. In other cases, the basis functions are not known a priori, and their choice can be carried out considering the numerous options available in the literature (e.g., Gaussian, sigmoidal, wavelet, polynomial, trigonometric). See [24] for a discussion on the main features of the most used basis functions and for indications for their choice.

Define the following *residue function*:

$$f_\Delta(x) \doteq f^o(x) - f^*(x). \tag{12.18}$$

From (12.11) and from the Lipschitz continuity of ϕ_i, it follows that f_Δ is Lipschitz continuous over the set $\mathcal{X}$:

$$f_\Delta \in \mathcal{F}(\Gamma_\Delta). \tag{12.19}$$

for some $\Gamma_\Delta < \infty$. Note that the Lipschitz constant Γ_Δ can be estimated by means of the algorithm presented in Section 12.6.

Remark 12.4. *The inclusion* (12.19) *corresponds to assume a* global *maximum rate of variation for f_Δ but a* local *maximum rate of variation for f^o. Indeed, for every $x, \widehat{x} \in \mathcal{X}$, the following inequalities hold*

$$-\Gamma_\Delta \leqslant \frac{f_\Delta(x) - f_\Delta(\widehat{x})}{\|x - \widehat{x}\|_2} \leqslant \Gamma_\Delta$$

$$\frac{f^*(x) - f^*(\widehat{x})}{\|x - \widehat{x}\|_2} - \Gamma_\Delta \leqslant \frac{f^o(x) - f^o(\widehat{x})}{\|x - \widehat{x}\|_2} \leqslant \frac{f^*(x) - f^*(\widehat{x})}{\|x - \widehat{x}\|_2} + \Gamma_\Delta.$$

We can observe that, as expected, the maximum rate of variation of f_Δ is constant and equal to Γ_Δ for any $\hat{x} \in \mathcal{X}$. Instead, the maximum rate of variation of f^o is Γ_Δ plus a quantity that depends locally on the point $\hat{x}$. For this reason, when $f^* = 0$ and thus $f_\Delta = f^o$, the approach is called global set membership approach *which was discussed in Section 12.3. Otherwise, the approach is called* local set membership approach.

Under the above assumptions and, in particular, under (12.19) and (12.7), we have that $f^o \in FFS$, where FFS is the feasible function set defined as follows:

Definition 12.7. *The* FFS *is*

$$FFS \doteq \{f : f = f^* + f_\Delta,\, f_\Delta \in \mathcal{F}(\Gamma_\Delta),\, \|\widetilde{y} - f(\widetilde{x})\|_q \leqslant \mu\}$$

where $\widetilde{y} = (\widetilde{y}_1, \ldots, \widetilde{y}_L)$ *and* $f(\widetilde{x}) \doteq (f(\widetilde{x}_1), \ldots, f(\widetilde{x}_L))$. □

According to this definition, FFS is the set of all functions consistent with the prior assumptions and data. In the set membership framework, the validation of prior assumptions is a fundamental step. It is usual to introduce the concept of prior assumption validation as consistency with the available data: the prior assumptions are considered validated if at least one estimate consistent with these assumptions and the data exists, i.e., if FFS is not empty, see, e.g., [21,23].

Definition 12.8. *The prior assumptions are validated if* $FFS \neq \varnothing$. □

The following theorem gives a necessary and sufficient condition for the validation of prior assumptions.

Theorem 12.4. $FFS \neq \varnothing$ *if and only if the optimization problem (12.30) is feasible.*

Proof. See [18]. □

The following theorem shows that the approximation f^* in (12.17) is interpolatory (and thus almost optimal). The theorem also provides an explicit bound on the worst case approximation error. Let us define the following functions:

$$\begin{aligned} \overline{f}_\Delta(x, \widehat{\varepsilon}) &\doteq \min_{k=1,\ldots,L} (\delta_k(a^*) + \widehat{\varepsilon}_k + \Gamma_\Delta \|x - \widetilde{x}_k\|_2) \\ \underline{f}_\Delta(x, \widehat{\varepsilon}) &\doteq \max_{k=1,\ldots,L} (\delta_k(a^*) - \widehat{\varepsilon}_k - \Gamma_\Delta \|x - \widetilde{x}_k\|_2) \end{aligned} \tag{12.20}$$

$$\begin{aligned} \overline{f}(x, \varepsilon) &= f^*(x) + \overline{f}_\Delta(x, \varepsilon) \\ \underline{f}(x, \varepsilon) &= f^*(x) + \underline{f}_\Delta(x, \varepsilon) \end{aligned} \tag{12.21}$$

where $\delta_k(a^*) = \widetilde{y}_k - f^*(\widetilde{x}_k)$ (see (12.31)) and $\widehat{\varepsilon}_k \geqslant 0$, $k = 1, \ldots, L$.

Theorem 12.5. *Assume the following:*

1. *The noise affecting the measurements* $\{\widetilde{x}_k, \widetilde{y}_k\}_{k=1}^L$ *is bounded according to (12.7).*
2. *The function* f^o *is Lipschitz continuous according to (12.11).*

Then, for $q = 2, \infty$ *and for any* $p \in [1, \infty]$*:*

1. *The approximation* f^* *defined in (12.17) is interpolatory (and thus almost-optimal).*
2. *The worst case approximation error of* f^* *is bounded as*

$$EN(f^*) \leqslant \max_{\substack{\|\widehat{\varepsilon}\|_q \leqslant \mu \\ \|\widehat{\epsilon}\|_q \leqslant \mu}} \left\| \overline{f}_\Delta(\cdot, \widehat{\varepsilon}) - \underline{f}_\Delta(\cdot, \widehat{\epsilon}) \right\|_p = 2 \inf_{\widehat{f}} EN(\widehat{f}). \tag{12.22}$$

Proof. See [18]. □

Suppose that $\widehat{\varepsilon}^* = \widehat{\epsilon}^* = \varepsilon$, where $\varepsilon = (\varepsilon_1, \ldots, \varepsilon_L)$ is obtained as described in Section 12.4.1. Now let the function f^c be defined as

$$f^c(x) \doteq f^*(x) + \frac{1}{2}\left[\overline{f}_\Delta(x, \varepsilon) + \underline{f}_\Delta(x, \varepsilon)\right]. \tag{12.23}$$

where $\overline{f}_\Delta, \underline{f}_\Delta$ are given in (12.20). The following result shows that this function f^c is an optimal approximation of f^o.

Theorem 12.6. *Assume the following:*

1. *The noise affecting the measurements $\{\widetilde{x}_k, \widetilde{y}_k\}_{k=1}^L$ is bounded according to (12.7).*
2. *The function f^o is Lipschitz continuous according to (12.11).*

Then, for $q = 2, \infty$ and for any $p \in [1, \infty]$:

1. *The approximation f^c defined in (12.23) is optimal.*
2. *The worst case approximation error of f^* is given by*

$$EN(f^c) = \frac{1}{2}\left\| \overline{f}_\Delta(\cdot, \varepsilon) - \underline{f}_\Delta(\cdot, \varepsilon) \right\|_p = \inf_{\widehat{f}} EN(\widehat{f}). \tag{12.24}$$

Proof. See [19]. □

In summary, we have shown that the function $f^*(x)$ is an almost optimal approximation, whereas the function $f^c(x) \doteq f^*(x) + [\overline{f}_\Delta(x, \varepsilon) + \underline{f}_\Delta(x, \varepsilon)]/2$ is an optimal approximation of f^o. The correction term $[\overline{f}_\Delta(x, \varepsilon) + \underline{f}_\Delta(x, \varepsilon)]/2$ can be useful to check how close is f^* to the optimum: The two approximations can be evaluated off-line on a set of data not used for identification. If the errors of the two approximations on these data are similar, it can be concluded that f^* is practically optimal. Otherwise, this comparison allows us to quantify the suboptimality level of f^* with respect to f^c.

Remark 12.5. *A subcase of the general theory presented here is when $f^* = 0$, in which we have the so-called global set membership approach, previously presented. Otherwise, if $f^* \neq 0$, we have the so-called local set membership approach. See also Remark 12.4 for an explanation of this terminology.*

12.4.1 Interval estimates

Interval estimate on the unknown function f^o are now derived. A general formulation is developed, allowing us to deal with the cases where the noise is bounded in ℓ_2 or ℓ_∞ norm (i.e., $q = 2$ or $q = \infty$ in (12.7)).

12.4.1.1 Noise bounded in ℓ_2 norm

Theorem 12.5 does not allow the evaluation of interval estimates, since the assumption that the noise sequence $d = (d_1, d_2, \ldots, d_L)$ is bounded in ℓ_2 norm gives no information on how the single elements d_k are bounded. In order to overcome this issue, some additional assumption has to be made on the element-wise boundedness of the noise sequence d. This kind of assumption can be obtained as given below.

Since f^* is an almost optimal approximation of f^o, we have that $f^*(\widetilde{x}_k) \cong f^o(\widetilde{x}_k)$ and, consequently, that $d_k = \widetilde{y}_k - f^o(\widetilde{x}_k) \simeq \widetilde{y}_k - f^*(\widetilde{x}_k) \doteq \delta_k(a^*)$. It is then reasonable to assume the following relative plus absolute error bound:

$$|d_k| \leqslant \varepsilon_k \doteq \varepsilon^r |\delta_k(a^*)| + \varepsilon^a, \quad k = 1, \ldots, L \tag{12.25}$$

where the term $\varepsilon^r |\delta_k(a^*)|$ accounts for the fact that $d_k \simeq \delta_k(a^*)$ and ε^a accounts for the fact that d_k and $\delta_k(a^*)$ are not exactly equal. The parameters $\varepsilon^r, \varepsilon^a \geqslant 0$ have to be taken such that $\varepsilon^r \mu + \varepsilon^a \sqrt{L} \leqslant \mu$. Indeed, if this inequality is satisfied, (12.25) is consistent with (12.7): $\|d\|_2 \leqslant \varepsilon^r \mu + \varepsilon^a \sqrt{L} \leqslant \mu$. Following this indication, ε^r and ε^a (together with Γ_Δ) can be chosen by means of the procedure presented in [19].

In order to satisfy the assumption (12.25), the following additional constraints have to be inserted in Algorithm 12.1 (in particular, in (12.30) and on line 5 of (12.32)):

$$|\widetilde{y}_k - \Phi_k(\widetilde{x}_k)\, a| \leqslant \varepsilon_k, \quad k = 1, \ldots, L \tag{12.26}$$

where $\Phi_k(\widetilde{x}_k)$ is the kth row of the matrix Φ.

□

12.4.1.2 Noise bounded in ℓ_∞ norm

If the noise is bounded in ℓ_∞ norm, the required interval estimates can be obtained without introducing any further assumption. Indeed, in this case,

$$|d_k| \leqslant \varepsilon_k \doteq \mu, \quad k = 1, \ldots, L. \tag{12.27}$$

□

The following theorem, holding for both the ℓ_2 and ℓ_∞ cases, provides tight point-wise interval estimates for $f^o(x)$ and gives an expression of the worst case error bound computable for any dimension n_x (a computationally tractable algorithm for this computation is presented in [27]).

Theorem 12.7. *Assume the following:*

1. *The noise affecting the measurements $\{\widetilde{x}_k, \widetilde{y}_k\}_{k=1}^{L}$ is bounded according to (12.25) or (12.27).*
2. *The function f^o is Lipschitz continuous according to (12.11).*

Then, for $q = 2, \infty$ and for any $p \in [1, \infty]$:

1. *The worst case approximation error of f^* is bounded as*

$$EN(f^*) \leqslant \left\| \overline{f}_\Delta(\cdot, \varepsilon) - \underline{f}_\Delta(\cdot, \varepsilon) \right\|_p = 2 \inf_{\widehat{f}} EN(\widehat{f}).$$

2. *For any $x \in \mathcal{X}$, $f^o(x)$ is tightly bounded as*

$$\underline{f}(x,\varepsilon) \leqslant f^o(x) \leqslant \overline{f}(x,\varepsilon). \tag{12.28}$$

Proof. See [18]. □

12.4.2 Local approach—identification algorithms

In this section, two algorithms for the identification of the parameters a_i in (12.17) are proposed.

In order to ensure suitable regularity properties of the approximation, limiting well-known issues such as overfitting and the curse of dimensionality, we require the vector $a = (a_1, a_2, \dots, a_N) \in \mathbb{R}^N$ of the coefficients in (12.17) to be sparse. Hence, under assumption (12.7), a solution to the identification Problem 12.1 could be found by solving the following optimization problem:

$$\begin{aligned} a^0 &= \arg\min_{a\in\mathbb{R}^N} \|a\|_0 \\ \text{subject to} &\quad \|\widetilde{y} - \Phi a\|_q \leqslant \mu. \end{aligned} \tag{12.29}$$

where

$$\begin{aligned} \widetilde{y} &\doteq (\widetilde{y}_1, \dots, \widetilde{y}_L) \\ \Phi &\doteq \begin{bmatrix} \phi_1(\widetilde{x}_1) & \cdots & \phi_N(\widetilde{x}_1) \\ \vdots & \ddots & \vdots \\ \phi_1(\widetilde{x}_L) & \cdots & \phi_N(\widetilde{x}_L) \end{bmatrix} \\ &= \begin{bmatrix} \phi_1(\widetilde{x}) & \cdots & \phi_N(\widetilde{x}) \end{bmatrix}, \end{aligned}$$

$\phi_1(\widetilde{x}) \doteq (\phi_1(\widetilde{x}_1), \dots, \phi_1(\widetilde{x}_L))$, and $\|a\|_0$ is the ℓ_0 quasi-norm of a, defined as the number of nonzero components of a. In fact, minimizing the ℓ_0 quasi-norm of a vector corresponds to minimizing the number of its nonzero elements, i.e., to maximizing its sparsity. On the other hand, the constraint $\|\widetilde{y} - \Phi a\|_q \leqslant \mu$ ensures that the identified coefficient vector is consistent with the measured data (12.6) and the prior assumption on noise (12.7).

However, the optimization problem (12.29) cannot be easily solved, since the ℓ_0 quasi-norm is a non-convex function, and its minimization is an nondeterministic polynomial NP-hard problem. The classical approach to overcome this issue is to replace the ℓ_0 quasi-norm with its convex envelope, i.e., the ℓ_1 norm, see, e.g., [28–30]. The identification Problem 12.1 can thus be solved efficiently by means of the following convex optimization problem.

Algorithm 12.1. *Function identification 1*

$$\begin{aligned} a^* &= \arg\min_{a\in\mathbb{R}^N} \|a\|_1 \\ \textit{subject to} &\quad \|\widetilde{y} - \Phi a\|_q \leqslant \mu \end{aligned} \tag{12.30}$$

where μ can be chosen as a positive number slightly larger than $\mu^{\min}$, the minimum value for which the problem is feasible (this choice is theoretically motivated by the validation Theorem 12.4). Provided that $\mu > \mu^{\min}$, the value of μ can be tuned to suitably manage the trade-off between accuracy and sparsity. □

Another interesting ℓ_1 algorithm, completely based on convex optimization, is now presented. As discussed below, this algorithm provides sparser solutions with respect to the standard algorithm (12.30), which is based on simple ℓ_1-norm minimization.

Without loss of generality, assume that the columns of Φ are normalized: $\|\phi_i(\widetilde{x})\|_2 = 1, i = 1, 2, \ldots, N$. Define the following quantity:

$$\xi(a) \doteq \frac{|\delta(a)|_1 + |\delta(a)|_{K_0}}{\underline{\sigma}^2(\Phi)}$$

where $K_0 \doteq 2\,\|a\|_0$, $\underline{\sigma}(\Phi)$ is the minimum nonzero singular value of Φ and

$$\begin{aligned} \delta(a) &\doteq \widetilde{y} - \Phi a \\ |w|_K &\doteq \sqrt{\sum_{i \in I_K} \left(w^T \phi_i(\widetilde{x})\right)^2}, \end{aligned} \tag{12.31}$$

being I_K the set of the K largest inner products $\left|w^T \phi_i(\widetilde{x})\right|$. Let card$(\cdot)$ denote the set cardinality.

Algorithm 12.2. *Function identification 2*

1. *Solve the optimization problem* (12.30) *and set* $a^1 := a^*$.
2. *Let* $r(a^1) \doteq \left\{i_1, \ldots, i_j : \xi\left(a^1\right) > |a^1_{i_1}| \geqslant \cdots \geqslant |a^1_{i_j}|\right\}$ *and let* $r_\lambda(a^1)$ *denote the subset of* $r(a^1)$ *with indices in* λ. *Compute the coefficient vector* a^* *as follows:*

$$\begin{aligned} &\textit{for } k = 1 : \textit{card}(r(a^1)) \\ &\qquad c^k = \arg\min_{a \in \mathbb{R}^N} \|\widetilde{y} - \Phi a\|_q \\ &\qquad \textit{subject to } a_i = 0, \ \forall i \in r_\lambda(a^1) \\ &\qquad\qquad\qquad \lambda = \{k, \ldots, \textit{card}(r(a^1))\} \\ &\qquad \textit{if } \left\|\widetilde{y} - \Phi c^k\right\|_q \leqslant \mu \\ &\qquad\qquad a^* := c^k \\ &\qquad\qquad \textit{break} \\ &\qquad \textit{end} \\ &\textit{end} \end{aligned} \tag{12.32}$$

□

The rationale behind Algorithm 12.2 can be explained as follows: In step 1, an optimization problem similar to (12.29) is solved, where the ℓ_0 quasi-norm is replaced by the ℓ_1 norm. The ℓ_1 norm is the convex envelope of the ℓ_0 quasi-norm, and its minimization yields a sparse vector a^1 [28–30]. However, it is not guaranteed that all the nonzero elements of a^1 are necessary to have $\|\widetilde{y} - \Phi a^1\|_q \leqslant \mu$. In step 2, only the elements of a^1 larger than $\xi(a^1)$ are kept (indeed, $\xi(a^1)$ discriminates between "important" and "less important" vector components [22]), while the remaining ones, ordered by decreasing amplitude, are progressively included to form the vector a^*. The algorithm stops when $\|\widetilde{y} - \Phi c^k\|_q \leqslant \mu$. The solution provided by step 2 is thus a vector a^* where the number of nonzero elements is further reduced with respect to the initial sparse solution a^1. The sparse approximation is given by (12.17), with $a_i = a_i^*, \forall i$.

12.5 Nonlinear set membership identification: quasi-local approach

In this section, the so-called quasi-local nonlinear set membership approach is presented, capturing the advantages of the global and local approaches of Sections 12.3 and 12.4 and avoiding some drawbacks of these two. On one hand, the quasi-local approach allows the derivation of significantly less conservative uncertainty bounds with respect to the global approach of Section 12.3. On the other hand, the quasi-local approach does not require to choose a suitable parametric form for the filter, as done in the local approach of Section 12.4. The filter is obtained directly from the data in a nonparametric closed form.

Based on assumption (12.11), we can define the following quantity, called the *quasi-local Lipschitz parameter*:

$$\gamma(x) \doteq \sup_{\widehat{x}\in\mathcal{X},\widehat{x}\neq x} \frac{|f_o(x) - f_o(\widehat{x})|}{\|x - \widehat{x}\|_2} \tag{12.33}$$

Obviously, the Lipschitz constant of f on $\mathcal{X}$ is given by

$$\Gamma = \sup_{x\in\mathcal{X}} \gamma(x). \tag{12.34}$$

Lemma 12.1. *For any $x \in \mathcal{X}$, a $\gamma(x)$ exists, such that*

$$|f_o(x) - f_o(\widehat{x})| \leqslant \gamma(x)\ \|x - \widehat{x}\|_2,\ \forall \widehat{x} \in \mathcal{X}.$$

Proof. The statement follows directly from (12.33). □

Let us now suppose that the quasi-local Lipschitz parameters $\gamma(\widetilde{x}_k)$, $k = 1, \ldots, L$, are known or can be estimated (a method for performing such an estimation is given in Section 12.6.1). Assume also that the noise is bounded in ℓ_∞ norm according to (12.7), with $q = \infty$ and $\varepsilon \doteq \mu$. On the basis of this information, we can define the following function set:

$$\mathcal{F}_{ql} \doteq \{f : |f(x) - f(\widehat{x}_k)| \leqslant \gamma(\widehat{x}_k)\ \|x - \widehat{x}_k\|_2, \forall x \in \mathcal{X},\ k = 1, \ldots, L\}. \tag{12.35}$$

This allows us to define the *FFS* as follows:

Definition 12.9. *The* FFS *is*

$$FFS \doteq \{f \in \mathcal{F}_{ql} : \|\widetilde{y} - f(\widetilde{x})\|_q \leqslant \mu\} \tag{12.36}$$

where $\widetilde{y} = (\widetilde{y}_1, \ldots, \widetilde{y}_L)$ *and* $f(\widetilde{x}) \doteq (f(\widetilde{x}_1), \ldots, f(\widetilde{x}_L))$. □

As discussed in Section 12.4, *FFS* is the set of all functions consistent with the prior assumptions and data. The prior assumptions are considered validated if at least one estimate consistent with these assumptions and the data exists, i.e., if *FFS* is not empty, see, e.g., [21,23].

Definition 12.10. *The prior assumptions are validated if* $FFS \neq \varnothing$. □

The following theorem gives a necessary and a sufficient condition for the validation of prior assumptions. Define the functions

$$\begin{aligned}\overline{f}(x) &\doteq \min_{k=1,\ldots,L} (\overline{h}_k + \gamma(\widetilde{x}_k)\,\|x - \widetilde{x}_k\|_2)\\ \underline{f}(x) &\doteq \max_{k=1,\ldots,L} (\underline{h}_k - \gamma(\widetilde{x}_k)\,\|x - \widetilde{x}_k\|_2)\end{aligned} \tag{12.37}$$

where $\overline{h}_k \doteq \widetilde{y}_k + \varepsilon_k$ and $\underline{h}_k \doteq \widetilde{y}_k - \varepsilon_k$.

Theorem 12.8.

1. *A necessary condition for prior assumptions to be validated is* $\overline{f}(\widetilde{x}_k) \geqslant \underline{h}_k, \underline{f}(\widetilde{x}_k) \leqslant \overline{h}_k$, $k = 1, \ldots, L$.
2. *A sufficient condition for prior assumptions to be validated is* $\overline{f}(\widetilde{x}_k) > \underline{h}_k, \underline{f}(\widetilde{x}_k) < \overline{h}_k$, $k = 1, \ldots, L$.

Proof. See [31]. □

Let us now define the function

$$f^c(x) \doteq \frac{1}{2}[\underline{f}(x) + \overline{f}(x)] \tag{12.38}$$

where $\underline{f}(x)$ and $\overline{f}(x)$ are given in (12.37). The next result shows that the approximation f^c is optimal for any L_p norm. The theorem also provides an explicit bound on the worst case approximation error.

Theorem 12.9. *Assume the following:*

1. *The noise affecting the measurements* $\{\widetilde{x}_k, \widetilde{y}_k\}_{k=1}^L$ *is bounded according to (12.7).*
2. *The function* f^o *is Lipschitz continuous according to (12.11).*

Then, for $q = \infty$ *and for any* $p \in [1, \infty]$*:*

1. *The approximation* f^c *defined in* (12.38) *is optimal.*
2. *The worst case approximation error of* f^c *is bounded as*

$$EN(f^c) = \frac{1}{2}\left\|\overline{f} - \underline{f}\right\|_p = \inf_{\widehat{f}} EN(\widehat{f}) \doteq \mathcal{R}_{\mathcal{I}}. \tag{12.39}$$

Proof. See [31]. □

12.5.1 Interval estimates

The interval estimates of the unknown function f^o are now given for the cases where the noise is bounded in ℓ_∞ norm (i.e., $q = \infty$ in (12.7)).

Theorem 12.10. *Assume the following:*

1. *The noise affecting the measurements $\{\widetilde{x}_k, \widetilde{y}_k\}_{k=1}^{L}$ is bounded according to (12.7).*
2. *The function f^o is Lipschitz continuous according to (12.11).*

Then, for $q = \infty$ and for any $x \in \mathcal{X}$, $f^o(x)$ is tightly bounded as

$$\underline{f}(x) \leqslant f^o(x) \leqslant \overline{f}(x). \tag{12.40}$$

Proof. See [18]. □

Remark 12.6. *The point-wise bounds (12.40) provide an interval estimate of the unknown value $f^o(x)$. Interval estimates allow us to quantify the uncertainty associated with the identification process and are thus important in system and control applications. Indeed, these estimates can be used, e.g., for robust control design, [32,33], prediction interval evaluation, [34], and fault detection, see the following sections.*

Remark 12.7. *A study on the relation between the representativeness and length of the available data and the quality of the approximation can be found in [27]. This study relies on the computation of the radius of information $\mathcal{R}_\mathcal{I}$ (see Definition 12.2), representing the minimum worst case error that can be achieved from the available prior and experimental information. Based on this concept, a methodology is proposed in [27], where $\mathcal{R}_\mathcal{I}$ is used to evaluate the "level of information" provided by a given data set and to obtain precise indications on the quality of the approximation that can be obtained.*

Figure 12.1 shows the comparison between the global and quasi-local set membership bounds for a nonlinear function. In Figure 12.1(a), a global Lipschitz constant Γ was assumed for the function f^o. In Figure 12.1(b), a quasi-local Lipschitz parameter $\gamma(x)$ was assumed. It can be noted that the resulting uncertainty bounds are clearly tighter in the quasi-local case, especially in regions where the function is relatively "flat."

12.6 Parameter estimation and adaptive set membership model

12.6.1 Parameter estimation

Estimates of the noise bound μ, Lipschitz constants Γ, Γ_Δ, and the quasi-local Lipschitz parameter $\gamma(x)$ such that the assumptions are validated can be obtained by means of two algorithms given in [35] and reported in the following. The first

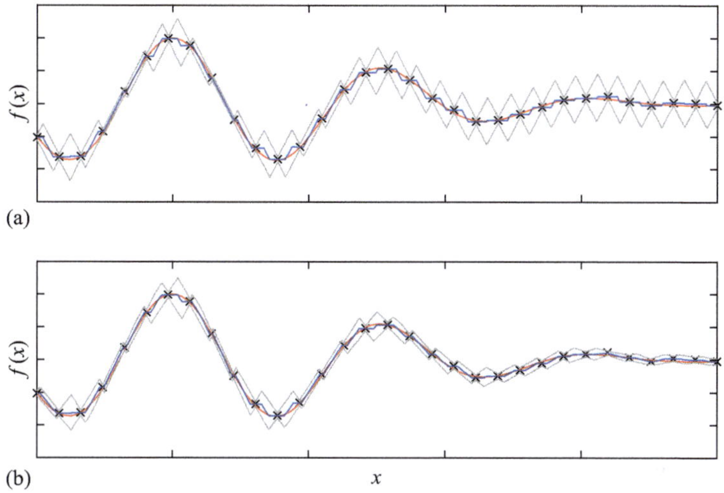

Figure 12.1 *(a) Global bounds; (b) quasi-local bounds; $f^o(x)$: red line; measurements: black cross; $f^c(x)$: blue line; $\overline{f}(x)$,$\underline{f}(x)$: gray line*

algorithm is directly taken from [35], while the second one is a generalization of the corresponding one in [35].

Algorithm 12.3. *Noise bound estimation* $\widehat{\mu}$

1. *Choose a "small"* $\rho > 0$. *For example,* $\rho = 0.01max_{k,t=1,\ldots,L-1} \|\widetilde{x}_k - \widetilde{x}_t\|$.
2. *Find the set of indexes:* $I_k \doteq \{k : \|\widetilde{x}_k - \widetilde{x}_t\| \leqslant \rho\}$. *If* $I_k = \emptyset$ *for all* $k = 1, \ldots, L-1$, *go to step 1 and choose a larger* ρ.
3. *For* $k = 1, \ldots, L-1$ *compute* $\delta\widetilde{y}_k = \max_{i \in I_k} |\widetilde{y}_k - \widetilde{y}_i|$. *If* $I_k = \emptyset$, *set* $\delta\widetilde{y}_k = \infty$.
4. *Obtain the estimate* $\widehat{\mu}$ *of the noise bound* μ *as* $\widehat{\mu} = \frac{1}{2N} \sum_{k \in Q} \delta\widetilde{y}_k$ *where* $Q \doteq \{k \in \{1, \ldots, L-1\} : \delta\widetilde{y}_k < \infty\}$ *and* $N \doteq \mathrm{card}(Q)$

□

Algorithm 12.4. *Lipschitz parameter estimation* $\widehat{\Gamma}$, $\widehat{\Gamma}_\Delta$, $\widehat{\gamma}(x)$

1. *For* $t, k = 1, \ldots, L-1$ *and* $\widetilde{x}_k \neq \widetilde{x}_t$, *compute*

$$\widehat{\Gamma} = \max_{k,t=1,\ldots,L-1} \begin{cases} \frac{|\widetilde{y}_k - \widetilde{y}_t| - 2\widehat{\mu}}{\|\widetilde{x}_k - \widetilde{x}_t\|_2} & \text{if} \quad |\widetilde{y}_k - \widetilde{y}_t| > 2\widehat{\mu} \\ 0 & \text{otherwise.} \end{cases} \tag{12.41}$$

2. *For* $t, k = 1, \ldots, L-1$ *and* $\widetilde{x}_k \neq \widetilde{x}_t$, *compute*

$$\widehat{\Gamma}_{\Delta} = \max_{k,t=1,\ldots,L-1} \begin{cases} \frac{|\widetilde{\delta}_k - \widetilde{\delta}_t| - 2\widehat{\mu}}{\|\widetilde{x}_k - \widetilde{x}_t\|_2} & \text{if } |\widetilde{\delta}_k - \widetilde{\delta}_t| > 2\widehat{\mu} \\ 0 & \text{otherwise.} \end{cases} \tag{12.42}$$

3. *For* $k = 1, \ldots, L-1$ *and* $\widetilde{x}_k \neq \widetilde{x}_t$, *compute*

$$\widehat{\gamma}(\widetilde{x}_k) = \max_{t=1,\ldots,L-1} \begin{cases} \frac{|\widetilde{y}_k - \widetilde{y}_t| + 2\widehat{\mu}}{\|\widetilde{x}_k - \widetilde{x}_t\|_2} & \text{if } |\widetilde{y}_k - \widetilde{y}_t| > 2\widehat{\mu} \\ 0 & \text{otherwise.} \end{cases} \tag{12.43}$$

□

The following theorems show that, under reasonable density conditions on the noise, the estimates given by these two algorithms converge to the corresponding true values.

Theorem 12.11. *(Theorem 2 of [35]) Let the set* $\{\widetilde{x}_k, d_k\}_{k=1}^{L}$ *appearing in (12.2) be dense on* $\mathcal{X} \times B_{\varepsilon}$ *as* $L \to \infty$. *Then,*

$$\lim_{L\to\infty} \widehat{\mu} = \mu. \quad \square$$

Theorem 12.12. *(Theorem 3 of [35]) Let the set* $\{\widetilde{x}_k, d_k\}_{k=1}^{L}$ *appearing in (12.2) be dense on* $\mathcal{X} \times B_{\varepsilon}$ *as* $L \to \infty$. *Then,*

$$\lim_{L\to\infty} \widehat{\Gamma} = \Gamma. \quad \square$$

12.6.2 Adaptive set membership model

In many applications, it may happen that the dynamics of the system changes over time or the model is not accurate enough in the whole domain of $\mathcal{X}$. One of the advantages of nonlinear set membership (global and quasi-local approach) is that it can be easily made adaptive since no optimization problem needs to be solved online.

As discussed in Section 12.2, in the set membership framework, the model accuracy is defined by the radius of information, and it can be computed in a deterministic way [27]. In the global and quasi-local approaches, the model is basically defined by the measurement set $\mathcal{D}$, noise bound μ and Lipschitz parameters Γ or $\gamma(x)$. Therefore, the model can be made adaptive by updating online the measurement set and the Lipschitz parameters.

Let us define the error function as follows:

$$f^e(x) = \frac{1}{2}[\overline{f}(x) - \underline{f}(x)] \tag{12.44}$$

This function allows us to write the radius of information as

$$\mathcal{R}_{\mathcal{I}} = \|f_e(\cdot)\|_p. \tag{12.45}$$

Suppose that a model with a desired radius of information $\mathcal{R}_d$ is looked for, where $\mathcal{R}_d < \mathcal{R}_{\mathcal{I}}$. If, at time instant $t, f^e(x_t) > \mathcal{R}_d$, it means that the model error in that point is larger than the desired radius. Therefore, the new measurement can be added to the set $\mathcal{D}$ in order to increase the model accuracy.

In the case where the dynamics of the system changes over time, a time label can be assigned to each element of the set $\mathcal{D}$ as $\mathcal{D}(L) = \{\widetilde{x}_k, \widetilde{y}_k, t_k\}_{k=1}^{L}$ where t_k indicates the time each measurement was taken. Then, since the model is running online, at each time instant t, we can eliminate the measurements which were taken at $t - \delta t$, where δt is a desired value which depends on the system dynamics variation.

Algorithm 12.5. *Online update of the set membership model*

1. *Define the measurement set* $\mathcal{D}$ *as* $\mathcal{D}(L) = \{\widetilde{x}_k, \widetilde{y}_k, t_k\}_{k=1}^{L}$.
2. *At time step t, compute the vector* x_t *and measure the system output* y_t.
3. *If* $f^e(x_t) > \mathcal{R}_d$. *Then,*

$$\begin{aligned} &\mathcal{D}(L+1) = \mathcal{D}(L) \cup \{\widetilde{x}_{L+1} = x_t, \widetilde{y}_{L+1} = y_t, t_{L+1} = t\} \\ &L = L+1. \end{aligned} \tag{12.46}$$

4. *Find the set of indexes* $I_k = \{k : t_k \in \mathcal{D}, t_k < t - \delta t\}$. *Then,*

$$\begin{aligned} &\mathcal{D}(L-N) = \mathcal{D}(L) \setminus \{\widetilde{x}_k, \widetilde{y}_k, t_k\}_{k \in I_k} \\ &L = L - N \end{aligned} \tag{12.47}$$

   where $N \doteq \mathrm{card}(I_k)$.
5. *If* $\mathcal{D}$ *is changed, update the model Lipschitz parameters according to Algorithm 12.4.*

□

12.7 Summary of set membership fault-detection procedure

The main steps of the proposed set membership fault-detection method are now summarized.

Off-line operations

1. *Define the measurement set* $\mathcal{D}$ *according to (12.2).*
2. *Estimate the noise bound* μ *according to Algorithm 12.3.*
3. *In the case of local approach, estimate a preliminary approximation* f^* *according to Algorithm 12.1 or 12.2.*
4. *Estimate the Lipschitz parameters according to Algorithm 12.4. In the case of global approach, compute* $\widehat{\Gamma}$*; in the case of quasi-local approach, compute* $\widehat{\gamma}(x)$*; in the case of local approach, compute* $\widehat{\Gamma}_{\Delta}$.

Online operations

1. *At each time step,*

$$\begin{aligned} &\textit{If} \quad \tilde{y}_k > \overline{f}(\tilde{x}_k) + \varepsilon_k \quad \textit{or} \quad \tilde{y}_k < \underline{f}(\tilde{x}_k) - \varepsilon_k \\ &\textit{Then} \quad \textit{Fault} = 1 \\ &\textit{Else} \quad \textit{Fault} = 0 \end{aligned}$$

2. *In the case of adaptive algorithm, update the set membership model according to Algorithm 12.5 if no fault has accrued in the system and the system is not recovering from a fault.*

Regarding the online operations, since the set membership model describes the normal dynamic behavior of the system, it has to be updated when no other dynamics are involved. In other words, during a fault or immediately after a fault, the measurements correspond to an abnormal behavior of the system, and therefore they are not suitable for updating the model.

12.8 Example: fault detection for a drone actuator

In this example, the proposed fault-detection algorithms was tested on a real drone actuator, in a laboratory experimental setup. The actuator is composed of three main components: a Brushless DC motor, a driver and a propeller. The motor makes the propeller rotate, and this rotation produces the required thrust and torque to drive a drone. The torque produced by the motor is given by

$$T = K_t(I - I_0) \tag{12.48}$$

where T is the motor torque, I is the input current, I_0 is the current when there is no load on the motor and K_t is the torque constant of the motor. The voltage across the motor is given by

$$V = R_m I + L_m \frac{\mathrm{d}I}{\mathrm{d}t} + K_e \omega \tag{12.49}$$

where V is the voltage drop across the motor, R_m and L_m are the motor resistance and inductance, respectively, and K_e is the motor speed constant. The motor is attached to a propeller, and therefore we have

$$T = B_m \omega + (J_m + J_p) \frac{\mathrm{d}\omega}{\mathrm{d}t} + T_D \tag{12.50}$$

where B_m is the motor friction, ω is the motor angular speed, J_m and J_p are the motor and propeller moments of inertia, respectively, and T_D is the propeller drag torque, given by

$$T_D = C_p \rho D^5 \omega^2 \tag{12.51}$$

Figure 12.2 Drone actuator

where C_p is the propeller power coefficient, ρ is air density and D is the propeller diameter. The power coefficient C_p is a nonlinear function of the propeller speed, which is difficult to derive from the geometric shape of the propeller [36].

12.8.1 Experimental setup

A Brushless DC motor (RIMFIRE.10) attached to a propeller (APC 10×4.7p) is connected to a driver and a 200 W power supply. The input command of the driver is a PWM signal with a duly cycle proportional to the voltage across the motor. An encoder is attached to the back of the motor to measure the angular speed of the propeller. The input command signal and the encoder are connected to a PC running MATLAB® through a National Instrument data acquisition device (PCI-6289). Data acquisition and online fault detection were carried out using the Simulink Desktop Real-Time. The drone actuator system is shown in Figure 12.2.

12.8.2 Nonlinear set membership fault detection

In order to identify a model, an experiment was carried out, where an amplitude modulated pseudorandom binary sequence (APRBS) command input with a duration of 100 s was applied to the actuator. From this experiment, a set of data was collected, using a sampling time of $T_s = 0.05$ s. The dataset was divided into an identification set, composed by the first 1,000 data and a validation set, composed by the remaining

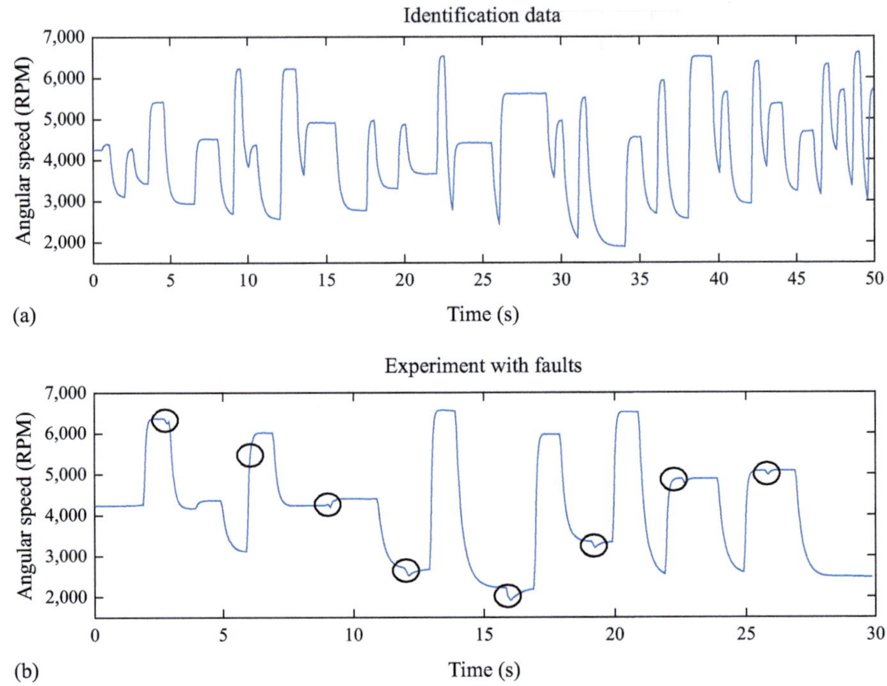

Figure 12.3 (a) First experiment used for identification, (b) second experiment online fault-detection test (black circles are where the faults has occurred)

1,000 data. See Figure 12.3(a). The measurement set was defined according to (12.2) as follows:

$$\begin{aligned} \mathcal{D} &= \{\widetilde{x}_k, \widetilde{y}_k\}_{k=4}^{1,000} \\ \widetilde{x}_k &= [\widetilde{y}_{k-1} \quad \widetilde{y}_{k-2} \quad \widetilde{u}_{k-2} \quad \widetilde{u}_{k-3}] \end{aligned} \tag{12.52}$$

To evaluate the performance of the fault-detection algorithm, a second experiment was carried out using another APRBS input command that was neither used for identification nor validation. This second experiment had a duration of 30 s and during this experiment, in order to introduce a fault scenario, a sheet of paper was placed between the blades, approximately every 3 s, for a total of eight times. In Figure 12.3(b), the fault occurrences are denoted by the circles. During this experiment, the following fault-detection algorithms were running online.

12.8.2.1 Global approach

A nonlinear set membership model was obtained assuming a global constant bound on the function gradient according to Section 12.3. A constant Lipschitz parameter $\Gamma = 0.7$ and a noise bound $\mu = 15$ were estimated according to Algorithms 12.3 and

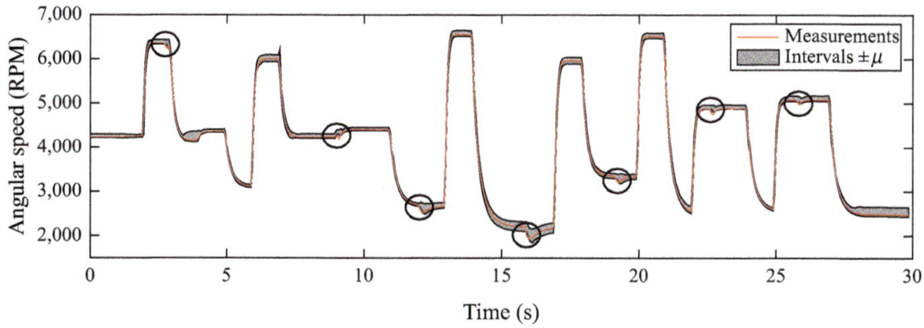

Figure 12.4 Global approach; black lines, $\overline{f} + \mu$,$\underline{f} - \mu$. Red line, $\widetilde{y}$. Circles, bound violation

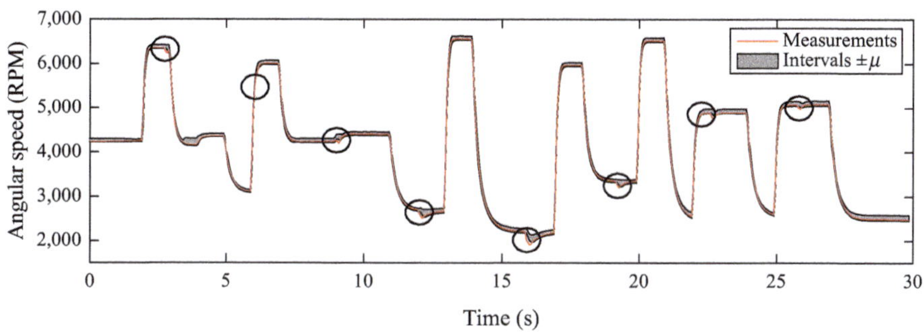

Figure 12.5 Quasi-local approach; black lines, $\overline{f} + \mu$,$\underline{f} - \mu$. Red line, $\widetilde{y}$. Circles, bound violation

12.4. The model was tested in simulation on the validation set and the root mean square simulation error (RMSE) was 42 RPM. Then, the fault-detection algorithm was applied online, as discussed above. Figure 12.4 shows the model intervals and the bound violations. The model was able to detect seven out of eight faults.

12.8.2.2 Quasi-local approach

A nonlinear set membership quasi-local model was obtained according to Section 12.5. A quasi-local Lipschitz parameter γ was derived according to Algorithm 12.4. The model was tested in simulation on the validation set and the RMSE was 30 RPM. Then, the fault-detection algorithm was applied online, as discussed above. Figure 12.5 shows the model intervals and the bound violations. The model was able to detect all the eight faults.

12.8.2.3 Local approach

A preliminary approximation f^* was derived according to Algorithm 12.1, using polynomial basis functions up to degree 5. No improvements were observed considering

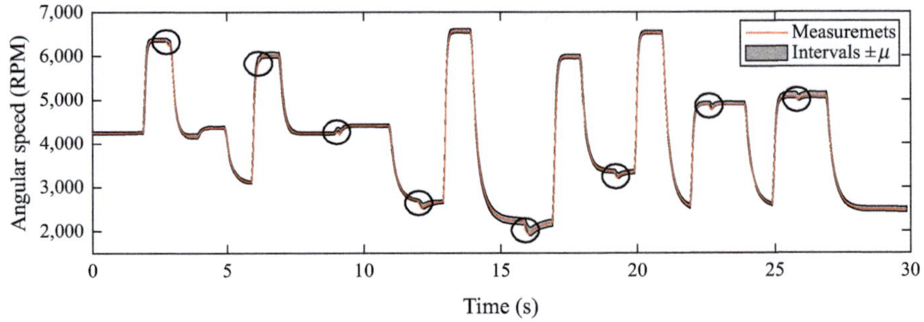

Figure 12.6 Local approach; Black lines, $\overline{f} + \mu, \underline{f} - \mu$. Red line, $\widetilde{y}$. Circles, bound violation

higher degrees. The number of basis functions in (12.30) is 81 where, due to ℓ_1 sparsification, only 33 have a non-null coefficient. The nonlinear set membership local model was obtained according to Section 12.4. A Lipschitz parameter $\Gamma_\Delta = 0.2$ was derived according to Algorithm 12.4. The model was tested in simulation on the validation set, and the RMSE was 23 RPM. Then, the fault-detection algorithm was applied online, as discussed above. Figure 12.6 shows the model intervals and the bound violations. The model was able to detect all eight faults.

12.9 Conclusions

A novel fault-detection approach, based on set membership interval estimates, has been presented, allowing us to overcome several problems of the standard techniques. Its effectiveness has been demonstrated in a laboratory study, related to fault detection for a real propeller. The possibility of extending the approach to fault isolation is currently under investigation.

References

[1] Zhang X, Polycarpou MM, Parisini T. A robust detection and isolation scheme for abrupt and incipient faults in nonlinear systems. IEEE Transactions on Automatic Control. 2002;47(4):576–593.

[2] Patton RJ, Frank PM. Fault Diagnosis in Dynamic Systems: Theory and Applications. Upper Saddle River, NJ: Prentice Hall; 1989.

[3] Gertler J. Fault Detection and Diagnosis in Engineering Systems. New York: Marcel Dekker; 1998.

[4] Persis CD, Isidori A. A geometric approach to nonlinear fault detection and isolation. IEEE Transactions on Automatic Control. 2001;46(6):853–865.

[5] Isermann R. Fault-Diagnosis Systems: An Introduction from Fault Detection to Fault Tolerance. Berlin, Germany: Springer; 2006.
[6] Ferrari RMG, Parisini T, Polycarpou MM. Distributed fault detection and isolation of large-scale discrete-time nonlinear systems: an adaptive approximation approach. IEEE Transactions on Automatic Control. 2012;57(2):275–290.
[7] Liu B, Si J. Fault isolation filter design for linear time-invariant systems. IEEE Transactions on Automatic Control. 1997;42:704–707.
[8] Simani S, Fantuzzi C, Patton RJ. Model-Based Fault Diagnosis in Dynamic Systems Using Identification Techniques. London: Springer Science & Business Media; 2013.
[9] Chen J, Patton RJ. Robust Model-Based Fault Diagnosis for Dynamic Systems. vol. 3. New York: Springer Science & Business Media; 2012.
[10] Patton RJ, Frank PM, Clark RN. Issues of Fault Diagnosis for Dynamic Systems. London: Springer Science & Business Media; 2013.
[11] Armengol J, Trave-Massuyes L, Vehi J, *et al.* A survey on interval model simulators and their properties related to fault detection. Annual Reviews in Control. 2000;24:31–39.
[12] Puig V, Quevedo J, Escobet T, *et al.* Passive robust fault detection of dynamic processes using interval models. IEEE Transactions on Control Systems Technology. 2008;16(5):1083–1089.
[13] Reppa V, Tzes A. Fault detection and diagnosis based on parameter set estimation. IET Control Theory and Applications. 2010;5(1):69–83.
[14] Blesa J, Puig V, Saludes J. Identification for passive robust fault detection using zonotope-based set-membership approaches. International Journal of Adaptive Control and Signal Processing. 2011;25(9):788–812.
[15] Raissi T, Videau G, Zolghadri A. Interval observer design for consistency checks of nonlinear continuous-time systems. Automatica. 2010;46(3): 518–527.
[16] Tornil-Sin S, Ocampo-Martinez C, Puig V, *et al.* Robust fault detection of non-linear systems using set-membership state estimation based on constraint satisfaction. Engineering Applications of Artificial Intelligence. 2012; 25(1):1–10.
[17] Tornil-Sin S, Ocampo-Martinez C, Puig V, *et al.* Robust fault diagnosis of nonlinear systems using interval constraint satisfaction and analytical redundancy relations. IEEE Transactions on Systems, Man, and Cybernetics: Systems. 2014;44(1):18–29.
[18] Novara C. Sparse set membership identification of nonlinear functions and application to fault detection. International Journal of Adaptive Control and Signal Processing. 2016;30(2):206–223.
[19] Milanese M, Novara C. Set membership identification of nonlinear systems. Automatica. 2004;40/6:957–975.
[20] Traub JF, Wasilkowski GW, Woźniakowski H. Information-Based Complexity. San Diego, CA: Academic Press, Inc.; 1988.
[21] Milanese M, Norton J, Lahanier HP, *et al.* Bounding Approaches to System Identification. New York: Plenum Press; 1996.

[22] Novara C. Sparse identification of nonlinear functions and parametric set membership optimality analysis. In: American Control Conference. San Francisco, USA; 2011.

[23] Chen J, Gu G. Control-Oriented System Identification: An H_∞ Approach. New York: John Wiley & Sons; 2000.

[24] Sjöberg J, Zhang Q, Ljung L, *et al.* Nonlinear black-box modeling in system identification: a unified overview. Automatica. 1995;31:1691–1723.

[25] Hsu K, Novara C, Vincent T, *et al.* Parametric and nonparametric curve fitting. Automatica. 2006;42/11:1869–1873.

[26] Novara C, Vincent T, Hsu K, *et al.* Parametric identification of structured nonlinear systems. Automatica. 2011;47(4):711–721.

[27] Milanese M, Novara C. Computation of local radius of information in SM-IBC identification of nonlinear systems. Journal of Complexity. 2007;23:937–951.

[28] Fuchs JJ. Recovery of exact sparse representations in the presence of bounded noise. IEEE Transactions on Information Theory. 2005;51(10):3601–3608.

[29] Tropp JA. Just relax: convex programming methods for identifying sparse signals in noise. IEEE Transactions on Information Theory. 2006;52(3): 1030–1051.

[30] Donoho DL, Elad M, Temlyakov VN. Stable recovery of sparse overcomplete representations in the presence of noise. IEEE Transactions on Information Theory. 2006;52(1):6–18.

[31] Karimshoushtari M, Novara C. Design of experiments for nonlinear dynamic systems based on set membership identification. Internal Report Politecnico di Torino. 2018.

[32] Freeman A, Kokotovic V. Robust Nonlinear Control Design. Boston: Birkhäuser; 1996.

[33] Qu Z. Robust Control of Nonlinear Uncertain Systems. Wiley Series in Nonlinear Science. New York: John Wiley & Sons, Inc.; 1998.

[34] Milanese M, Novara C. Set membership prediction of nonlinear time series. IEEE Transactions on Automatic Control. 2005;50(11):1655–1669.

[35] Fagiano L, Novara C. Learning a nonlinear controller from data: theory, computation and experimental results. IEEE Transactions on Automatic Control. 2016;61(7):1854–1868.

[36] Deters R, Selig M. Static testing of micro propellers. In: 26th AIAA Applied Aerodynamics Conference; 2008. p. 6246.

Chapter 13

Robust data-driven control of systems with nonlinear distortions

Achille Nicoletti[1], Christoph Kammer[2], and Alireza Karimi[2]

The frequency-domain methods that exist for investigating the behaviour of linear systems have become fundamental tools for the control systems engineer. However, due to the increased performance demands on today's industrial systems, the effects of certain nonlinearities can no longer be neglected in modern control applications; for such systems, direct application of these frequency-domain tools is not possible. In the current literature, however, frequency-domain methods exist where the underlying linear dynamics of a nonlinear system can be captured in an identification experiment; in this manner, the nonlinear system is replaced by a linear model with a noise source where a best linear approximation of the nonlinear system is obtained with an associated frequency-dependent uncertainty. With the frequency-domain data and uncertainty obtained from an identification experiment, robust control algorithms can then be used to ensure performance for the underlying linear system. This chapter presents a data-driven robust control strategy which implements a convex optimization algorithm to ensure the performance and closed-loop stability of a linear system that is subject to nonlinear distortions (by considering a model-reference objective). The effectiveness of the proposed data-driven method is illustrated by designing a controller for an inertial positioning system that possesses nonlinear torsional dynamics.

13.1 Introduction

A common issue encountered in control systems is the problem of unmodelled dynamics in parametric models. Systems are usually approximated with low-order models in order to reduce the complexity of a control design strategy. However, this approximation can lead to stability and performance degradations since these low-order models are subject to model uncertainty. The data-driven control strategy mitigates the problems with model-based controller designs since the data-driven scheme avoids

[1]Technology Department, European Organization for Nuclear Research (CERN), Geneva, Switzerland
[2]Automatic Control Laboratory at Ecole Polytechnique Fédérale de Lausanne (EPFL), Lausanne, Switzerland

the problem of unmodelled dynamics associated with low-order parametric models. A survey on the differences between the model-based control and data-driven control schemes has been addressed in [1,2].

Frequency response data have been predominantly used for analysis and design of linear systems. The tools developed by Bode and Nyquist continue to be indispensable for the controls engineer since, to an extent, the effects of nonlinearities could be ignored because they did not impair system performance. However, due to the increased performance demands on today's industrial systems, the effects of certain nonlinearities can impact the behaviour of these systems. For many of today's systems, the effects of nonlinearities can no longer be neglected (see [3,4]). Due to the extensive use of frequency-domain techniques for linear systems within the control systems community, and given the need for analysing the effects of nonlinear systems, it is thus natural to extend the frequency-domain analysis and control schemes for linear systems where nonlinear distortions can occur. A comparative study of frequency-domain methods for nonlinear systems has recently been addressed in [5].

Data-driven methods for controlling systems with nonlinearities is a field which continues to spark the interests of many researchers. The authors in [6] present a model-free approach to design controllers that guarantee stability for a class of nonlinear discrete-time systems; in [7], this method is extended to the multiple-input–multiple-output (MIMO) nonlinear systems. A virtual reference feedback tuning (VRFT) method is proposed in [8] to design controllers for nonlinear plant models using a direct "one-shot" data-driven method. This method implements a nonlinear controller and minimises a model-reference criterion using time-domain data. The authors in [9] build an iterative data-driven algorithm to design controllers for a class of nonlinear autoregressive exogenous models. A method for designing controllers in a data-driven setting for constrained linear systems is presented in [10]. The work in [11] extends on the concept of the VRFT method and implements a data-driven scheme to design linear parameter-varying model-reference controllers.

The objective of this paper is to combine the ideas presented in [12,13] and develop a data-driven controller design methodology that guarantees $\mathcal{H}_\infty$ performance and closed-loop stability for linear systems that are subject to nonlinear distortions. In [13], the frequency response function (FRF) of a nonlinear system is modelled as a best-linear-approximation (BLA) with an associated frequency-dependent uncertainty. By performing a set of identification experiments on the nonlinear system, the dynamics of the underlying linear system are guaranteed to lie in the set of these uncertainties. A model-reference problem is then formulated to achieve the desired performance and guarantee the closed-loop stability; this problem is devised based on the ideas presented in [12] where $\mathcal{H}_\infty$ performance was achieved for a linear plant model that was subjected to frequency-dependent uncertainties. Thus, by considering the BLA of the nonlinear system as the nominal model, and by designing a controller which accounts for the frequency-dependent uncertainties obtained from an identification experiment of the nonlinear system, the closed-loop stability and performance is guaranteed for the underlying linear system.

This paper is organised as follows: In Section 13.2, the class of controllers and nonlinearities is defined. A data-driven frequency-domain approach is recalled in

Section 13.3 where the FRF measurement of a nonlinear system is modelled as a BLA with an associated frequency-dependent uncertainty. Section 13.4 will address the control objectives and the conditions required for obtaining $\mathcal{H}_\infty$ performance and closed-loop stability for the underlying linear model of a nonlinear system. Section 13.5 will demonstrate the effectiveness of the proposed method by designing a controller that ensures robust stability and performance for the underlying linear dynamics of a torsional plant system (which possesses a considerable amount of static friction). Finally, the concluding remarks are given in Section 13.6.

Notation: In order to avoid the risk of any confusion, the notation for the symbols employed in this chapter will be defined here. $\mathbb{R}$ and $\mathbb{R}_+$ define the sets of all real numbers and real numbers greater than zero, respectively. $\Re\{\cdot\}$ denotes the real part of the argument. $(\cdot)^\star$ denotes the complex conjugate of the argument. The variables s and z are the complex frequency variables used to represent continuous-time and discrete-time systems, respectively.

13.2 Preliminaries

13.2.1 Class of nonlinearities

The class of nonlinearities discussed in this work is now addressed with the following definition:

Definition 13.1. *Class $\mathcal{N}$ of nonlinear systems is the set of nonlinear systems for which the following properties hold [13]*

- *The influence of the initial conditions vanishes asymptotically.*
- *The steady state response to a periodic input is a periodic signal with the same period as the input. Nonlinearities such as bifurcation, chaos and subharmonics are excluded; however, strongly nonlinear phenomena such as saturation and discontinuities are permitted.*
- *Only a point-wise approximation of the output is obtained.*

The nonlinear systems described by $\mathcal{N}$ include a class of nonlinearities known as the so-called Wiener systems (should not be confused with Hammerstein–Wiener models) [14]. A nonlinear system which abides by the above definition will be denoted as $G_{\mathcal{N}}(\cdot)$ (i.e. $G_{\mathcal{N}}(\cdot) \in \mathcal{N}$).

13.2.2 Class of controllers

A fixed-order one-degree-of-freedom polynomial control structure is considered. The general structure of this control system is shown in Figure 13.1. The functions $C(z^{-1}, \boldsymbol{\rho})$ and $D(z^{-1}, \boldsymbol{\rho})$ each represent polynomials in z^{-1}, i.e.

$$C(z^{-1}, \boldsymbol{\rho}) = c_0 + c_1 z^{-1} + \cdots + c_{n_c} z^{-n_c} \tag{13.1}$$

$$D(z^{-1}, \boldsymbol{\rho}) = 1 + d_1 z^{-1} + \cdots + d_{n_d} z^{-n_d} \tag{13.2}$$

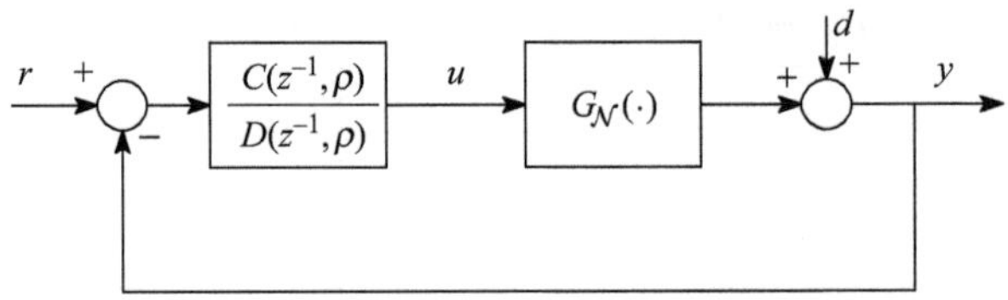

Figure 13.1 Discrete-time controller structure

where c_i and d_i are the controller parameters and $\{n_c, n_d\}$ are the orders of the polynomials C and D, respectively. The vector of controller parameters $\boldsymbol{\rho}$ is defined as

$$\boldsymbol{\rho}^\top = [c_0, c_1, \ldots, c_{n_c}, d_1, d_2, \ldots, d_{n_d}] \tag{13.3}$$

where $\boldsymbol{\rho} \in \mathbb{R}^n$ with $n = n_c + n_d + 1$.

13.3 Frequency-domain identification

In this section, a data-driven method is implemented such that the underlying linear dynamics of a system $G_{\mathcal{N}}(\cdot)$ are captured in the FRF during a set of experiments. Moreover, the distortion of the frequency function originating from system nonlinearity is modelled as a stochastic frequency-domain uncertainty.

Suppose that the signals u and y are measurable. Let the frequency spectrum of the signals u and y be denoted as $U(e^{-j\omega})$ and $Y(e^{-j\omega})$, respectively. According to [13], for a certain class of reference signals and nonlinear systems, the FRF obtained during an experiment with a nonlinear plant can be described by a linear system plus an error term $Y_S(e^{-j\omega})$ (see Figure 13.2). The class of nonlinear systems that can be considered with this approach are those in $\mathcal{N}$.

The idea asserted in [13] is to perform multiple experiments with full or random-phase multi-sine signals as the reference input. Averaging of the FRFs over the consecutive periods quantifies the noise level. Averaging of these mean FRFs over multiple experiments quantifies the level of the stochastic nonlinear distortions (with the sum of the remaining noise level).

Definition 13.2. *A random-phase multi-sine signal $u(t)$ is defined as*

$$u(t) = \sum_{k=-K/2+1}^{K/2-1} U_k e^{j2\pi f_s kt/K} \tag{13.4}$$

where $U_k = U^\star_{-k} = |U_k| e^{j\phi_k}$, f_s is the clock frequency of the waveform generator, K is the number of samples in the signal period, and the phases ϕ_k are a realisation of an independent distributed random process in $[0, 2\pi)$ where the expected value of $e^{j\phi_k}$ is equal to zero.

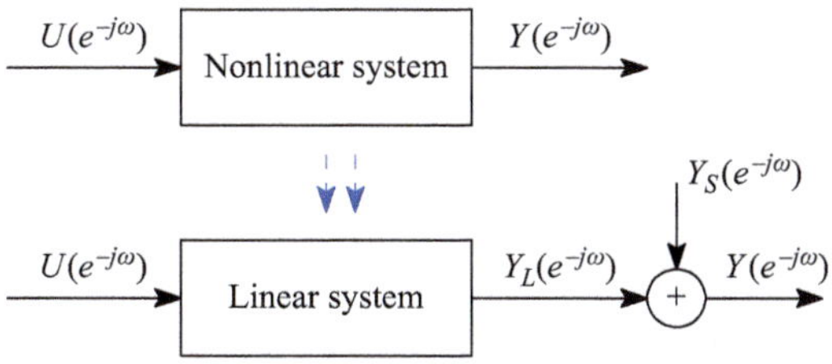

Figure 13.2 Representation of a nonlinear system by a linear system for a certain class of inputs

13.3.1 Stable plant

The case of stable plant models will first be considered. For a given known input signal, a set of open-loop experiments can be performed to obtain the FRF of the BLA and the noise and distortion variances. Let $G^{[q,p]}(e^{-j\omega})$ be defined as the FRF estimate of $G_{\mathcal{N}}(\cdot)$ for the pth period of a qth experiment (with P denoting the total number of periods in each experiment and Q being the total number of experiments):

$$\begin{aligned} G^{[q,p]}(e^{-j\omega}) &= \frac{Y^{[q,p]}(e^{-j\omega})}{U^{[q]}(e^{-j\omega})} \\ &= G(e^{-j\omega}) + G_S^{[q]}(e^{-j\omega}) + E_G^{[q,p]}(e^{-j\omega}) \end{aligned} \tag{13.5}$$

where G is the FRF of the BLA, $G_S^{[q]} = Y_S^{[q]}/U^{[q]}$ is the stochastic nonlinear contributions and $E_G^{[q,p]}$ is the error due to the output noise. The sample mean and the sample variance of the FRF estimates over P periods are determined as follows:

$$\begin{aligned} G^{[q]}(e^{-j\omega_k}) &= \frac{1}{P}\sum_{p=1}^{P} G^{[q,p]}(e^{-j\omega_k}) \\ \sigma_n^{2[q]}(k) &= \frac{1}{P(P-1)}\sum_{p=1}^{P} \left|G^{[q,p]}(e^{-j\omega_k}) - G^{[q]}(e^{-j\omega_k})\right|^2 \end{aligned} \tag{13.6}$$

where $\sigma_n^{2[q]}$ is the sample noise variance of the sample mean $G^{[q]}$. The BLA of the plant G with the associated sample total variance σ_G^2 can then be determined with the following relations [13]:

$$\begin{aligned} G(e^{-j\omega_k}) &= \frac{1}{Q}\sum_{q=1}^{Q} G^{[q]}(e^{-j\omega_k}) \\ \sigma_G^2(k) &= \frac{1}{Q(Q-1)}\sum_{q=1}^{Q} \left|G^{[q]}(e^{-j\omega_k}) - G(e^{-j\omega_k})\right|^2 \end{aligned} \tag{13.7}$$

13.3.2 Unstable plant

For the case of unstable plant models, an open-loop experiment cannot be performed to obtain the FRFs. A stabilising controller would first need to be implemented in order to stabilise the closed-loop system (i.e. all outputs remain bounded for a bounded reference signal). Now, suppose that the signal r is measurable where the frequency spectrum of r is denoted as $R(e^{-j\omega})$. Additionally, define $N(e^{-j\omega})$ as the FRF of the BLA between the signals r to y and $M(e^{-j\omega})$ as the FRF of the BLA between the signals r to u. Since the nonlinear system is described by a linear system plus an error term $Y_S(e^{-j\omega})$, then it is evident that the FRF of the BLA of the plant model is $G(e^{-j\omega}) = N(e^{-j\omega})M^{-1}(e^{-j\omega})$. This is known as a *coprime factorisation* of the FRF G where N and M are the FRFs of stable transfer functions [15].

According to [13], the sample means and total (co)variances can be determined as follows:

$$
\begin{aligned}
Y^{[q]}(e^{-j\omega_k}) &= \frac{1}{P}\sum_{p=1}^{P} Y^{[q,p]}(e^{-j\omega_k}) \\
Y(e^{-j\omega_k}) &= \frac{1}{Q}\sum_{q=1}^{Q} Y^{[q]}(e^{-j\omega_k}) \\
\sigma_Y^2(k) &= \frac{1}{Q(Q-1)}\sum_{q=1}^{Q} |Y^{[q]}(e^{-j\omega_k}) - Y(e^{-j\omega_k})|^2 \\
\sigma_{YR}^2(k) &= \frac{1}{Q(Q-1)}\sum_{q=1}^{Q} [Y^{[q]}(e^{-j\omega_k}) - Y(e^{-j\omega_k})]\cdot \\
&\qquad [R^{[q]}(e^{-j\omega_k}) - R(e^{-j\omega_k})]^\star \\
\sigma_{UR}^2(k) &= \frac{1}{Q(Q-1)}\sum_{q=1}^{Q} [U^{[q]}(e^{-j\omega_k}) - U(e^{-j\omega_k})]\cdot \\
&\qquad [R^{[q]}(e^{-j\omega_k}) - R(e^{-j\omega_k})]^\star
\end{aligned}
\tag{13.8}
$$

where the FRFs and variances for the signals u (i.e. $U(e^{-j\omega_k})$ and $\sigma_U^2(k)$) and r (i.e. $R(e^{-j\omega_k})$ and $\sigma_R^2(k)$) are computed in the same manner as $Y(e^{-j\omega_k})$ and $\sigma_Y^2(k)$, respectively. For notation purposes, the dependency in $e^{-j\omega}$ will be omitted and will only be reiterated when deemed necessary. Finally, the FRF of the BLA for each coprime can then be obtained as $N = YR^{-1}$ and $M = UR^{-1}$ where the associated total variance for each coprime is calculated as follows:

$$
\begin{aligned}
\sigma_N^2(k) &= |N|^2\left(\frac{\sigma_Y^2(k)}{|Y|^2} + \frac{\sigma_R^2(k)}{|R|^2} - 2\Re\left\{\frac{\sigma_{YR}^2(k)}{YR^\star}\right\}\right) \\
\sigma_M^2(k) &= |M|^2\left(\frac{\sigma_U^2(k)}{|U|^2} + \frac{\sigma_R^2(k)}{|R|^2} - 2\Re\left\{\frac{\sigma_{UR}^2(k)}{UR^\star}\right\}\right)
\end{aligned}
\tag{13.9}
$$

Remark. *Note that in [13], the FRF estimate of G (and the associated uncertainty) can be obtained from the signals u and y directly. However, the coprime formulation was needed in this paper in order to apply the proposed controller design schemes (which are asserted in Section 13.4).*

13.3.3 Uncertainty filters for coprime factorisation

Suppose that the uncertainty associated with a given FRF is described by an additive uncertainty:

$$\begin{aligned}\tilde{N}(e^{-j\omega}) &= N(e^{-j\omega}) + |W_n(e^{-j\omega})|\delta_n e^{j\theta_n} \\ \tilde{M}(e^{-j\omega}) &= M(e^{-j\omega}) + |W_m(e^{-j\omega})|\delta_m e^{j\theta_m}\end{aligned} \tag{13.10}$$

where $\tilde{N}$ and $\tilde{M}$ are the *true* FRFs of coprime factorisation and $|\delta_n| \leq 1$, $|\delta_m| \leq 1$ with $\{\theta_n, \theta_m\} \in [0, 2\pi]$ represent the frequency-domain uncertainty disks. Finally, $|W_n|$ and $|W_m|$ are the size of uncertainties around N and M, respectively, which can be determined from the covariance of the estimates for a given confidence interval. Given the frequency spectrums of Y, U and R, the estimates of the real and the imaginary part of $N(e^{-j\omega})$ and $M(e^{-j\omega})$ can be formulated; these estimates are asymptotically uncorrelated and normally distributed [16]. For any given ω, the additive uncertainty for $\tilde{N}$ can be described in the complex plane with a disk centred at N having a radius of $|W_n|$ (when the worst case scenario is considered with $\delta_n = 1$). A similar representation can be made with $\tilde{M}$. The uncertainty is characterised by the Rayleigh distribution and can be determined for any probability level. For example, if it is desired to construct an uncertainty disk such that the true-frequency response lies within the disk with a probability level of 0.95, then the radius of this disk(s) will be

$$|W_n(e^{-j\omega_k})| = \sqrt{5.99\sigma_{\tilde{N}}^2(k)}; \quad |W_m(e^{-j\omega_k})| = \sqrt{5.99\sigma_{\tilde{M}}^2(k)} \tag{13.11}$$

13.4 Robust controller design

In this section, a robust control design method is introduced to compute a stabilising controller with $\mathcal{H}_\infty$ performance using only the frequency response data. A convex optimisation problem is formulated by minimising the norm of a weighted sensitivity function or the difference between a sensitivity function and a desired one (model reference control problem). The nonlinear distortions are considered as additive frequency-domain uncertainties and taken into account for the controller design.

13.4.1 Control performance

Since the design techniques introduced in this paper belong to the $\mathcal{H}_\infty$ framework, it is appropriate to consider the various sensitivity functions associated with the unit feedback controller structure. The four sensitivity functions for the closed-loop system can be asserted as follows:

$$\mathcal{S}(z^{-1},\boldsymbol{\rho}) = \frac{M(z^{-1})D(z^{-1},\boldsymbol{\rho})}{\psi(z^{-1},\boldsymbol{\rho})} \tag{13.12}$$

$$\mathcal{T}(z^{-1},\boldsymbol{\rho}) = \frac{N(z^{-1})C(z^{-1},\boldsymbol{\rho})}{\psi(z^{-1},\boldsymbol{\rho})} \tag{13.13}$$

$$\mathcal{U}(z^{-1},\boldsymbol{\rho}) = \frac{M(z^{-1})C(z^{-1},\boldsymbol{\rho})}{\psi(z^{-1},\boldsymbol{\rho})} \tag{13.14}$$

$$\mathcal{V}(z^{-1},\boldsymbol{\rho}) = \frac{N(z^{-1})D(z^{-1},\boldsymbol{\rho})}{\psi(z^{-1},\boldsymbol{\rho})} \tag{13.15}$$

where $\psi(z^{-1},\boldsymbol{\rho}) = N(z^{-1})C(z^{-1},\boldsymbol{\rho}) + M(z^{-1})D(z^{-1},\boldsymbol{\rho})$. For notation purposes, the dependency in z^{-1} will be omitted and will only be reiterated when deemed necessary. In the general model-reference $\mathcal{H}_\infty$ control problem for *linear systems*, the objective is to minimise an upper bound γ to find the stabilising controller parameter vector $\boldsymbol{\rho}$ such that

$$\sup_{\omega\in\Omega} \left| W_2(e^{-j\omega})\left[\mathcal{T}(e^{-j\omega},\boldsymbol{\rho}) - \mathcal{T}_d(e^{-j\omega})\right]\right| < \gamma \tag{13.16}$$

where $\Omega := [-\pi,\pi]$, $\gamma \in \mathbb{R}_+$, $\mathcal{T}_d(e^{-j\omega})$ is the desired FRF, and $W_2(e^{-j\omega})$ is a weighting FRF.

Remark. *Note that the performance specification is defined for the shaping of the complementary sensitivity function $\mathcal{T}$; however, other sensitivity functions can be considered in this problem. Moreover, the constraint in (13.16) is general enough to cover the constraints on the weighted sensitivity function (where $\mathcal{T}_d \equiv 0$) or the case of model reference control (where $W_2 \equiv 1$) or a combination where the model reference error is minimised in a frequency region of interest.*

13.4.2 Convex formulation for robust performance

For the nominal plant model represented as $G = NM^{-1}$ and the definition of the sensitivity function $\mathcal{T}$ in (13.13), the condition in (13.16) can be expressed as follows:

$$\gamma^{-1}\left|W_2\left(NC(\boldsymbol{\rho}) - \mathcal{T}_d\psi(\boldsymbol{\rho})\right)\right| < |\psi(\boldsymbol{\rho})|, \quad \forall\omega\in\Omega \tag{13.17}$$

For any given frequency in Ω, the condition in (13.17) represents a disk in the complex plane centred at $\psi(\boldsymbol{\rho})$ with a radius of $\gamma^{-1}|W_2[NR(\boldsymbol{\rho}) - \mathcal{T}_d\psi(\boldsymbol{\rho})]|$, which does not include the origin. In [12], it is shown that there exists a complex function $f(e^{-j\omega})$ which can rotate this disk such that it lies on the right-hand side of the imaginary axis. This geometrical construction is used to formulate a necessary and sufficient condition for (13.16), which is recalled in the following Theorem:

Theorem 13.1. *Given the FRF $G = NM^{-1}$, the frequency response of a weighting filter W_2, and the frequency response of the desired sensitivity $\mathcal{T}_d$, then the following statements are equivalent:*

1. *There exists a controller that stabilises G and*

$$\sup_{\omega\in\Omega} |W_2\,(\mathcal{T}(\boldsymbol{\rho}) - \mathcal{T}_d)| < \gamma \tag{13.18}$$

2. *There exists a controller CD^{-1} such that*

$$\gamma^{-1}\,|W_2[NC(\boldsymbol{\rho}) - \mathcal{T}_d\psi(\boldsymbol{\rho})]| < \Re\,\{\psi(\boldsymbol{\rho})\} \quad \forall\omega\in\Omega \tag{13.19}$$

Proof: The proof is similar to the proof of Theorem 1 in [12] and is omitted. ■

Given the additive uncertainty in (13.10), the desired performance condition in (13.16) can be satisfied for all models in the uncertain set (13.10) if

$$\sup_{\omega\in\Omega} |W_2[\tilde{\mathcal{T}}(\boldsymbol{\rho}) - \mathcal{T}_d]| < \gamma \tag{13.20}$$

where $\tilde{\mathcal{T}}(\boldsymbol{\rho}) = \tilde{N}C(\boldsymbol{\rho})\tilde{\psi}^{-1}(\boldsymbol{\rho})$ and $\tilde{\psi}(\boldsymbol{\rho}) = \tilde{N}C(\boldsymbol{\rho}) + \tilde{M}D(\boldsymbol{\rho})$. As a worst case consideration, δ_m and δ_n can be selected to be equal to one in (13.10) (which ensures that the uncertainty in the entire disk is taken into account). By substituting the expressions in (13.10) into this condition, the following constraint can be devised:

$$\left|W_2\left[C(\boldsymbol{\rho})(N + |W_n|e^{j\theta_n}) - \mathcal{T}_d\left(\psi(\boldsymbol{\rho}) + \Gamma(\boldsymbol{\rho},\theta_n,\theta_m)\right)\right]\right| < \gamma|\psi(\boldsymbol{\rho}) + \Gamma(\boldsymbol{\rho},\theta_n,\theta_m)|$$
$$\forall\omega\in\Omega, \forall\{\theta_n,\theta_m\}\in[0,2\pi] \tag{13.21}$$

where

$$\Gamma(\boldsymbol{\rho},\theta_n,\theta_m) = D(\boldsymbol{\rho})|W_m|e^{j\theta_m} + C(\boldsymbol{\rho})|W_n|e^{j\theta_n} \tag{13.22}$$

For a given $\{\omega,\theta_n,\theta_m\}$, (13.21) represents a disk centred at $\psi(\boldsymbol{\rho}) + \Gamma(\boldsymbol{\rho},\theta_n,\theta_m)$ with a radius of

$$x_r'(\boldsymbol{\rho},\theta_n,\theta_m) = \gamma^{-1}\left|W_2\left[C(\boldsymbol{\rho})(N + |W_n|e^{j\theta_n}) - \mathcal{T}_d\left(\psi(\boldsymbol{\rho}) + \Gamma(\boldsymbol{\rho},\theta_n,\theta_m)\right)\right]\right|$$

that does not include the origin.

Given the results from Theorem 13.1, a necessary and sufficient condition for (13.21) can be formulated as follows:

$$x_r'(\boldsymbol{\rho},\theta_n,\theta_m) < \Re\{\psi(\boldsymbol{\rho}) + \Gamma(\boldsymbol{\rho},\theta_n,\theta_m)\}, \quad \forall\omega\in\Omega, \forall\{\theta_m,\theta_n\}\in[0,2\pi] \tag{13.23}$$

By gridding in ω, θ_m and θ_n, (13.23) then becomes a convex constraint (with respect to $\boldsymbol{\rho}$); however, gridding in all of these variables can be computationally expensive. Therefore, a sufficient condition for (13.21) can be devised as follows:

$$\sup_{\omega\in\Omega} \frac{|W_2|[|NC(\boldsymbol{\rho}) - \mathcal{T}_d\psi(\boldsymbol{\rho})| + |C(\boldsymbol{\rho})W_n| + |\mathcal{T}_d|\Gamma_s(\boldsymbol{\rho})]}{|\psi(\boldsymbol{\rho})| - \Gamma_s(\boldsymbol{\rho})} < \gamma \tag{13.24}$$

where $\Gamma_s(\boldsymbol{\rho}) = |D(\boldsymbol{\rho})W_m| + |C(\boldsymbol{\rho})W_n|$. With this condition, the dependency in θ_m and θ_n has been removed, and gridding in only one variable (i.e. ω) is required. The condition in (13.24) can be represented as a disk in the complex plane which is centred at $\psi(\boldsymbol{\rho})$ and has radius

$$x_r(\boldsymbol{\rho}) = \gamma^{-1}|W_2|[|NC(\boldsymbol{\rho}) - \mathcal{T}_d\psi(\boldsymbol{\rho})| + |C(\boldsymbol{\rho})W_n| + |\mathcal{T}_d|\Gamma_s(\boldsymbol{\rho})] + \Gamma_s(\boldsymbol{\rho}) \tag{13.25}$$

Therefore, a set of convex constraints (with respect to $\boldsymbol{\rho}$) can be devised with the following condition:

$$x_r(\boldsymbol{\rho}) < \Re\{\psi(\boldsymbol{\rho})\}, \quad \forall\omega \in \Omega \tag{13.26}$$

Note that (13.24) (and hence (13.26)) introduces some conservatism; however, this conservatism can always be reduced by imposing (13.23) (at the price of a larger computation time).

13.4.3 Controller design by convex optimisation

With the constraints developed in the previous section, an optimisation problem can be formulated to guarantee $\mathcal{H}_\infty$ performance and closed-loop stability for the underlying linear system. For the control performance objective (i.e. $\|W_2[\tilde{\mathcal{T}} - \mathcal{T}_d]\|_\infty < \gamma$), the following optimisation problem is considered:

$$\begin{aligned} &\underset{\gamma,\boldsymbol{\rho}}{\text{minimise}} \quad \gamma \\ &\text{subject to:} \quad x_r(\boldsymbol{\rho}) < \Re\{\psi(\boldsymbol{\rho})\} \\ &\qquad\qquad\quad \forall\omega \in \Omega_l = \{\omega_1, \ldots, \omega_l\} \end{aligned} \tag{13.27}$$

This optimisation problem is quasi-convex; to solve such a problem, a bisection algorithm can be realised where an iterative approach is implemented in order to obtain an asymptotically convergent solution for γ. In the above optimisation problem, the infinite number of constraints is converted to a finite number of constraints by gridding in ω. Thus, the problem has been transformed from a semi-infinite programming (SIP) problem to a semi-definite programming (SDP) problem. The frequency grid can be predefined in a variety of manners (see [17,18]).

13.5 Case study

This case study will demonstrate the effectiveness of the proposed robust design method by applying a random-phase multi-sine signal that excites the nonlinearity in the system. The FRF obtained from this identification will then be used to model the BLA with an associated uncertainty and design a robust controller that achieves a desired performance. It will also be desired to investigate the response of a controller when the uncertainties in the design are neglected and the FRF of the coprimes are obtained from a given standard identification experiment.

13.5.1 System description

In this case study, the system to be controlled is a torsional plant built by Quanser and consists of three disks connected by a metallic rod (see Figure 13.3). The lowest disk is actuated, and the control objective is to regulate the angular position of the upper disk. High resolution encoders (16,000 counts/rev) are used for feedback where the regulation sampling time is $T_s = 0.04$ s. As the diameter of the connecting rod is small compared to its length, there are strong torsional dynamics that need to be

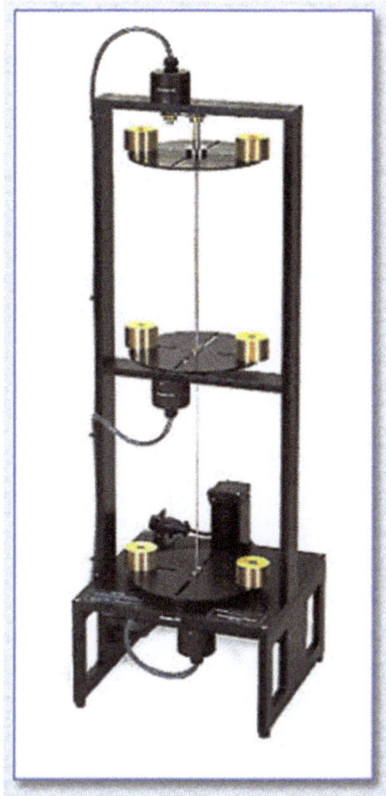

Figure 13.3 Torsional plant by Quanser. Three plates with variable masses are suspended on a vertical rod (i.e. coupled dynamics)

damped. Furthermore, the system exhibits notable static friction; thus, the system can be characterised as having a dead-zone nonlinearity and belongs in $\mathcal{N}$.

13.5.2 Identification experiments

It is known that the plant contains an integrator and so is unstable in open-loop operation. Thus, a closed-loop structure was used to capture the dynamics of the system via the coprime models. For this case study, the closed-loop system was stabilised with a simple proportional-integral controller. The closed-loop system was then excited with a periodic random-phase multi-sine (with an amplitude range of ± 1.2); 16 experiments were performed where the system was excited with 10 periods of this signal where the period length was 2,000 samples, and each period contained 500 sinusoids with random phases.

For comparative purposes, it was desired to compare the identified models when the nonlinear distortions were taken into account with the identified FRF using a standard identification method. The FRF of BLAs with the associated uncertainties for

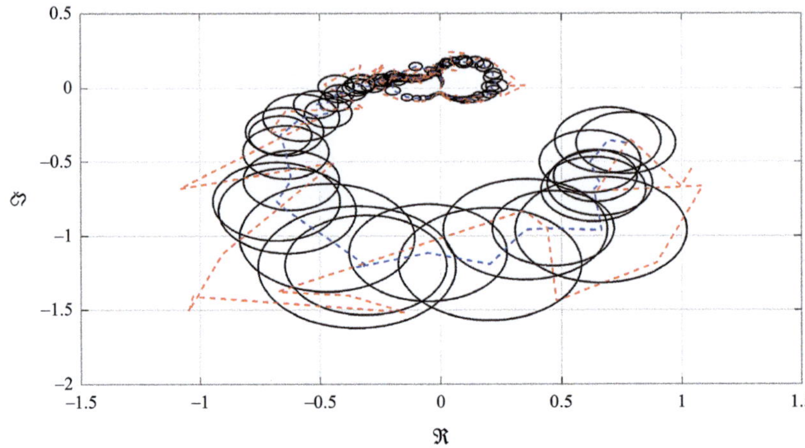

Figure 13.4 N (dashed-blue line) with the nonlinear distortion uncertainties at each frequency (black circles). The FRF obtained between r(t) and y(t) for a given experiment using standard spectral analysis (dashed-red line)

$\tilde{N}$ and $\tilde{M}$ are shown in Figures 13.4 and 13.5, respectively. The radii of the uncertainty disks for each coprime were computed using (13.11). It can be observed that the uncertainty disks related to the nonlinear distortions are quite large (particularly at lower frequencies). This can be explained by the effect of static friction in the torsional system. Moreover, the error of the FRF estimate between the standard identified models and those considering nonlinear distortions is large at low frequencies.

13.5.3 Performance specification

The desired performance will be achieved by satisfying the model-reference objective (i.e. $W_2 \equiv 1$). Since the system is limited in the amount of voltage that can be applied to the motor (i.e. saturation limits), a constraint on the uncertain input sensitivity function $\tilde{\mathcal{U}}$ will be added to limit the control signal for all models in the uncertainty set (including nonlinear distortions). Additionally, a constraint on the infinity norm of the uncertain sensitivity function $\tilde{\mathcal{S}}$ will be added to ensure sufficient modulus margin (for increased robustness) [19]. Therefore, the optimisation problem to solve for this case study can be expressed as follows:

$$\begin{aligned} &\underset{\gamma,\rho}{\text{minimise}} && \gamma \\ &\text{subject to:} && \left\| \tilde{\mathcal{T}}(\rho) - \mathcal{T}_d \right\|_\infty < \gamma \\ & && \left\| W_1 \tilde{\mathcal{S}}(\rho) \right\|_\infty < 1 \\ & && \left\| W_3 \tilde{\mathcal{U}}(\rho) \right\|_\infty < 1 \end{aligned} \tag{13.28}$$

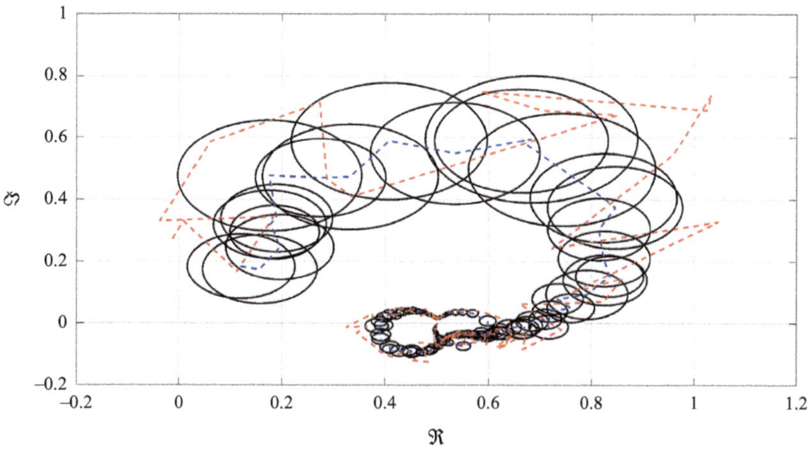

Figure 13.5 M (dashed-blue line) with the nonlinear distortion uncertainties at each frequency (black circles). The FRF obtained between r(t) to u(t) for a given experiment using standard spectral analysis (dashed-red line)

This problem as stated is not convex and requires a parametric model. Thus, the proposed approach is used to convexify the problem and guarantee the closed-loop stability whilst considering the uncertainty from the nonlinear distortions; this problem can be formulated as follows:

$$\begin{aligned} \underset{\gamma,\rho}{\text{minimise}} \quad & \gamma \\ \text{subject to} \quad & x_r(\rho) < \Re\{\psi(\rho)\} \\ & x_m(\rho) < \Re\{\psi(\rho)\} \\ & x_u(\rho) < \Re\{\psi(\rho)\} \\ & \forall \omega \in \Omega_l \end{aligned} \tag{13.29}$$

where x_r is given in (13.25), and

$$\begin{aligned} x_m &= |W_1 M D(\rho)| + |W_1 W_m D(\rho)| + \Gamma_s(\rho) \\ x_u &= |W_3 M C(\rho)| + |W_3 W_m C(\rho)| + \Gamma_s(\rho) \end{aligned} \tag{13.30}$$

The weighting filters W_n and W_m for the uncertainties in $\tilde{N}$ and $\tilde{M}$ were calculated using (13.11). A modulus margin of 0.4 was desired for sufficient robustness; therefore, $W_1 = 0.4$ was selected. Finally, to limit the control signal to the plant and avoid saturation, $W_3 = 0.01$ was selected.

The desired reference model $\mathcal{T}_d$ was chosen as a standard second order model

$$\mathcal{T}_d(s) = \frac{\omega_d^2}{s^2 + 2\zeta\omega_d s + \omega_d^2}$$

where ζ is the damping factor and

$$\omega_d = 2\pi f_d \left[1 - 2\zeta^2 + \sqrt{2 - 4\zeta^2 + 4\zeta^4}\right]^{-0.5}$$

and f_d (Hz) is the desired closed-loop bandwidth. For this case study, $\zeta = 0.8$ and $f_d = 2$ Hz was selected.

13.5.4 Experimental results

The problem in (13.29) was solved in SDP form with the 500 frequency points obtained from the random-phase multi-sine experiments and with a seventh-order controller. The YALMIP interface [20] was used together with MATLAB in order to obtain a solution to the problem. To invoke integral action, the controller was prefixed with an integrator. Three design schemes were considered:

- **(A)** A design in which the FRF of the BLA with the nonlinear distortion uncertainties were considered.
- **(B)** A design where no uncertainties were considered (i.e. $|W_n| = |W_m| = 0$) and the FRF of the coprimes was obtained from a standard spectral analysis method.
- **(C)** A design in which the minimum value of γ obtained from **(A)** (which will be denoted as γ_A^*) was fixed and was used to solve (13.29) where no uncertainties were considered (i.e. $|W_n| = |W_m| = 0$), and the FRF of the coprimes was obtained from a standard spectral analysis method.

The last design scheme was studied in order to qualitatively compare the results when the same γ and weighting filters were considered.

The optimal solution to the proposed convex problem for case (**A**) was computed as $\gamma_A^* = 4.335$ (using a tolerance of 10^{-3} for the bisection algorithm) with an optimisation time of approximately 220 s. With the same tolerance, the optimal solution for case (**B**) was computed as $\gamma_B^* = 1.274$ with an optimisation time of approximately 100 s. This optimisation time was calculated based on a computer having the following hardware specifications: Intel-Core i7, 3.4 GHz CPU, 8 GB RAM. The optimisation algorithms were run using MATLAB version (R2017a) on a Windows 7 platform (64-bit). The solution obtained from the proposed method (i.e. case (**A**)) produced a controller with no unstable poles. Figure 13.6 shows the Nyquist plot of the open-loop FRF with the associated uncertainties. It can be observed that the Nyquist criterion for closed-loop stability is satisfied while ensuring good robustness (by satisfying the modulus margin constraint).

A step reference with low amplitude was used (i.e. where the effect of the nonlinearity was more significant) in order to investigate the closed-loop response. The closed-loop step responses of the nonlinear system are shown in Figure 13.7; it can be observed that when the frequency-dependent uncertainties were considered in the design, good performance and stability was achieved. When the uncertainties were neglected in the design, the effect of the nonlinearity was more significant and the settling time was significantly larger. This was caused by the modelling error from the closed-loop experiment (which can be seen in Figures 13.4 and 13.5 where the FRF lies outside the uncertainty disks at various frequency points).

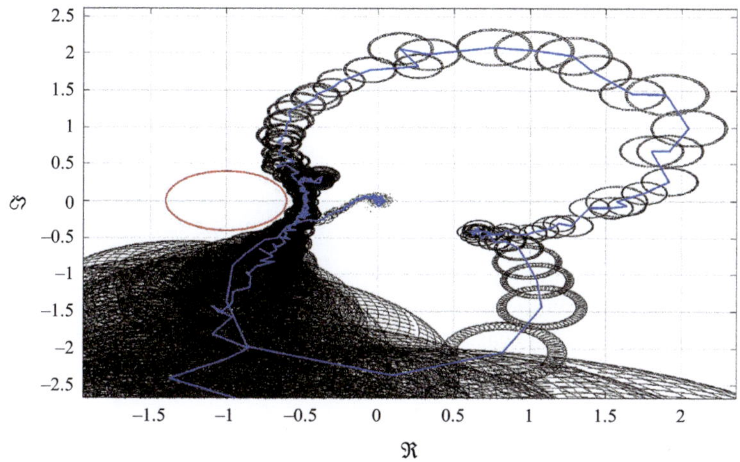

Figure 13.6 Nyquist plot of the open-loop FRF of the BLA (blue line) with the associated uncertainties (black disks). Modulus margin constraint of 0.4 (red disk)

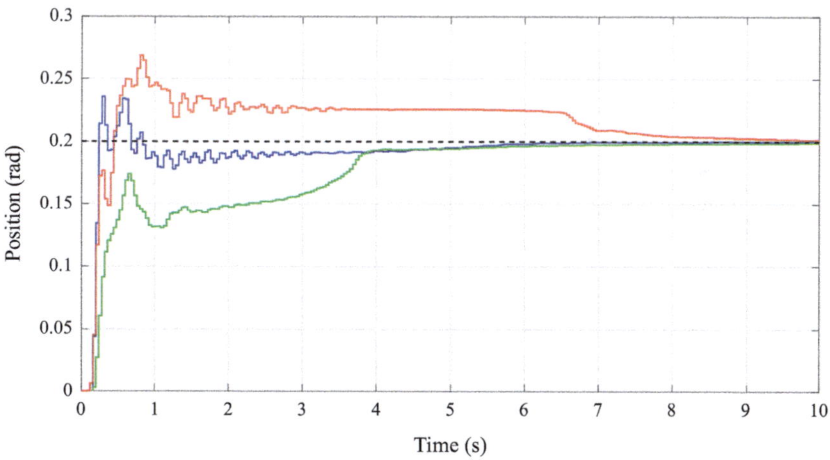

*Figure 13.7 Step response of the nonlinear system. The desired closed-loop response (black line); the response with the proposed method (case (**A**), blue line); the response with no uncertainties considered (case (**B**), red line); the response with no uncertainties considered and same specifications as case (**A**) (case (**C**), green line)*

The control signals from the step responses are shown in Figure 13.8. The applied voltage for case (**C**) was significantly lower than the other two cases. The reason for this is due to the fact that the controller gain for case (**C**) is lower than the other cases (as shown in Figure 13.9), particularly at lower frequencies where the

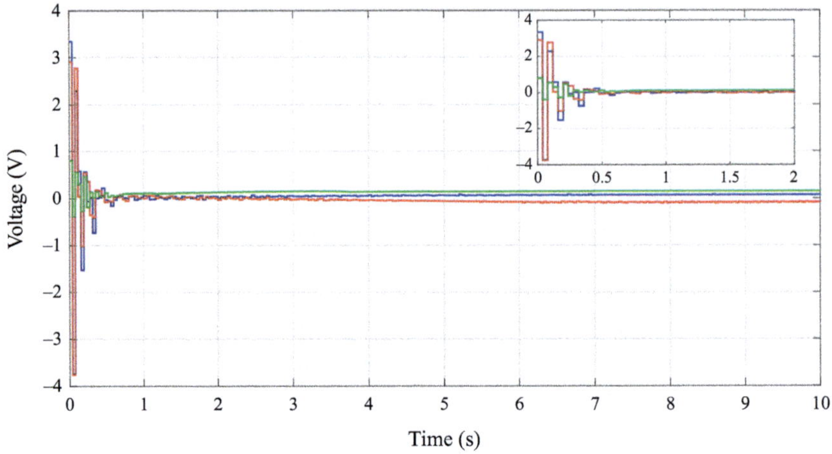

*Figure 13.8 Control signal produced from the step response. The signal produced with the proposed method (case (**A**), blue line); the signal produced with no uncertainties considered (case (**B**), red line); the signal produced with no uncertainties considered and same specifications as case (**A**) (case (**C**), green line)*

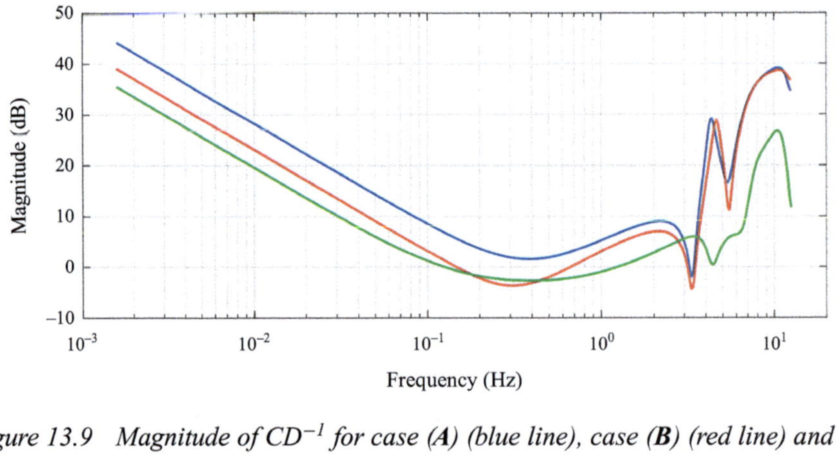

*Figure 13.9 Magnitude of CD^{-1} for case (**A**) (blue line), case (**B**) (red line) and case (**C**) (green line)*

uncertainties are larger. Thus, the effect of the nonlinearity was more significant, and as a result, the performance was degraded. The controller obtained with the proposed method produced the largest applied voltage during the transient where the plant was behaving more linearly for larger control signals. Thus, with the proposed method, the performance and stability of the underlying linear system was guaranteed by considering the frequency-dependent uncertainties obtained from the random-phase multi-sine identification experiments performed on a nonlinear system.

13.6 Conclusion

In this paper, a data-driven method has been proposed for designing controllers for linear systems with nonlinear distortions. In this method, a robust design was implemented where the FRF obtained from an identification experiment of the nonlinear system was modelled as a BLA (to capture the dynamics of the underlying linear system) with an associated uncertainty (representing the system nonlinearities). A convex optimisation algorithm was then devised to guarantee $\mathcal{H}_\infty$ performance and stability for this underlying linear uncertain system. The case study has confirmed the effectiveness of the proposed method by designing a controller for a torsional plant which possessed significant static friction. The step responses showed that the effect of the nonlinearity in the system was more significant when the uncertainties were neglected in the design process. For future work, it will be desired to extend the proposed robust control design methodology for MIMO systems.

References

[1] Hou ZS, Wang Z. From model-based control to data-driven control: Survey, classification and perspective. Information Sciences. 2013;235:3–35.

[2] Formentin S, Van Heusden K, Karimi A. A comparison of model-based and data-driven controller tuning. International Journal of Adaptive Control and Signal Processing. 2013;28(10):882–897.

[3] Meyer Y, Cumunel G. Active vibration isolation with a MEMS device: Effects of nonlinearities on control efficiency. Smart Materials and Structures. 2015;24(8):085004. Available from: http://stacks.iop.org/0964-1726/24/i=8/a=085004.

[4] Barboza E, Silva M, Coelho L, *et al.* Impact of nonlinear effects on the performance of 120 Gb/s 64 QAM optical system using adaptive control of cascade of amplifiers. In: 2015 SBMO/IEEE MTT-S International Microwave and Optoelectronics Conference (IMOC); 2015. p. 1–5.

[5] Rijlaarsdam D, Nuij P, Schoukens J, *et al.* A comparative overview of frequency domain methods for nonlinear systems. Mechatronics. 2017; 42 (Supplement C):11–24. Available from: http://www.sciencedirect.com/science/article/pii/S0957415816301568.

[6] Hou Z, Jin S. A novel data-driven control approach for a class of discrete-time nonlinear systems. IEEE Transactions on Control Systems Technology. 2011;19(6):1549–1558.

[7] Hou Z, Jin S. Data-driven model-free adaptive control for a class of MIMO nonlinear discrete-time systems. IEEE Transactions on Neural Networks. 2011;22(12):2173–2188.

[8] Campi MC, Savaresi SM. Direct nonlinear control design: The virtual reference feedback tuning (VRFT) approach. IEEE Transactions on Automatic Control. 2006;51(1):14–27.

[9] Radac MB, Precup RE, Petriu EM, *et al.* Iterative data-driven tuning of controllers for nonlinear systems with constraints. IEEE Transactions on Industrial Electronics. 2014;61(11):6360–6368.

[10] Piga D, Formentin S, Bemporad A. Direct data-driven control of constrained systems. IEEE Transactions on Control Systems Technology. 2017; 26(4):1–8.

[11] Formentin S, Piga D, Tóth R, *et al.* Direct learning of LPV controllers from data. Automatica. 2016;65:98–110.

[12] Karimi A, Nicoletti A, Zhu Y. Robust H_∞ controller design using frequency-domain data via convex optimization. International Journal of Robust and Nonlinear Control. 2018;25(12):3766–3783.

[13] Pintelon R, Schoukens J. System identification: A frequency domain approach. New York: John Wiley & Sons; 2012.

[14] Schetzen M. The Volterra and Wiener theories of nonlinear systems. New York: Wiley; 1980. Available from: http://www.amazon.com/exec/obidos/redirect?tag=citeulike07-20\&path=ASIN/0471044555.

[15] Zhou K, Doyle JC. Essentials of robust control. NY: Prentice-Hall; 1998.

[16] Ljung L. System identification – Theory for the user. 2nd ed. Upper Saddle River, NJ: Prentice Hall; 1999.

[17] Seiler P, Vanek B, Bokor J, *et al.* Robust H_∞ filter design using frequency gridding. In: American Control Conference (ACC), 2011. IEEE; 2011. p. 1801–1806.

[18] Galdos G, Karimi A, Longchamp R. H_∞ Controller design for spectral MIMO models by convex optimization. Journal of Process Control. 2010; 20(10):1175–1182.

[19] Landau ID, Zito G. Digital control systems: Design, identification and implementation. London: Springer; 2006.

[20] Löfberg J. YALMIP: A toolbox for modeling and optimization in MATLAB. In: CACSD Conference; 2004. http://control.ee.ethz.ch/joloef/yalmip.php.

Index